Optical Properties
and
Remote Sensing
of
Inland and Coastal Waters

Optical Properties
and
Remote Sensing
of
Inland and Coastal Waters

Robert P. Bukata
John H. Jerome
Kirill Ya. Kondratyev
Dimitry V. Pozdnyakov

Environment Canada
Aquatic Ecosystem Conservation Branch
National Water Research Institute
Burlington, Ontario, Canada

CRC Press
Boca Raton   London   New York   Washington, D.C.

Library of Congress Cataloging-in-Publication Data

Optical properties and remote sensing of inland and coastal waters /
   Robert P. Bukata...[et al.].
      p. cm.
   Includes bibliographical references and index.
   ISBN 0-8493-4754-8 (alk. paper)
   1. Optical oceanography. 2. Lakes—Optical properties.
   3. Oceanography—Remote sensing. 4. Lakes—Remote sensing.
   I. Bukata, R. P.
   GC178.2.066 1995
   551.465′01—dc20   95-15157
                                                                   CIP

No claim to original U.S. Government works
International Standard Book Number 0-8493-4754-8
Library of Congress Card Number 95-15157
Printed in the United States of America   2 3 4 5 6 7 8 9 0
Printed on acid-free paper

PREFACE

The interaction of downwelling solar and sky radiation with natural water bodies has fascinated earthbound observers for centuries. Excellent treatises on photon propagation through attenuating media have been written from the perspectives of geometric optics, physical optics, and quantum dynamics. The radiative transfer mechanisms governing the establishment of the light regime beneath the air-water interface of oceans likewise have been considered through avenues of dialogue and mathematical formulism and/or simulation. The era of environmental remote sensing is generally considered as having begun with the launch of ERTS-1 in 1972. Satellite monitoring and remote sensing research in oceanography have very rapidly developed thanks to (a) the wealth of research in optical oceanography and atmospheric physics that had been performed in many countries for decades prior to the inception of global monitoring from remote platforms and (b) the teams of talented researchers in several oceanographic institutes who were able to transform the optics of oceanic photon propagation into the development of algorithms that would give a scientifically justifiable meaning to the optical data collected above the ocean. As a testimony to their talents and creativity, satellite mapping of the near-surface chlorophyll concentrations and primary production of the global oceans is now being performed on a routine basis using commercially available computer algorithms with an accuracy and reproducibility generally deemed acceptable for most purposes. Excellent textbooks chronicling the development of this and other applications of remotely sensed data to oceanography also exist, many authored by workers who continue to contribute to the current status enjoyed by both optical oceanography and remote sensing.

Within such a scenario, the writing of a monograph/textbook dealing with the optical properties of natural water bodies and the remote sensing of such aquatic resources might seem to be at worst presumptuous and at best redundant. Such a book, however, is certainly not our intent. The natural water bodies that we have for years been considering under the general umbrella of aquatic optics and remote sensing are non-mid-oceanic. While such inland and coastal water regions are a source of delight for scientific query and investigation, they are also a source of frustration for would-be remote sensors. These frustrations emerge from the fact that, in general, the optical complexity of a natural water body rapidly escalates as its proximity to a land mass increases. In essence, our book is a consequence of these frustrations. It certainly would be most gratifying if limnologists could directly apply the inversion algorithms and remote sensing methods that are serving the oceanographers in such good stead. Such, however, is not the case. The cumulative surface area of water bodies totally encapsulated by a single continent is essentially negligible when compared to the cumulative surface area of intercontinental waters. The limnological frustrations could, therefore, even be reluctantly tolerated if the non-mid-oceanic waters were environmentally inconsequential when compared (e.g., in total productivity) to the global oceans. However, this too, is not the case. Anthropogenic invasions upon the Earth's ecosystems (e.g., acid rain, greenhouse gases, atmospheric ozone depletion, pollutant injection and transport) have manifested as fluctuations in the hydrological cycle, climatic patterns, terrestrial and aquatic productivity, biological species adaptation and transformation, and population dynamics on regional as well as global scales. Inland and oceanic coastal water regimes play major roles in global bioproductivity, and these roles are continuing to broaden with time. Remote sensing of inland and coastal waters is being integrated into national and international programs dedicated to monitoring and assessing environmental impacts. Thus, our frustrations are now being shared among an ever-expanding group of colleagues that includes environmental scientists, resource managers, undergraduate and graduate students in a variety of disciplines, as well as limnologists, climatologists, and remote sensing specialists. We hope that our colleagues will find some value in this book.

We have attempted to write *Optical Properties and Remote Sensing of Inland and Coastal Waters* from the perspective of a physicist with a background in optics and mathematics, some knowledge of and interest in physical limnology and oceanography, an awareness of biological limnology and oceanography, an appreciation of but not necessarily extensive experience in remote sensing of natural resources, and a realization that, despite the emphasis on physics in this book, the remote sensing of optically complex water bodies is a multidisciplinary venture. Whether or not such a perspective adequately serves our intended colleagues is, of course, moot. Admittedly, it is hazardous to write a book from the vantage point of optical physics and assume it be appropriate for an audience whose interests may transcend or circumvent optical physics. Nevertheless, it is the subsurface physical light field that must be directly compared to and interpreted in terms of the above-surface light field recorded at a remote platform. To address this problem we have generated a logical progression of discussions in which mathematical formulations are appropriately interwoven with responsible narrative. The theory of radiative transfer processes and photon propagation through attenuating media is briefly reviewed, as is the propagation of downwelling solar and sky radiation through the air-water interface. The light regimes of Case I and non-Case I waters (i.e., mid-oceanic and non-mid-oceanic) are discussed in terms of co-existing organic and inorganic aquatic component concentrations and the spectrally selective specific scattering and absorption coefficients that may be ascribed to these components. The roles of these optical properties in the remote sensing of water color, water quality, and water bioproductivity are considered and analyzed. In an attempt to maintain continuity within the text, a certain amount of intentional redundancy is included, largely in the form of reiterations. Illustrative examples from both published and unpublished data sets obtained from optically complex waters are frequently employed. In this regard we have unashamedly drawn quite extensively from our own research activities and experiences with the non-Case I waters comprising the Laurentian Great Lakes.

Optics and radiative transfer theory are well established fields of physics, while the remote sensing of natural aquatic resources is a relatively new field of scientific endeavor. The remote sensing of inland and coastal waters is even newer and less established. Thus, we have attempted to write an instructional text that is simultaneously timeless and timely. The mathematical formulations of radiative transfer theory can very quickly become intimidating to an audience of varied background. Such intimidation rarely serves a useful purpose to any but the most theoretically erudite. Similarly, no useful purpose is achieved by reducing such mathematical intimidation to trite approximations of reality. In an effort to achieve some middle ground between trite and intimidating, however, we may have succeeded in satisfying no one. To those whose optical knowledge and expertise exceed ours we apologize for being perceived as ineffectual and boring. To those whose enthusiasm for limnological optics does not parallel ours, we apologize for being perceived as pompous and self-serving. To those who want quick and reliable solutions to the problems of remotely sensing optically complex waters, we apologize for our shortcomings as researchers. To those who have either accepted or waived their rights to our apologies, we bid you welcome to *Optical Properties and Remote Sensing of Inland and Coastal Waters*.

R. P. Bukata
J. H. Jerome
K. Ya. Kondratyev
D. V. Pozdnyakov

AUTHORS

Robert Peter Bukata, Ph.D., received his doctorate in physics and mathematical physics in 1964 from the University of Manitoba, where he designed and utilized a pair of mutually perpendicular high energy particle telescopes rotating in opposition to the Earth's rotation to discover a galactic source of cosmic radiation in the region of the constellation Aquila. He spent seven years on the faculty of the Southwest Center for Advanced Studies in Dallas, Texas, where he was Co-Principal Investigator of cosmic ray particle studies aboard the Pioneers 6, 7, 8, 9, and 10 deep-space probes. He and his colleagues executed *in situ* studies of solar flare effects and were the first to observe solar flares directly during solar minimum, filamentary interplanetary magnetic field structures, solar M-region magnetic storm modulations of the cosmic ray flux, co-rotating Forbush decreases, and energetic storm particle (ESP) events. He was also active in studies involving the Earth-orbiting satellites Explorers 35 and 41, the worldwide network of ground-based neutron monitors, and atmospheric balloon measurements of the secondary cosmic radiation. In 1971 he returned to Canada, where he designed remote sensing projects for the Faculty of Agriculture at the University of Manitoba. In 1973 he joined the National Water Research Institute (NWRI) in Burlington, Ontario, as Head of the Environmental Spectro-Optics Section. His research interests there have included lake optics, limnology, and the development of models that relate apparent and inherent optical properties of natural water to the bio-physical activities that control them. He is currently Chief of Atmospheric Change Impacts at NWRI, where his research activities focus upon impacts to aquatic ecosystems of climate change and stratospheric ozone depletion. He has authored over 110 articles and book segments in cosmic ray physics, aquatic optics, remote sensing, limnology, hydrology, and climate change.

John Harvey Jerome, B. Eng., received his Bachelor of Engineering in Engineering Physics from McMaster University in Hamilton, Ontario, in 1971. Upon graduating he joined the Aquatic Physics and Systems Division of the Canada Centre for Inland Waters in Burlington, Ontario. From 1971 to 1978 he carried out the first extensive program of the measurement and characterization of the optical properties of the Laurentian Great Lakes for the Canadian Department of the Environment. With the launch of Nimbus-

7 carrying the Coastal Zone Color Scanner, his work became focused on the determination of the relationships among the optical properties of natural waters, the natural constituents of these waters, and the optical spectra recorded by remote sensing devices over these waters. He devised a Monte Carlo simulation of radiative transfer in natural waters to aid in the quantification of such inter-relationships. While the work in oceanographic remote sensing concentrated on a single component concentration algorithm, his work on limnological remote sensing concentrated upon the development of a bio-optical model that would enable the simultaneous extraction of several aquatic components from the measurement of *in situ* volume reflectance. He continued to work on various *in situ* and remote sensing techniques employing optical devices to monitor the natural components of fresh waters. He has authored over 50 scientific papers. He is currently a Study Leader in the Atmospheric Change Impacts Project at the National Water Research Centre in Burlington, where his research activities continue on the refinement of bio-optical models in addition to the investigation of UV-B levels in inland waters and the use of remote sensing to assess the impacts of climate change on aquatic ecosystems.

Kirill Yakovlevich Kondratyev, Ph.D., longstanding senior member of the Russian (formerly USSR) Academy of Sciences, received his doctorate in physical-mathematical science from Leningrad State University in 1946. He remained associated with the University until 1978, serving in the capacities of Professor, Department Head, Vice-Rector for Science and Research, and Rector of LSU. From 1958 to 1982 he was Department Head at the Main Geophysical Observatory at the Institute for Lakes Research of the USSR Academy of Sciences. He remained at the Institute for Lakes Research until 1992, when he assumed the responsibilities of Counsellor of the Russian Academy of Sciences at the Research Centre for Ecological Safety in St. Petersburg, Russia, where he currently is engaged in the development of remote sensing techniques for environmental studies. An internationally recognized expert in atmospheric radiation processes and physical principles of climate evolution, he has authored over 1000 scientific papers and 93 monographs. His monographs have included "Meteorological Satellites" (1963), "Radiative Heat Exchange in the Atmosphere" (1965), "Weather and Climate on Planets" (1982), "Global Climate and its Changes" (1987), "Climate Shocks: Natural and Anthropogenic" (1988), and "Ecology and Politics" (1993). His numerous scientific awards include the State Prize, the World Meteorological Organization Gold Medal, and the Symons Medal. He holds honorary doctorates from the Universities of Lille (France) and Budapest (Hungary). He is a member of the American Meteorological Society, the German Natural Science Academy, the Royal Meteorological Society, and the International Academy of Astronautics. A member of editorial boards of several international scientific journals, he is currently Editor-in-Chief of Earth Observation and Remote Sensing.

Dimitry Victorich Pozdnyakov, Ph.D., received his doctorate in physics in 1972 from the State University of St. Petersburg, Russia, where he conducted infrared studies simulating gas/aerosol interactions. His work revealed new sink mechanisms of climate-controlling trace gases, including the stratospheric ozone-depleting flourocarbons. He then accepted a lecturing post in the Department of Atmospheric Physics at the State University. As a Visiting Professor he lectured in physics for five years at the University of Conarky, Guinea. In 1983 he joined the Institute for Lakes Research of the Russian (then USSR) Academy of Sciences, where his research interests focused upon limnological ecology and hydro-optics. He has developed bio-optical algorithms for the remote sensing of water quality parameters utilizing passive spectrometric and active lidar techniqies. His scientific team has remotely investigated the trophic status of nearly all the large lakes and water storage reservoirs of the former USSR. He was awarded a D.Sc. degree in 1992. He has authored more than 75 scientific papers, brochures, and books. He currently holds the position of Leading Scientist at the Nansen International Environmental and Remote Sensing Centre (NIERSC), as well as Invited Professor at the Electrotechnical University, both in St. Petersburg. His research activities continue to be directed toward the optical properties of inland and coastal waters.

In 1992 the authors were recipients of the Chandler-Misener Award presented by the International Association for Great Lakes Research for their work documented in the companion publications:

Bukata, R. P., Jerome, J. H., Kondratyev, K. Ya., and Pozdnyakov, D. V., Estimation of Organic and Inorganic Matter in Inland Waters: Optical Cross Sections of Lakes Ontario and Ladoga, *Journal of Great Lakes Research*, 17, 461–469, 1991.

Bukata, R. P., Jerome, J. H., Kondratyev, K. Ya., and Pozdnyakov, D. V., Satellite Monitoring of Optically-Active Components of Inland Waters: An Essential Input to Regional Climate Change Impact Studies, *Journal of Great Lakes Research*, 17, 470–478, 1991.

TABLE OF CONTENTS

Chapter 1

INTRODUCTORY THEORY

1.1 AQUATIC OPTICS

In its broadest terms, *optics* is generally defined as that branch of physics concerned with the interactions of light with its containment medium as the light propagates through that medium. Aquatic optics generally restricts its definition of containment media to those found in naturally occurring water bodies. Since, however, the point of origin and/or final destination of the light propagation under study are very often in media other than those comprising the naturally occurring water bodies, aquatic optics must also contend with the transference of light among these adjacent media. The propagation of light through the atmosphere and the interaction of light with the solid matter containing the natural water body must, therefore, be appropriately incorporated into the principles of aquatic optics.

Aquatic optics can be subdivided according to whether the natural water body is salty (oceanographic), inland or fresh (limnological), or coastal. Oceanographic optics has long been recognized as an integral component of oceanography, and has benefitted from rapid advances in the development of sophisticated submersible spectro-optical devices and computer-based interpretative techniques. Limnological optics has not enjoyed the same longstanding recognition as has oceanographic optics, although the sophisticated spectro-optical devices and computer-based interpretative techniques offer comparable benefits. It is not the purpose of this book to dwell on the possible reasons for such an historical lack of recognition, but rather to address the nature of the limnological light field and the behavior of photons within such an aquatic environment. The fundamental theory of radiative transfer developed for atmospheric and oceanographic optics is directly applicable to the study of both the limnological and the coastal environments. However, due to the basic differences amongst oceanographic and limnological aquatic compositions, much of the interpretive techniques are not directly applicable. These similarities and differences will be discussed throughout this book.

The launch of ERTS-1 (the Earth Resources Technology Satellite, later renamed Landsat-1) in August 1972 introduced remote sensing as a potential tool to the protocols of environmental monitoring. Mounted onboard ERTS-1 was a broadband multispectral scanning device which recorded the radiation upwelling from the target contained within the field-of-view of the downward-looking multispectral scanning device. This target was an environmental ecosystem comprised of atmospheric, terrestrial, and/or aquatic regimes. Consequently, the satellite sensor recorded an integrated spectral signature representing the reflected portion of the solar and sky radiation impinging upon the earth (later sensors recorded a second signature representing the thermal energy generated by the ecosystem itself). Since the parameter recorded at the orbiting satellite was optical in nature, it logically followed that such optical data should be interpreted in terms of the optical properties of the environmental targets. Remote sensing and its potential role in environmental monitoring activities became, therefore, intimately linked with aquatic optics, providing, as it did, a need to understand the interactive nature of electromagnetic radiation and natural water bodies.

As subsequent satellites were launched (some such as Nimbus-7 and Seasat were dedicated predominantly to the study of oceans and ocean color), techniques to extract oceanographic information from the continuous streams of remotely-sensed data were rapidly developed. Understandably, these environmental satellites contained payloads of optical devices sensitive to various energy regions of the electromagnetic spectrum. It is the intention of this book to review the philosophy and scientific theory governing such information extraction within a restricted region of the electromagnetic spectrum, namely the visible wavelength region. While other regions of the electromagnetic spectrum (most notably microwave, radar, and thermal infrared) have been justifiably represented by the complements of sensors comprising both aircraft and satellite remote sensing missions dedicated to aquatic studies, measurement of water colour continues to comprise a major component of such remote sensing activities. For this reason, the major thrust of this book will be directed towards an interpretation of visible light (that is, radiation contained within the boundaries of the color wavelengths) emanating from inland and coastal natural water bodies (that is, natural waters influenced by their proximity to land masses).

To accommodate this thrust, the basic theory and general principles of aquatic optics and light propagation will first be briefly reviewed. A glossary is included at the end of this book defining the terminology that will be used throughout the text. Remote sensing will be treated as a parallel topic with aquatic optics, and the progression of the chapters of this book begins with an introduction to the general problem of extracting aquatic information from spectral data collected remotely over inland and coastal waters. This will be followed by the development of theoretical and experimental approaches that have constituted the quest for solving this problem.

1.2 THE NATURE OF ELECTROMAGNETIC RADIATION AND LIGHT

Electromagnetic radiation exists as discrete energy packets known as *quanta* or *photons*. While the term *photon* appears to define the particulate aspect of electromagnetic radiation, the more fundamental physical properties of electromagnetic radiation are defined in terms of its wave-like aspects. Every photon is characterized by a specific energy ξ (in units of joules), a specific wave frequency ν (in units of cycles s^{-1}), a specific wavelength λ (in units of m), the speed of light in a vacuum c (3.00×10^8 m s^{-1}), and Planck's Constant h ($h = 6.625 \times 10^{-34}$ joule s) according to the relationships:

$$\lambda = c/\nu \tag{1.1}$$

and

$$\xi = h\nu = hc/\lambda \tag{1.2}$$

The standard unit of measurement for wavelength λ is meters. However, for some regions of the electromagnetic spectrum it may be more convenient to express wavelength either in terms of nanometers [a nanometer (nm) being equivalent to 10^{-9} m] or in terms of micrometers [a micrometer (μm) being equivalent to 10^{-6} m or 10^3 nm]. The inverse of the wavelength is termed the wavenumber κ and is measured in units of inverse meters (m^{-1}).

$$\kappa = 1/\lambda \tag{1.3}$$

The term *light* is generally understood to refer to radiation in that portion of the electromagnetic spectrum to which the human eye is sensitive (that is, the visible region of the spectrum very broadly considered to lie within the wavelength interval 390 nm to 740 nm). It is the integration of radiation in this spectral band with the sensitivity response function of the human eye that results in the neurophysical sensation of perceived color. It logically follows, therefore, that aquatic color would be intimately related to an observer's critique of the aesthetic value of a natural water body. The color of naturally occurring water is, thus, often considered as an indicator of water quality. Such a consideration strains the scientifically dispassionate definitions of both water quality and water color. However, the passionate direct viewing of water color coupled with the considerably less passionate viewing accomplished with the aid of optical sensors has played a major role in the monitoring as well as the interpretation of water quality in terms of its spectral signatures in the visible wavelength region of the electromagnetic spectrum. Such a role is anticipated to continue well into the foreseeable future and is reflected in the payloads of existing and planned environmental satellites.

The totality of light emanating from a natural water body in a particular wavelength interval $\lambda_1 \leq \lambda \leq \lambda_2$ would, therefore possess an energy ξ_T given by the integration of equation (1.2) over the wavelength interval, namely:

$$\xi_T = hc \int_{\lambda_1}^{\lambda_2} \frac{p(\lambda)d\lambda}{\lambda} \tag{1.4}$$

where $p(\lambda)$ represents the number of photons characterized by wavelength λ.

In this discussion light is being considered as a flux of photons possessing a composite energy ξ_T which is shared among photons of differing wavelengths. The *photon flux N* is defined as the number of photons arriving at a unit area per unit time. The *energy flux* Φ_N associated with this photon flux at each wavelength λ would then be the product of the individual photon energies and the number of such photons per unit area and time. That is:

$$\Phi_N = \frac{energy}{photon} \times \frac{number\ of\ photons}{area \times time}$$

$$= h\nu \times N \text{ joules m}^{-2} \text{ s}^{-1} \tag{1.5}$$

Thus, if a monochromatic flux of light (that is, light comprised of a single wavelength λ, numerically expressed in nanometers) were emanating from the aquatic surface, the totality of light energy ξ_T would simply be the product of the number of monochromatic photons p, the wavenumber κ, Planck's Constant h, and the speed of light c. Substituting the numerical values of h and c into equation (1.2), this monochromatic energy becomes:

$$\xi_T = (1988p\kappa) \times 10^{-19} \text{ joules} \tag{1.6}$$

If a polychromatic flux of light (light comprised of a variety of wavelengths λ, again numerically expressed in nm) were emanating from the aquatic surface, but the flux were comprised of the same number of photons p at each wavelength, then the integration of equation (1.4) yields:

$$\xi_T = (1988p)[\ln(740) - \ln(390)] \times 10^{-19} \text{ joules}$$

$$= (1273p) \times 10^{-19} \text{ joules} \tag{1.7}$$

If, however, a non-"white" polychromatic flux of light (light comprised of not necessarily identical numbers of photons at each of the visible wavelengths λ) were emanating from the aquatic surface, the number of discrete photons of each wavelength would be required before an estimate of total energy could be determined from equation (1.4). Thus, to evaluate the color of natural water bodies, it is first essential to possess knowledge of the spectral distribution of the photons comprising the broad band visible radiant energy emanating from the aquatic surface. Relating this upwelling energy spectrum to the aquatic constituents responsible for that upwelling energy spectrum has been the goal of bio-optical modelling as well as *in situ* and remote sensing research activities performed at oceanographic and limnological institutes on a global scale for many years.

A commonly used and more convenient symbol for ξ_T is Q, which represents the quantity of energy (in joules, J) transferred by radiation. The *radiant flux* Φ is defined as the time rate of flow of radiant energy Q, namely:

$$\Phi = \frac{dQ}{dt} \quad \text{J s}^{-1} \text{ or watts (W)} \tag{1.8}$$

In any medium light travels more slowly than it does in a vacuum. The index of refraction, $n(\lambda)$, of an optically transparent medium is defined as the ratio of the speed of light at wavelength λ in a vacuum to the speed of light at wavelength λ in that medium. [The index of refraction is a complex number, i.e., it is expressed as the sum of a real n_R and an imaginary in_I part, where i is $\sqrt{-1}$. That is, $n(\lambda) = n(\lambda)_R + in(\lambda)_I$]. For the restricted band of visible wavelengths considered herein, the real part of the index of refraction can be considered as spectrally invariant, and thus represented as a dimensionless number > 1. (The imaginary part of the index of refraction over the visible spectrum is in the range 10^{-8} to 10^{-6} and is related to the amount of absorption that occurs in the medium at each wavelength). For example, the index of refraction of air (under conditions of standard atmospheric composition, temperature, and pressure) is taken to be 1.00028. That is, the speed of propagation of light in air does not significantly differ from the speed of propagation of light in a vacuum. The index of refraction of sea water,[198] although displaying a small dependence on salt content, temperature, and pressure, is generally taken to be 1.333. The speed of light in most natural waters, therefore, is roughly three-quarters the speed of light in a vacuum (2.25×10^8 m s^{-1}). Thus, while the wave frequency v of the light in water remains the same as in a vacuum (or, equivalently for our purposes, as in air), its wavelength diminishes in proportion to its diminished velocity. Since the velocity of the light and its wavelength undergo simultaneous changes in *non vacuo* media, the energy per photon does not change, and the only variable to be considered as light travels from one medium to another is the photon flux N (Conserving the energy flux across the media interface, however, gives rise to an n^2 law governing this photon flux). Consequently, equations (1.1) through (1.6) apply as equally to water as they do to air and to a vacuum. Throughout this book we shall refer to the wavelength of light as the wavelength it would possess if measured in a vacuum, and the analyses techniques and theory will be primarily concerned with the changes in photon flux as the photons propagate through or transfer among media of various indices of refraction.

From the perspective of aquatic optics, this change in photon flux is a direct consequence of the composite nature of the water mass under survey and, consequently, directly influences the manner in which remotely sensed data collected over that water body are to be interpreted. This introductory section has treated light as a flux of photons of varying wavelengths. While such a simplistic treatment allows for a basic description of the general energy balances and propagation of photons through an aquatic system, it obviously falls considerably short of yielding precise information concerning the radiative transfer mechanisms governing the optical interactions between the propagating photons and their containment media. It is not the purpose of this book to provide a detailed treatise regarding such radiative transfer mechanisms, which cover the historical development of optical physics dating from the early geometrical/physical work of Alhazen, Huygens, and Snell through the electromagnetic formulations of Fraunhofer, Maxwell, Kirchhoff, Hertz, Fresnel, and Brewster and into the quantum dynamical formulations of Einstein, Rayleigh, Planck, Raman, and Chandrasekhar, to name but a minuscule number of the pioneers of modern radiative transfer theory. Many excellent books exist covering various aspects of this continuing adventure in scientific thought, and the reader is strongly encouraged to capitalize upon their existence. This book will assume such reference material is readily available to the reader, and will concentrate on those aspects of aquatic optics that are directly applicable to the interpretation of upwelling light from inland and coastal waters.

1.3 RADIANCE AND IRRADIANCE

Radiance and *irradiance* are arguably the two most important terms required for understanding the behavior of photon flux. This should become very evident as the development of both *in situ* and remote sensing optical methods and algorithms unfold in the upcoming chapters. Both terms refer to a measure of radiant flux Φ. Each, however, while representing the number of photons at a particular site within the containment medium, represents a different subset of the totality of photons, these subsets being a function of the directions of photon arrival at the site and the manner in which these photons are observed and recorded.

Light that originates from a point source (the fixed stars would qualify as such point sources when viewed from a great distance) is observed to propagate radially outward in all directions. *Radiant intensity, I,* is a measure of the radiant flux Φ per unit solid angle in a particular direction. Imagine a sphere of radius r with the point source of light at its center. Further imagine an infinitesimal cone, its apex at the point source, extending from

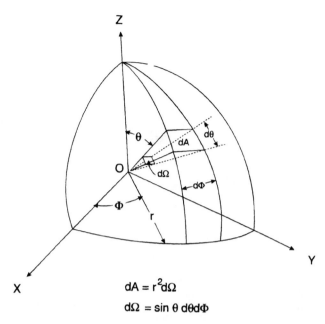

$$dA = r^2 d\Omega$$

$$d\Omega = \sin\theta \, d\theta d\Phi$$

FIGURE 1.1
Spherical coordinate representation of the geometry of point-source photon progagation.

the center to intersect the surface of the sphere. This infinitesimal cone subtends an infinitesimal solid angle $d\Omega$ (in steradians, sr). The intersection of the cone with the spherical surface defines a surface area dA. The radiant intensity, I, of light emanating from the point source of light as contained by and measured within the direction defined by the infinitesimal cone is then given by:

$$I = \frac{d\Phi}{d\Omega} \text{ J s}^{-1} \text{ sr}^{-1} \text{ or W sr}^{-1} \tag{1.9}$$

Figure 1.1 illustrates the geometry of this point source propagation in spherical polar coordinates (polar angle θ, azimuth angle ϕ, and radius vector of magnitude r). The infinitesimal solid angle $d\Omega$ is defined as $\sin\theta d\theta d\phi$. Expressed in polar coordinates, equation (1.9) becomes:

$$I = \frac{d\Phi}{\sin\theta d\theta d\phi} \tag{1.10}$$

Radiant intensity may also be used to describe incoming radiation at a point in space which is removed from the source of radiation. At such a point, the infinitesimal solid angle is used to define the containment of incoming,

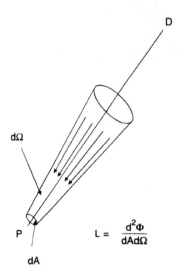

$$L = \frac{d^2\Phi}{dAd\Omega}$$

FIGURE 1.2
Concept of radiance at a point P in space.

rather than outgoing, photons. Thus, *radiant intensity* may be readily defined in a comparable manner as the radiant flux incident at that point per unit solid angle.

If, in addition to the photons contained either within an infinitesimal exit or within an infinitesimal arrival solid angle, the flux of photons imping-ing upon a unit area perpendicular to the direction of photon flow is also considered, then the concept of *radiance* emerges. The radiance, L, in a speci-fied direction at a point in the radiation field is then defined as the radiant flux at that point per unit solid angle per unit area at right angles to the direction of photon propagation. Expressed mathematically, radiance is given by:

$$L = \frac{d^2\Phi}{dAd\Omega} \ \text{J s}^{-1} \ \text{m}^{-2} \ \text{sr}^{-1} \ \text{or W m}^{-2} \ \text{sr}^{-1} \qquad (1.11)$$

where dA represents the infinitesimal area perpendicular to the specified direction of photon propagation. The standard unit of radiance will be taken to be W m^{-2} sr^{-1} throughout the duration of these discussions.

Figure 1.2 schematically illustrates the concept of radiance as defined by equation (1.11) at a point in space. In this instance the radiance at the point P is defined as the radiant flux contained within the infinitesimal solid angle $d\Omega$ surrounding the selected direction of photon propagation DP and passing through the infinitesimal surface area dA which is perpendicular to the direction DP. If a point source of radiation were located at the point P, and the directions of propagating photons were reversed, equation (1.11)

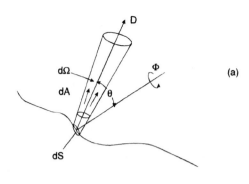

(a)

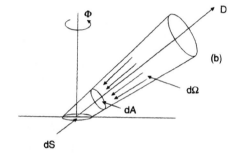

(b)

FIGURE 1.3
(a) Radiance associated with radiation emitted from an extended source. (b) Radiance associated with radiation impinging upon an extended surface.

would readily define the radiance of the point source along the selected direction *PD*.

Remote sensing of natural waters, however, must contend with light emanating from areas of extended sources of radiation as well as with light impinging upon extended surface areas which are not perpendicular to the selected direction of photon propagation. Figure 1.3(a) schematically illustrates the situation in which photons are emanating from an extended source at a selected direction *D* which is not perpendicular to the surface of the extended source. Figure 1.3(b) schematically illustrates the radiance at a point on an extended surface. In both these instances the photon propagation is considered for a direction which is not at right angles to the infinitesimal surface area *dS*. $L(\theta,\phi)$ is the radiance either emanating from or impinging upon the infinitesimal surface area dS at an angle θ (measured relative to the normal at the surface) and an azimuthal angle ϕ (measured in the plane containing *dS*). The radiant flux is contained within the solid angle *dΩ* centred on the selected direction of photon propagation *D* (defined by θ and ϕ). The radiant flux passes perpendicularly across the area *dA* which is the projected area of dS as would be presented to a viewing direction aligned with *D*. This projected surface area is defined as $dS\cos\theta$. Thus, the radiance impinging upon an extended surface or the radiance emitted from an extended source along direction D is given by:

$$L(\theta, \phi) = \frac{d^2\Phi}{dS\cos\theta d\Omega} \qquad (1.12)$$

The term *irradiance* E refers to the radiant flux per unit area at a point within the radiative field or at a point on an extended surface. Irradiance is defined as the radiant flux impinging upon an infinitesimal surface area dS (containing the point in question) divided by that infinitesimal area. That is:

$$E = \frac{d\Phi}{dS} \text{ W m}^{-2} \qquad (1.13)$$

The directional aspects attributable to radiance are not attributable to irradiance, since irradiance is comprised of all the radiant flux impinging upon a selected point within the radiative field, i.e., all the directions associated with the photon propagation. The relationship between radiance and irradiance may be readily appreciated in terms of the following considerations:

If the radiance from all directions within a hemisphere were measured with a radiometer, the total radiant flux Φ would be determined (from equation 1.12) as:

$$\Phi = \int_S \int_\Omega L(\theta, \phi) \, dS\cos\theta d\Omega \qquad (1.14)$$

where the angular integration occurs over 2π steradian.

From equation (1.13), the total radiant flux Φ in terms of irradiance would be determined as:

$$\Phi = \int_S E \, dS \qquad (1.15)$$

Comparing equations (1.14) and (1.15), radiance $L(\theta,\phi)$ and irradiance E are seen to be inter-related according to:

$$E = \int_\Omega L(\theta, \phi) \cos\theta d\Omega \qquad (1.16)$$

Expressed in polar coordinates (from Figure 1.1 $d\Omega$ may be expressed as $\sin\theta d\theta d\phi$) equation (1.16) becomes:

$$E = \int_0^{2\pi} \int_0^{\pi/2} L(\theta, \phi) \cos\theta\sin\theta d\theta d\phi \tag{1.17}$$

Equation (1.17) considers the radiance impinging upon a point from the θ and ϕ directions comprising the hemisphere immediately above the horizontal plane containing the point. Consequently, it represents the downwelling irradiance resulting from the totality of directionally dependent radiances $L(\theta,\phi)$.

If the downwelling radiation received at the point under consideration is *isotropic* (that is, the radiance $L(\theta,\phi)$ is independent of viewing direction and can be considered to be a constant value L for all values of θ and ϕ in the downwelling hemisphere), then equation (1.17) reduces to:

$$E = 2\pi L \times (1/2) \int_0^{\pi/2} \sin2\theta d\theta = \pi L \tag{1.18}$$

Similarly, if isotropic photon propagation were considered to be occurring in a full sphere around a point in the radiative field, then the irradiance would be double that of one hemisphere, namely:

$$E = 2\pi L \tag{1.19}$$

The simplistic relationships of equations (1.18) and (1.19) hold true for many aspects of radiative field configurations (for example, the radiance of a perfect Planck blackbody radiator is obtained by dividing its hemispherical exit irradiance by π). However, the anisotropies defining the photon propagation paths in natural water bodies (a direct consequence, as will be discussed later, of the variable scattering and absorption centre densities comprising these water bodies) generally prohibit the use of such simplistic expressions and the more general angularly dependent descriptions of $L(\theta,\phi)$ are required.

Technically, however, irradiance can be treated as a directional parameter with upwelling and downwelling irradiances being considered as distinguishable entities in both atmospheric and aquatic optics.

Downwelling irradiance E_d is defined as the irradiance at a point due to the stream of downwelling light. It is the irradiance that would be recorded by an upward-looking radiometer (equipped with a cosine response) and represents the irradiance attributable to the hemisphere above the horizontal plane containing the point of interest. Similarly, the *upwelling irradiance E_u* is defined as the irradiance at a point due to the stream of upwelling light. It would be recorded by a downward-looking radiometer (equipped with a cosine response), and represents the irradiance attributable to the hemisphere below the horizontal plane containing the point of interest.

Thus, as was the case for equation (1.17), downwelling irradiance is obtained by integrating equation (1.16) over the upper hemisphere:

$$E_d = \int_0^{2\pi} \int_0^{\pi/2} L(\theta, \phi) \cos\theta\sin\theta d\theta d\phi \ W \ m^{-2} \qquad (1.20)$$

Upwelling irradiance is obtained by integrating equation (1.16) over the lower hemisphere (adjusting for the fact that $\cos\theta$ is negative for $\pi/2 \leq \theta \leq \pi$):

$$E_u = -\int_0^{2\pi} \int_{\pi/2}^{\pi} L(\theta, \phi) \cos\theta\sin\theta d\theta d\phi \ W \ m^{-2} \qquad (1.21)$$

The ratio of the upwelling irradiance at a point to the downwelling irradiance at that point is termed the *irradiance reflectance*. Within a water body this ratio is known as the *subsurface volume reflectance*, or simply the *volume reflectance*. The volume reflectance, R, is one of the singularly most important parameters in the interpretation of remotely sensed spectral data in terms of the compositions of natural water bodies. Considerable use will be made of this optical parameter throughout the remainder of this book.

$$R = \frac{E_u}{E_d}$$

$$= -\frac{\int_0^{2\pi} \int_{\pi/2}^{\pi} L(\theta, \phi)\cos\theta\sin\theta d\theta d\phi}{\int_0^{2\pi} \int_0^{\pi/2} L(\theta, \phi)\cos\theta\sin\theta d\theta d\phi} \qquad (1.22)$$

At any point within the water column, the *net downward irradiance, $E\downarrow$*, is simply the difference between the downwelling and upwelling irradiances at that point, namely:

$$E\downarrow = E_d - E_u \qquad (1.23)$$

Equation (1.20) readily yields a positive numerical value for E_d by integrating over the angular ranges ($0 \leq \theta \leq \pi/2$) and ($0 \leq \phi \leq 2\pi$). Equation (1.21) readily yields a positive numerical value of E_u by integrating over the angular ranges ($\pi/2 \leq \theta \leq \pi$) and ($0 \leq \phi \leq 2\pi$). Subtracting equation (1.21) from equation (1.20) is directionally consistent with integrating equation (1.17) over a full sphere (4π space).

Therefore, the net downward irradiance at a point is given by:

$$E\downarrow = \int_0^{2\pi} \int_0^{\pi/2} L(\theta, \phi)\cos\theta\sin\theta d\theta d\phi$$

$$+ \int_0^{2\pi} \int_{\pi/2}^{\pi} L(\theta, \phi)\cos\theta\sin\theta d\theta d\phi$$

$$= \int_0^{2\pi} \int_0^{\pi} L(\theta, \phi) \cos\theta\sin\theta d\theta d\phi \tag{1.24}$$

Equation (1.24) integrates the product of each individual directional radiance $L(\theta,\phi)$ with the cosine of its corresponding polar angle θ over all the directions comprising both the upper and lower hemispheres. Downwelling irradiance, when considered as a function of depth in an aquatic medium, forms the basis for discussions of the concepts of light extinction, photic depths, scattering, absorption, and primary production, among other manifestations of aquatic optics to be encountered as this book progresses.

It is often appropriate (and this will be discussed more fully in our treatment of photosynthesis and primary production) to consider the total radiant intensity at a point in space independent of its arrival directions. Such a consideration regards radiance from each direction equally. This is equivalent to equating the infinitesimal area dS [the actual area upon which the directional radiance $L(\theta,\phi)$ is impinging] to the infinitesimal area dA [the area of dS that would be projected to a radiant flux passing perpendicularly through it]. This, in turn, is equivalent to the removal of the $\cos\theta$ term from equation (1.24). When such an equal treatment of radiances is considered in the integration, the resulting irradiance at that point is termed the *scalar irradiance* E_0 and is defined as the integral of the radiance distribution over all directions (4π space) at that point. That is:

$$E_0 = \int_0^{2\pi} \int_0^{\pi} L(\theta, \phi)\sin\theta d\theta d\phi \ \text{W m}^{-2} \tag{1.25}$$

Equation (1.25) integrates radiance $L(\theta,\phi)$ over all the directions comprising the upper and lower hemispheres. Although each direction in the radiance distribution is treated equally (represented by the removal of the $\cos\theta$ term from the $E\downarrow$ expression), the directional nature of $L(\theta,\phi)$ is maintained.

Submersible irradiance sensors are designed, depending upon their purpose, to measure upwelling, downwelling, or scalar irradiance at a point in the aquatic medium. As illustrated by equations (1.21) and (1.25), those sensors responsive to irradiance (as opposed to scalar irradiance) are required to possess a cosine response.

As was the case for irradiance, a quasi-directionality can be assigned to scalar irradiance (a "vector" property of a "scalar" commodity). Similar to the development of equations (1.20) and (1.21), the *downwelling scalar irradiance*, E_{0d}, and the *upwelling scalar irradiance*, E_{0u}, are defined as follows:

The downwelling scalar irradiance at a point is defined as the integrated radiance distribution over the hemisphere above the horizontal plane containing the point, namely:

$$E_{0d} = \int_0^{2\pi} \int_0^{\pi/2} L(\theta, \phi)\sin\theta d\theta d\phi \qquad (1.26)$$

The upwelling scalar irradiance at a point is defined as the integrated radiance distribution over the hemisphere below the horizontal plane containing the point, namely:

$$E_{0u} = \int_0^{2\pi} \int_{\pi/2}^{\pi} L(\theta, \phi)\sin\theta d\theta d\phi \qquad (1.27)$$

Since the $\cos\theta$ term is absent from the scalar irradiance expressions, no mathematical compensation is required to obtain E_{0u} in the lower hemisphere, and equation (1.27) as written results in a value of scalar irradiance that is numerically positive. The *net downwelling scalar irradiance, $E_0\downarrow$*, is then obtained as the difference between equations (1.26) and (1.27):

$$E_0\downarrow = E_{0d} - E_{0u}$$

$$= \int_0^{2\pi} \int_0^{\pi/2} L(\theta, \phi)\sin\theta d\theta d\phi$$

$$- \int_0^{2\pi} \int_{\pi/2}^{\pi} L(\theta, \phi)\sin\theta d\theta d\phi \qquad (1.28)$$

The terms *downwelling light* and *upwelling light* usually, although not always, refer to downwelling and upwelling irradiance as opposed to downwelling and upwelling scalar irradiance fluxes. Throughout the text we will adhere to the irradiance implication of these terms unless it is obvious from the context of the discussion that some other radiance distributions are being considered.

The weighted mean value of a parameter x, ($\bar{x}$), with respect to another parameter y is mathematically defined by:

$$\bar{x} = \sum_{i=1}^{i=n} x_i y_i \bigg/ \sum_{i=1}^{i=n} y_i \qquad (1.29)$$

The forms of equations (1.20) and (1.26) are amenable to the application of equation (1.29) and enable a weighted average value of the cosine ($\bar{\mu}_d$) of the polar angle, θ, to be determined for a downwelling radiance distribution. Thus, the average cosine, $\bar{\mu}_d$, of a downwelling radiance distribution may be obtained by first determining the downwelling irradiance, E_d, at a point in the radiation field [equation (1.20)], then dividing this irradiance by the downwelling scalar irradiance, E_{0d} [equation (1.26)]. That is:

$$\bar{\mu}_d = \frac{E_d}{E_{0d}} \tag{1.30}$$

Similarly, the average value of the cosine, $\bar{\mu}_u$, of an upwelling radiance distribution may be obtained by first determining the upwelling irradiance, E_u, at a point in the radiation field [equation (1.21)], then dividing this irradiance by the upwelling scalar irradiance, E_{0u} [equation (1.27)].

$$\bar{\mu}_u = \frac{E_u}{E_{0u}} \tag{1.31}$$

To determine the average value of the cosine, $\bar{\mu}$, of the polar angle θ for the total 4π (i.e., upwelling and downwelling) radiance distribution at a point, it is essential to first integrate the product of each radiance and its corresponding $\cos\theta$ over all directions [which from equation (1.24) yields the net downwelling irradiance $E\downarrow$] and then to divide by the total radiance from all directions at that point (i.e., the scalar irradiance). The average value of the cosine μ is thus given by:

$$\bar{\mu} = \frac{E\downarrow}{E_0}$$

$$= \frac{E_d - E_u}{E_0} \tag{1.32}$$

To this stage of our discussions of radiance and irradiance, we have been emphasizing the directional aspects of the photon fluxes either as they exit from or arrive at a point within an optical medium. Radiance and irradiance, in addition to displaying dependencies upon directions of propagation (that is, upon θ and ϕ), display dependencies upon wavelength λ, which may vary markedly over the visible wavelength region considered herein. Knowledge of the variations of these wavelength-dependent radiant fluxes with aquatic depth, time, and physical location is essential to understanding and evaluating the physical aspects of the scattering and absorption processes that will be discussed in upcoming chapters. The wavelength and depth dependencies of radiance and irradiance are also determining factors of the photosynthetic capabilities of natural water bodies. Obtaining reliable relationships between the subsurface radiance and irradiance fields and the organic and inorganic materials comprising the water column (as a function of wavelength, time, and global location) is obligatory to the interpretation of remotely sensed data collected over inland and coastal waters. Similarly, as will be discussed later, obtaining reliable relationships between the subsurface aquatic and above-surface atmospheric radiance and irradiance fields is also obligatory.

1.4   ATTENUATION OF LIGHT IN AN AQUATIC MEDIUM

Photons entering and propagating within a natural water body will undergo scattering or absorption interactions with the materials comprising the natural water body. Both scattering and absorption interactions result in changes to the original subsurface radiance distribution as the photon flux propagates through the aquatic medium. Absorption and scattering processes combine to reduce the intensity of the radiance distribution, while the scattering processes also change the directional character of the radiance distribution. As the photons continue downwelling propagation, the number of optical interactions also continue. At some depth the intensity of the photons reaches zero. $L(\theta,\phi)$, therefore, displays a relationship with aquatic depth z (measured in meters positively vertically downward from the air-water interface). $L(\theta,\phi)$, as alluded to earlier, also displays a dependence upon wavelength λ. This is a consequence of the spectral nature of the original source of photons (for example, solar or sky radiation above the air-water interface), as well as the spectral dependencies of the optical properties of the scattering and absorption centers comprising the aquatic medium. Consequently, $L(\theta,\phi)$ is more appropriately written as $L(\theta,\phi,z,\lambda)$.

Recall from equation (1.5) that the energy flux Φ_N is defined as the product of the energy per photon and the number of photons per unit area per unit time (that is, $\Phi_N = h\nu N$). In order to describe the attenuation of a downwelling photon flux after it penetrates a natural water body, consider the particulate nature of the photon flux and the probabilistic nature of its fate. To do so note that (a) the probability that a photon of specific energy $h\nu$ will be absorbed is proportional to the change in energy flux Φ_N (since $h\nu$ of the photon flux is kept fixed, this is equivalent to the probability being proportional to a change in the photon flux N), and (b) the probability of absorption increases as the thickness of the absorbing medium increases. Consider an incident photon flux N_{inc} upon a unit area with thickness Δr in the absorbing medium. Consider further that this incident flux is reduced to a transmitted flux, N_{trans}, subsequent to its exit from this volume. The difference between N_{inc} and N_{trans} would then be proportional to the product of N_{inc} and Δr. That is:

$$N_{\text{trans}} - N_{\text{inc}} = -\alpha N_{\text{inc}} \Delta r$$

or

$$\Delta N = -\alpha N_{\text{inc}} \Delta r \tag{1.33}$$

where the constant of proportionality α is the absorption coefficient (in units of m^{-1}) appropriate to the fixed value of λ.

In the limit as both ΔN and Δr approach zero, equation (1.33) becomes:

$$\frac{dN}{N} = -\alpha dr \tag{1.34}$$

which is known as *Beer's Law.*

An alternate, and perhaps more familiar, form of Beer's Law is obtained by integrating equation (1.34) from an initial origin ($r = 0$) to a distance r in the absorbing medium. That is:

$$N(r) = N_0 e^{-\alpha r} \tag{1.35}$$

Note that in this development of Beer's Law, the absorbing medium (aquatic water body) has been considered to be spatially homogeneous (i.e., α is considered invariant with respect to r).

A convenient form of Beer's Law results from the consideration of the loss of energy from a beam of light (photons comprised of a spectrum of energy values $h\nu$). For such a beam the operative parameter is the radiant flux Φ (time rate of flow of radiant energy) rather than the energy flux Φ_N or the photon flux N. The absorption loss of beam energy from an initial radiant flux value of Φ_{inc} to a final radiant flux value of Φ_{trans} subsequent to passing through an attenuating medium of thickness Δr would then be given [comparable to equation (1.33)] by

$$\Phi_{\text{trans}} - \Phi_{\text{inc}} = -a\Phi_{\text{inc}} \Delta r \tag{1.36}$$

where the constant of proportionality a is defined as the *absorption coefficient.* Since the beam attenuation coefficient is a function of wavelength λ, it follows from equation (1.36) that:

$$\Phi(r, \lambda) = \Phi(0, \lambda) e^{-a(\lambda)r} \tag{1.37}$$

from which

$$a(\lambda) = -\frac{[\partial \Phi(r, \lambda)]_{abs}}{\Phi(r, \lambda)\partial r} \, m^{-1} \tag{1.38}$$

which is a variation of equation (1.37) with the subscript abs added to the partial derivative of Φ to indicate the process of absorption.

Equation (1.38) then provides the general definition of *absorption coefficient* $a(\lambda)$ as the fraction of radiant energy absorbed from a beam as it traverses an infinitesimal distance ∂r divided by ∂r. To this point of the discussion we have considered attenuation of a photon flux or a radiant flux as due solely to absorption processes and independent of any attenuation due to scattering processes.

The radiant flux, however, is also subject to attenuation due to scattering, and in a manner similar to the development of equations (1.33) to (1.38), the *scattering coefficient* $b(\lambda)$, defined as the fraction of radiant energy scattered from a beam per unit distance as it traverses an infinitesimal distance ∂r, is mathematically expressed as:

$$b(\lambda) = -\frac{[\partial \Phi(r, \lambda)]_{scatt}}{\Phi(r, \lambda)\partial r} \, m^{-1} \tag{1.39}$$

where the subscript *scatt* denotes that the diminution of radiant energy as it traverses an infinitesimal distance dr is due entirely to scattering processes. Equation (1.39) describes the attenuation of a radiant flux of wavelength λ in the absence of absorption processes in the same manner as equation (1.38) describes the attenuation of a radiant flux of wavelength λ in the absence of scattering processes. In a natural medium such as air and water where both absorption and scattering processes are responsible for attenuation, the *beam attenuation coefficient,* $c(\lambda)$, is defined as the fraction of radiant energy removed from an incident beam per unit distance as it traverses an infinitesimal distance ∂r due to the combined processes of absorption and scattering. The beam attenuation coefficient is then *mathematically defined* as the sum of the absorption coefficient and the scattering coefficient.

$$c(\lambda) \equiv a(\lambda) + b(\lambda) \, m^{-1} \tag{1.40}$$

or, equivalently,

$$c(\lambda) \equiv -\left[\frac{(\partial \Phi(r, \lambda))_{abs} + (\partial \Phi(r, \lambda))_{scatt}}{\Phi(r,\lambda)\partial r}\right] \tag{1.41}$$

The units of $a(\lambda)$, $b(\lambda)$, and $c(\lambda)$ are all m^{-1}. The terms *absorption coefficient, scattering coefficient,* and *beam attenuation coefficient* have, over the years, acquired the common usage terms *absorption coefficient, scattering coefficient,* and *total attenuation coefficient,* respectively. Consequently, throughout this book, these common usage terms will be adopted whenever $a(\lambda)$, $b(\lambda)$, and $c(\lambda)$ are discussed.

In its most simplistic concept, *radiative transfer* is the dynamics of the energy changes associated with the propagation of radiation through media which absorb, scatter, and/or emit photons. It would, therefore, be most advantageous if radiative transfer could be considered in terms of appropriate manipulations of the absorption, scattering, and total attenuation coefficients pertinent to the media [that is, the optical properties, $a(\lambda)$, $b(\lambda)$, and $c(\lambda)$]. In order to do so, such optical properties should be independent of the manner in which they are being measured as well as independent of the manner in which the medium under observation is being illuminated. Such optical properties, therefore, should be *inherent* optical properties of the medium, that is, independent of the radiation distribution within that medium. All three of the optical properties $a(\lambda)$, $b(\lambda)$, and $c(\lambda)$ qualify as inherent optical properties of the medium. Under *in situ* conditions, the total attenuation coefficient $c(\lambda)$ for natural waters is generally directly measured by means of submersible transmissometers. The scattering coefficient $b(\lambda)$ may be inferred through indirect measurements [e.g., as discussed below, from measurements of the volume scattering function $\beta(\theta)$]. The absorption coefficient $a(\lambda)$ is then obtained as the difference between $c(\lambda)$ and $b(\lambda)$.

Chapter 3 of this book will expand upon the inherent optical properties of water bodies. Suffice at this point to reiterate that *inherent optical properties* of a water mass are those that are totally independent of the spatial distribution of the impinging radiation. The so-called *apparent optical properties* of a water mass are those that *are* dependent upon the spatial distribution of the impinging radiation. Remote sensing, by the very nature of its data gathering devices, its varying viewing positions above the targeted water body, and its varying atmospheric conditions is not granted the luxury of directly recording the inherent properties of the aquatic medium. The interpretation of data remotely sensed above an aquatic body must, therefore, rely heavily upon the relationships that link the apparent optical properties of that aquatic body with its inherent optical properties.

The inherent optical properties $a(\lambda)$, $b(\lambda)$, and $c(\lambda)$ as defined above refer to the removal of energy from a light beam as it propagates through the aquatic medium. While absorption may be considered as a rectilinear loss of radiant energy, scattering possesses an additional associated directional dependence. The angular distribution of the scattered flux resulting from such interactions is specified in terms of a *volume scattering function,* $\beta(\theta,\phi)$ where θ is the polar angle of scattering and ϕ is the azimuthal angle of scattering. Figure 1.4 schematically illustrates an irradiance E_{inc} incident upon an infinitesimal volume dV within an attenuating medium. The scat-

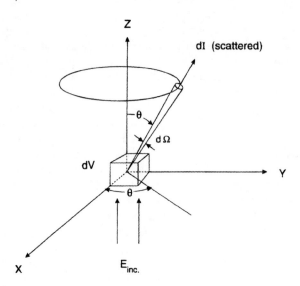

FIGURE 1.4
Illustration of volume scattering and the volume scattering function $\beta(\theta,\phi)$.

tered radiant intensity, dI, resulting from this volumetric encounter is shown as being contained within the cone defined by the solid angle $d\Omega$ at a location (reckoned from the origin of the scattering interaction) defined by the angular coordinates θ and ϕ. The *volume scattering function*, $\beta(\theta,\phi)$, is then defined as the scattered radiant intensity dI in a direction (θ,ϕ) per unit scattering volume dV normalized to the value of the incident irradiance E_{inc}:

$$\beta(\theta, \phi) = \frac{dI(\theta, \phi)}{E_{inc}dV} \tag{1.42}$$

Since $dI(\theta,\phi)$ is measured in W sr^{-1} and E_{inc} is measured in W m^{-2}, the units of $\beta(\theta,\phi)$ become m^{-1} sr^{-1}.

It is routinely observed for single particle scattering that the scattered intensity $dI(\theta,\phi)$ possesses symmetry about the z-axis. Thus, for a constant value of polar angle θ, $dI(\theta,\phi)$, is invariant with respect to ϕ. Equation (1.42) may be then simply written as:

$$\beta(\theta) = \frac{dI(\theta)}{E_{inc}dV} \text{ m}^{-1} \text{ sr}^{-1} \tag{1.43}$$

Equation (1.43) represents the characteristic shape of the distribution of the scattered flux resulting from the scattering processes occurring within an infinitesimal volume of an attenuating medium. Each attenuating medium is, therefore, characterized by its own particular $\beta(\theta)$ function, this function

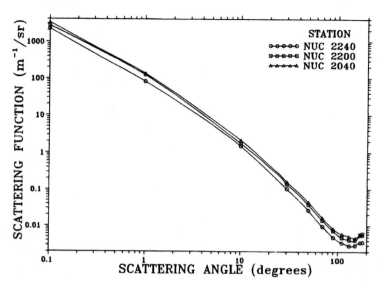

FIGURE 1.5
Volume scattering function for selected stations in San Diego Harbor. (From Petzold, T. J., *Volume Scattering Functions for Selected Ocean Waters*, Scripps Institute of Oceanography, LaJolla, CA, 1972. With permission.)

being a direct consequence of the nature and type of the particular scattering centers comprising the medium through which the irradiance is propagating. Since naturally occurring water bodies display spatial and temporal variations in their organic and inorganic compositions, it is to be expected that their associated $\beta(\theta)$ functions will reflect these spatial and temporal variations. Obtaining directly measured $\beta(\theta)$ functions for natural waters is not a simple task, a consequence of both the instrumental requirement of sophisticated small angle and large angle scattering meter configurations. Among others, small angle scattering meters developed by Bauer and Morel,[20] Kullenberg,[240] and Petzold[308] and high angle scattering meters developed by Tyler and Richardson,[402] Jerlov,[194] and Petzold[308] have contributed to the obtaining of volume scattering functions $\beta(\theta)$ for selected natural waters under a limited number of environmental conditions. For example, Figure 1.5, taken from Figure 13 of Petzold,[308] illustrates the $\beta(\theta)$ function obtained by Scripps Institute of Oceanography for stations in San Diego Harbor.

Since, from equation (1.9), the radiant intensity I in a particular direction is defined as the rate of change of radiant flux Φ with respect to solid angle $d\Omega$ (that is, $I = d\Phi/d\Omega$), and, from equation (1.13), the irradiance E is defined as the radiant flux per unit area (that is, $E = d\Phi/dS$), the volume scattering coefficient may be expressed as:

$$\beta(\theta) = \frac{d^2\Phi}{d\Phi_{inc} dr d\Omega} \tag{1.44}$$

where the infinitesimal volume dV has been replaced by $dSdr$. The infinitesimal solid angle, $d\Omega$, into which the photons are scattered is equal to $\sin\theta d\theta d\phi$.

If $d\Phi_{scatt}$ is the radiant flux scattered into $d\Omega$, then the fraction of the incident radiant energy flux Φ_{inc} scattered into this solid angle (per unit pathlength dr within the attenuating medium) would be expressed as:

$$\frac{d\Phi_{scatt}}{\Phi_{inc}} = \beta(\theta)\sin\theta d\theta d\phi \tag{1.45}$$

To obtain the fraction of the incident radiant energy flux Φ_{inc} scattered into all 4π directions (per unit pathlength dr within the attenuating medium) integration of equation (1.45) must be performed over the angular ranges $(0 \leq \theta \leq \pi)$ and $(0 \leq \phi \leq 2\pi)$. Thus, for a solid angle of 4π:

$$\frac{\Phi_{Scatt}}{\Phi_{inc}} = \int_0^{2\pi} \int_0^{\pi} \beta(\theta)\sin\theta d\theta d\phi$$

$$= 2\pi \int_0^{\pi} \beta(\theta)\sin\theta d\theta \tag{1.46}$$

Recall, however, from equation (1.39) that the scattering coefficient $b(\lambda)$ is defined as the fraction of radiant energy scattered from an incident flux as it traverses an infinitesimal distance within an attenuating medium. This is the same parameter that results from the integration of $\beta(\theta)$ shown in equation (1.46). Thus, the inherent optical property $b(\lambda)$ is the integral of the volume scattering coefficient $\beta(\theta)$ over all directions.

$$b = 2\pi \int_0^{\pi} \beta(\theta)\sin\theta d\theta \tag{1.47}$$

where the wavelength λ dependency of the parameters b and β, while not explicitly written, is nonetheless understood. For isotropic scattering, β is a constant independent of θ, and equation (1.47) reduces to:

$$b = 2\pi\beta \int_0^{\pi} \sin\theta d\theta = 4\pi\beta \tag{1.48}$$

The *scattering phase function*, $P(\theta,\phi)$, or more simply the *scattering phase func-*

tion, $P(\theta)$, is a form of the volume scattering function, $\beta(\theta)$, normalized to the scattering coefficient, b, and is given by:

$$P(\theta, \phi) = \frac{4\pi\beta(\theta, \phi)}{b} \tag{1.49}$$

where the θ and ϕ dependencies of P and β have been included for completeness. The phase function P for isotropic scattering has a value of unity.

In the study of photon propagation through aquatic media, it is often essential to distinguish among total scattering [as considered by equation (1.47)], directional scattering into the hemisphere ahead of the incident flux (*forwardscattering*), and directional scattering into the hemisphere trailing the incident flux (*backscattering*). The scattering coefficient (also oftimes referred to as the *total scattering coefficient*), b, may be compartmentalized into a *forwardscattering coefficient*, b_F, and a *backscattering coefficient*, b_B:

$$b = b_F + b_B \tag{1.50}$$

where, once again, the wavelength λ dependency is implicit.

Similar to equation (1.47), we may write:

$$b_F = 2\pi \int_0^{\pi/2} \beta(\theta)\sin\theta d\theta \tag{1.51}$$

and

$$b_B = 2\pi \int_{\pi/2}^{\pi} \beta(\theta)\sin\theta d\theta \tag{1.52}$$

The *forwardscattering coefficient*, b_F, is, therefore, the integral of the volume scattering function, $\beta(\theta)$, over the hemisphere ahead of the incident flux, which is defined by the angular ranges ($0 \leq \theta \leq \pi/2$) and ($0 \leq \phi \leq 2\pi$). The *backscattering coefficient*, b_B, is the integral of the volume scattering function, $\beta(\theta)$, over the hemisphere trailing the incident flux which is defined by the angular ranges ($\pi/2 \leq \theta \leq \pi$) and ($0 \leq \phi \leq 2\pi$).

Scattering and absorption phenomena that occur in natural water bodies, as in all attenuating media, are probabilistic in nature (a reality which, as we shall discuss in more detail in Chapter 3, enables the use of Monte Carlo techniques to be most effective when describing the propagation of photons through the media). Consequently, photons scattered into forward or backward directions are spoken of in terms of probabilities. Such probabilities may be simply described as the fractions of the total scattered flux that are directed into each hemisphere. Thus:

The *forwardscattering probability*, F, is defined as the ratio of the scattering into the forward hemisphere (the one ahead of the incident flux) to the total scattering occurring in all directions.

$$F = \frac{b_F}{b} \qquad (1.53)$$

The *backscattering probability*, B, is defined as the ratio of the scattering into the backward hemisphere (the one trailing the incident flux) to the total scattering occurring in all directions.

$$B = \frac{b_B}{b} \qquad (1.54)$$

Rewriting equations (1.53) and (1.54) results in convenient expressions for the forwardscattering coefficient, b_F, and the backscattering coefficient, b_B, respectively:

$$b_F = Fb \qquad (1.55)$$

and

$$b_B = Bb \qquad (1.56)$$

Clearly, all of the scattered photon flux is contained within the forward and backward hemispheres (that is, $F + B = 1$). Thus:

$$b_F + b_B = Fb + Bb$$
$$= b(F + B)$$
$$= b \qquad (1.57)$$

consistent with equation (1.50).

Briefly recapping, light propagating in a natural water body will undergo attenuation. This attenuation can be treated as an additive consequence of the absorption and scattering processes that occur among the photons and the organic and inorganic materials present in the natural water body (including the "pure" water molecules as well). This attenuation is described in terms of total attenuation, total absorption, and total scattering coefficients which are themselves inter-related ($a + b = c$), as well as the volume scattering function, directional scattering coefficients, and scattering probabilities that describe the directionality associated with the scattering encounters. This attenuation forms the basis for the interpretation of remote measurements in the visible region of the electromagnetic spectrum, and will be referred to extensively in the text to follow, as will be the concept of volume reflectance mentioned earlier. Before leaving this introduction to attenuation, however, four additional important optical parameters will be discussed, namely, the *irradiance attenuation coefficient*, the *optical depth*, the *attenuation length*, and the *scattering albedo*.

The *irradiance attenuation coefficient*, $K(\lambda,z)$, is defined as the logarithmic depth derivative of the spectral irradiance at subsurface depth z.

$$K(\lambda, z) = -\frac{1}{E(\lambda, z)} \left[\frac{\partial E(\lambda, z)}{\partial z} \right] m^{-1} \tag{1.58}$$

The definition of $K(\lambda,z)$ in equation (1.58) results from Beer's Law [equations (1.36), (1.37), and (1.41)] being used to describe the attenuation of spectral irradiance with depth z. That is:

$$E(\lambda, z) = E(\lambda, 0^-)\exp[-K(\lambda, z)z] \tag{1.59}$$

where $E(\lambda,z)$ and $E(\lambda,0^-)$ are the values of the irradiance at depth z and just below the air-water interface, respectively.

Since irradiance, as discussed earlier, is conveniently divided into its downwelling and upwelling components, two values of K, namely K_d and K_u may be used to refer to the attenuation of the downwelling, E_d and upwelling E_u irradiances at depth z, respectively:

$$K_d(\lambda, z) = -\frac{1}{E_d(\lambda, z)} \left[\frac{\partial E_d(\lambda, z)}{\partial z} \right] m^{-1} \tag{1.60}$$

and

$$K_u(\lambda, z) = -\frac{1}{E_u(\lambda, z)} \left[\frac{\partial E_u(\lambda, z)}{\partial z} \right] m^{-1} \tag{1.61}$$

The scalar irradiance $E_o(\lambda,z)$ also has its associated $K_o(\lambda,z)$ which is expressed as:

$$K_o(\lambda, z) = -\frac{1}{E_o(\lambda, z)} \left[\frac{\partial E_o(\lambda, z)}{\partial z} \right]. \tag{1.62}$$

In equations (1.60), (1.61), and (1.62), $K_d(\lambda,z)$, $K_u(\lambda,z)$, and $K_o(\lambda,z)$ are the values of the *irradiance attenuation coefficient*, for the downwelling, upwelling, and scalar irradiances, respectively. The logarithmic depth dependencies of

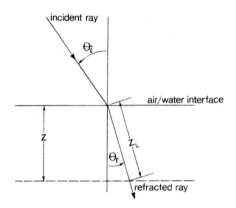

incident ray

θ_i

air/water interface

z

z_L

θ_r

refracted ray

FIGURE 1.6
Ray diagram illustrating Snell's Law: An incident ray of angle θ_i (measured from the normal to the air-water interface) is refracted to an angle θ_r (measured from the normal to the air-water interface).

the irradiance as defined here are reckoned along the downward z direction. Thus, equations (1.60), (1.61), and (1.62) consider that the downwelling flux incident upon the air-water interface is from the zenith direction (that is, $\theta = 0°$). Vertical incidence, of course, is not universally encountered, and an angular dependence must be incorporated into the use of irradiance attenuation coefficients to accommodate the non-vertical fluxes impinging upon the surface of natural water bodies. Since the incident radiation of principal interest in this book is the downwelling solar and sky radiation, and since the radiant flux directionality associated with these radiations are, among other determinants, a function of solar position, the θ dependencies of the irradiance attenuation coefficients can be expressed as time dependencies. The situation is schematically illustrated in Figure 1.6. The path of an incident beam (solar zenith angle θ_i and refracted angle θ_r) entering the water is shown in ray form. The path length along the principal direction of subsurface propagation to a particular depth level z is indicated as z_L. z_L is equal to z only during that time of day at which the sun is vertically overhead (that is, the time for which $\theta = 0°$). For all other times of the day, Beer's Law is appropriately written:

$$E(\lambda, z) = E(\lambda, 0)\exp[-K(\lambda)z_L(\theta_r)] \qquad (1.63)$$

where the in-water refracted angle θ_r is given (from *Snell's Law*) as:

$$\theta_r = \arcsin\left\{\sin\theta_i\left[\frac{\text{index of refraction of air}}{\text{index of refraction of water}}\right]\right\}$$

$$= \arcsin\,(0.75\,\sin\theta_i) \qquad (1.64)$$

Equation (1.64) indicates that the actual value of the irradiance attenuation coefficient $K(\lambda,z,\theta_i)$ varies with the time of day (that is, with the value of the solar zenith angle θ). The use of a temporally invariant value of K (the irradiance attenuation coefficient) throughout the day is inappropriate when evaluating diurnal variations in subsurface attenuation. Such time-dependent attenuation coefficients can be significant parameters when determining subsurface irradiation and primary production. We shall return to these considerations in Chapter 3.

A comparison of equations (1.58) and (1.41) shows that the irradiance attenuation coefficient K is to the downwelling irradiance E what the total attenuation coefficient c is to the downwelling radiant flux Φ. Both c and K represent total subsurface attenuation and consequently define the removal of beam energy due to the combined processes of absorption and scattering. However, c is not constrained to a pre-selected direction of downwelling flux, nor, in fact, is the property c dependent upon the presence of photons in the optical medium. Consequently, c is an *inherent optical property* of the aquatic medium, while K, being dependent upon the directionality of the radiance distribution comprising the downwelling irradiance, is an *apparent optical property* of the aquatic medium. Baker and Smith[14] have referred to K as a "quasi-inherent" optical property. However, K is generally considered to be an apparent optical property and will be so considered here.

In a similar manner, a *radiance attenuation coefficient*, $K(\lambda,z,\theta,\phi)$ and a *scalar irradiance attenuation coefficient*, $K_o(\lambda,z)$ may also be defined from expressions of Beer's Law. The angular dependencies of $K(\lambda,z,\theta,\phi)$ logically render the radiance attenuation coefficient an apparent optical property, while the scalar irradiance attenuation coefficient qualifies as an inherent optical property.

The downwelling irradiance, $E_d(\lambda,z)$, in a natural water mass has been considered to attenuate exponentially with depth z.

$$E_d(\lambda, z) = E_d(\lambda, 0)\exp[-K_d(\lambda)z] \tag{1.65}$$

where $K_d(\lambda)$ is the average value of the downwelling irradiance attenuation coefficient over the aquatic depth interval 0 to z. The downwelling irradiance exponentially approaches zero, and the depth at which this limit is effectively reached will, of course, depend upon the degree of clarity possessed by the water body under observation.

The integration of the downwelling irradiance attenuation coefficient $K_d(\lambda,z)$ over the subsurface depth z is defined as the *optical depth*, $\zeta(\lambda,z)$:

$$\zeta(\lambda, z) = \int_0^z K_d(\lambda, z)z \tag{1.66}$$

If K_d is the average value of the irradiance attenuation coefficient over the depth interval 0 to z, then the optical depth ζ is given as the product:

$$\zeta(\lambda, z) = K_d z \qquad (1.67)$$

A variety of optical depths can, of course, be associated with the same physical aquatic depth z. Very clear waters (waters with minimal concentrations of organic and inorganic materials in addition to the scattering and absorption centers of the water molecules themselves) are characterized by low values of K_d. Very turbid waters (waters with substantial concentrations of organic and/or inorganic materials) are characterized by higher values of K_d. Thus, for a given physical aquatic depth z, the optical depth ζ of turbid waters will be numerically greater than the optical depth ζ of clear waters.

Optical depths are a convenient means of referring to the depths of irradiance levels in a natural water body by defining the attenuation experienced by the downwelling irradiance at that level. This attenuation is defined in terms of the percentage of original subsurface irradiance $E(\lambda,0^-)$ remaining at the optical depth. For example, the 10% subsurface irradiance level refers to the depth at which the ratio $E(\lambda,z)/E(\lambda,0^-) = 0.1$. From equation (1.65), this is equivalent to:

$$0.1 = \exp(-K_d z) \qquad \text{from which}$$

$$2.303 = K_d z$$

Thus, the 10% subsurface irradiance level corresponds to an optical depth ζ of 2.303. Similarly, the 1% subsurface irradiance level corresponds to a ζ of 4.605, the 50% subsurface irradiance level to a ζ of 0.693, and the 100% subsurface irradiance level to a ζ of 0. As will be discussed later, the subsurface layer bounded by the 100% and 1% irradiance levels (4.605 optical depths) is referred to as the *photic zone* or *euphotic zone* and represents the region in which most of the aquatic photosynthesis occurs.

The total radiation in the 400 nm to 700 nm wavelength interval is defined as the *photosynthetic available Radiation*, (PAR) and may be expressed in units of energy or quanta (i.e., W m^{-2}, or μeinsteins m^{-2} sec^{-1}). Because of the importance of PAR to the overall considerations of aquatic photosynthesis and primary productivity, great emphasis is placed upon optical measurements in this wavelength interval, and the downwelling irradiance attenuation coefficient for PAR will be herein designated as K_{PAR}.

Quite distinct from the optical depth $\zeta(\lambda,z)$ is the parameter *attenuation length*, $\tau(\lambda,z)$. The attenuation length τ is the path distance in the attenuating medium that is required to reduce the radiant energy of a light beam by a factor of $1/e$ (i.e., to 0.367879 of its value). Thus, $\tau(\lambda,z)$ is defined as the inverse of the total attenuation coefficient $c(\lambda,z)$:

$$\tau(\lambda, z) = 1/c(\lambda, z) \text{ m} \qquad (1.68)$$

As was the case for the optical depths ζ, a variety of attenuation lengths τ

can be associated with a given physical aquatic depth z depending upon the degree of clarity or turbidity characterizing the water body. Very clear natural waters are characterized by numerically low values of c. Very turbid natural water bodies are characterized by numerically high values of c. Thus, mid-oceanic waters could be characterized by attenuation lengths of 20 m or more, while inland and coastal waters could be characterized by attenuation lengths of a few meters to a few cm.

The *scattering albedo*, ω_0, is defined as the number of scattering interactions that occur within a fixed volume of the aquatic medium expressed as a fraction of the total number of optical interactions (both scattering and absorption) that occur within that fixed volume. Since $a + b = c$, the mathematical definition of ω_0 is readily seen to be:

$$\omega_0 = \frac{b}{c} \tag{1.69}$$

Notice that the scattering albedo, being a ratio, provides information *solely* on the *relative* amounts of scattering and absorption occurring within a natural water mass. High numerical values of ω_0 indicate that the scattering phenomena substantially outnumber the absorption phenomena in a water column, just as low numerical values of ω_0 indicate that the absorption phenomena substantially outnumber the scattering phenomena. The scattering albedo provides no information whatsoever regarding the *absolute* numbers of scattering and absorption interactions that are occurring and, consequently, provides no information regarding the degree of clarity characterizing the water column. High numerical values of ω_0, as well as low numerical values of ω_0, could refer to the entire range of water types from very clear to very turbid.

The scattering albedo, ω_0, will be used throughout this text as a most convenient method of classifying inland and coastal water masses.

1.5 SPECTRAL ALBEDO

The term *albedo, $A(\lambda)$*, which is extensively used in remote sensing terminology, is formally defined as the ratio of energy returning from a point or a surface to the energy incident upon that point or surface. The circumstances dictating the albedo associated with the air-water interface will be discussed in Section 2.6 and Section 3.2. However, if $E_{au}(\lambda,0^+)$ is the upwelling irradiance in air just above the air-water interface, and $E_{ad}(\lambda,0^+)$ is the downwelling irradiance in air just above the air-water interface, then the albedo $A(\lambda)$ of the water body would be defined as:

$$A(\lambda) = \frac{E_{au}(\lambda, 0^+)}{E_{ad}(\lambda, 0^+)} \qquad (1.70)$$

As will be discussed in Sections 2.6 and 3.2, $A(\lambda)$ of the water body is a function of not only the energy reflected from the air-water interface, but also from the upwelling subsurface energy that exits through the air-water interface.

Spectral albedo, therefore, refers to the radiation return from a target normalized to the radiation impinging upon that target. The visible wavelength return from an optically opaque target comprises reflected light, while the visible wavelength return from an optically penetrable target (such as clouds, water, and snow) comprises surface reflection plus volume reflectance. Technically, therefore, every data bit recorded by an environmental satellite or aircraft, once normalized to the downwelling radiation at the target, is albedo, whether that albedo is associated with land, air, or water.

1.6   ENVIRONMENTAL REMOTE SENSING AND SPACE SCIENCE

Remote sensing refers to the study of objects or scenes at remote distances from those objects or scenes. In its broadest terms it includes the collection of data or the execution of experimental activities in which the detector system is not directly linked to the target. The indirect linkages are invariably the electromagnetic radiations emanating from the target in the form of both self-emissive and reflected radiations (there are also remote sensing techniques that utilize force fields as the indirect linkages between target and sensor, but these techniques will not be discussed here). Thus, monitoring and/or research activities involving the use of remote sensing techniques would comprise the collection of data by means of all contactless detectors responsive to any range of electromagnetic emanations from the very long wavelength radio waves to the very short wavelength cosmic radiation (that is, radio waves, radar, microwaves, infrared, visible, ultraviolet, x-rays, γ-rays, and high energy cosmic rays). The detector systems employed in the remote sensing of targets vary from photographic imaging sensors such as aerial cameras to dipole sensors such as microwave radiometers. The contactless sensors may be separated from the target by distances which vary from a fraction of a centimeter (as in the case of laboratory bench experiments) to several astronomical units (as for the telescopic observations of the interplanetary medium). The sensors may be mounted on platforms located on the surface of the earth, beneath the surface of the earth or its waters, or above the surface of the earth in aircraft, balloons, rockets, earth-orbiting satellites, or artificial planets that orbit the sun.

According to the above definition, remote sensing has been practiced since at least the times of Galileo, Copernicus, Kepler, and other pioneers of astronomy. Modern environmental remote sensing, however, is a relatively young discipline, probably having its roots in the military reconnaissance missions of World War I and receiving a major impetus from the space program of the 1950s and 1960s. Environmental remote sensing has unquestionably benefitted from the technological advances that emerged from the space program, particularly in the development of sophisticated sensor packages and high-speed computer systems to record, store, and rapidly disseminate the copious streams of digital data generated by these sensor packages. However, it is informative to briefly focus on some of the differences between the scenario of the space program and the scenario of environmental remote sensing.

Environmental remote sensing is designed to study and/or monitor the Earth's biospheric ecosystem. The unmanned component of the space program was designed to study and/or monitor the geophysical, the electromagnetic, and the solar and intergalactic cosmic ray particle propagation processes impacting and modulating the earth and the solar system. These activities can be more appropriately considered to comprise *space science.* The satellite payloads of these earth-orbiting and deep-space probe missions generally included sensors which could directly measure magnetic field strengths and directions; unambiguously identify atomic and nucleonic particle types, their flux densities and directions of propagation; directly measure plasma densities and velocities; and obtain a variety of inner solar cavity physical parameters that could be fused into models that would successfully describe the *physics* of the interplanetary medium and the interactions of solar emissions with planetary atmospheres. Thus, while measurements of the physical parameters were indeed measured *remotely,* they were performed *in situ,* and *not* in a *contactless* manner. Further, measurements of *physical* parameters were used to interpret the *physical* properties of the terrestrial and interplanetary media. With the exception of direct measurements of some physical and chemical properties of the terrestrial atmosphere performed from aircraft, balloons, rockets, and satellites (which are comparable to the direct measurements that formed the major thrust of the unmanned space exploration programs of the 1950s and 1960s), the environmental remote sensing missions are designed to acquire information on the *biological* status of the various components of the earth's ecosystem. No instruments currently exist that can directly measure biological parameters from a remote platform. As with interplanetary payloads, the terrestrial remote sensors are designed to measure physical parameters. This requirement to use *physics* (*specifically optics*) to infer *biology* generates the basic problem that remote sensing scientists and resource managers must overcome. This is the problem that we will address in the upcoming chapters.

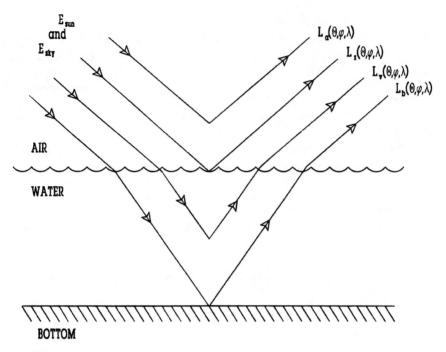

FIGURE 1.7
Components of the radiance recorded by a remote sensor over a natural water body.

1.7   REMOTE MEASUREMENTS ABOVE WATER

Figure 1.7 illustrates the familiar configuration of solar and terrestrial conditions confronting the worker concerned with collecting upwelling radiance from an aquatic medium. The figure is restricted to reflected solar and sky radiation, and schematically depicts (in ray form) downwelling light which interacts with the atmospheric and aquatic media. No consideration is given in Figure 1.7 to self-emissive radiation originating either in the atmosphere or in the natural water body. These downwelling solar and sky irradiances, labelled E_{sun} and E_{sky}, respectively, will be discussed at greater length in Chapter 2.

Optical sensors may be constructed to respond to either radiance or irradiance. The optical sensing devices generally mounted on remote platforms are responsive to photons arriving at the platform within a field-of-view (FOV) defined by the geometric configuration of the optical device. The photon fluxes arriving at the sensor are thus generally recorded as radiances.

The radiance $L(\theta,\phi,\lambda)$ recorded at the remote sensing platform (be it aircraft or satellite) can comprise as many as four components:

$L_a(\theta,\phi,\lambda)$:—that portion of the recorded radiance resulting from the downwelling solar and sky radiation that never reaches the air-water interface. It therefore represents a return from the atmosphere.

$L_s(\theta,\phi,\lambda)$:—that portion of the recorded radiance resulting from the downwelling solar and sky radiation that reaches but does not penetrate the air-water interface. It therefore represents a reflection from the aquatic surface.

$L_v(\theta,\phi,\lambda)$:—that portion of the recorded radiance resulting from the downwelling solar and sky radiation that penetrates the air-water interface and re-emerges from the water column without encountering the bottom of the natural water body. It therefore represents a return from the volume of the water column.

$L_b(\theta,\phi,\lambda)$:—that portion of the recorded radiance resulting from the downwelling solar and sky radiation that penetrates the air-water interface, reaches the bottom of the natural water body, and re-emerges from the water column. It therefore represents a return from the bed of the natural water body.

The above demarcation of the recorded radiance $L(\theta,\phi,\lambda)$ into an atmospheric, a surface, a subsurface volumetric, and a bottom return is, of course, somewhat artificial since the four radiances are not necessarily as distinct as the demarcation would imply. To reach the bottom of the water body the downwelling photon flux must first interact with the atmosphere, then the air-water interface, and then the water column. To be recorded at the remote sensing platform the photons must then retrace this path. Nevertheless, this demarcation is very often conceptually quite useful. Thus, $L(\theta,\phi,\lambda)$ may be represented as:

$$L(\theta,\,\phi,\,\lambda) = L_a(\theta,\,\phi,\,\lambda) + L_s(\theta,\,\phi,\,\lambda) + L_v(\theta,\,\phi,\,\lambda) + L_b(\theta,\,\phi,\,\lambda) \quad (1.71)$$

The component of equation (1.71) which is of principal interest to the aquatic research scientist and/or resource manager is the term $L_v(\theta,\phi,\lambda)$. Since it is a direct consequence of the aquatic absorption and scattering processes impacting the photon flux in its subsurface propagation, $L_v(\theta,\phi,\lambda)$ contains the most direct information on the water column. Extracting the volumetric component of $L(\theta,\phi,\lambda)$ and relating $L_v(\theta,\phi,\lambda)$ to the organic and inorganic absorption and scattering centers comprising the water column becomes the problem to be solved if the upwelling radiance spectrum recorded at a remote platform is to provide an estimate of natural water quality. That is, the basic starting point becomes:

$$L_v = L - (L_a + L_s + L_b) \quad (1.72)$$

As represented in equation (1.72) the atmosphere, the air-water interface (the first subsurface millimeter or so which is often referred to as the *free surface layer* or *boundary layer*), and the water bed are considered as obfuscations to the aquatic return signal recorded at the remote platform. For the water researcher this is indeed the case, and as a consequence, the past two decades have seen substantial effort dedicated towards the removal of these obfuscations through the development of atmospheric correction algorithms, surface reflection correction algorithms, and, where necessary, bottom reflectance compensation methodologies.

The atmosphere, however, like the aquatic medium, is comprised of a spatially and temporally varying composite of scattering and absorption centers, the studies of which parallel the studies of their aquatic counterparts. Thus, a remote sensing device pointing downward over water is viewing one attenuating medium through another attenuating medium.

The free surface layer (boundary layer) is a major target in climatology, meteorology, hydrology, and oceanography. Projects attempting to utilize remote sensing to provide information on the nature and composition of lake and river beds are being initiated. In these and similar interests, therefore, $L_v(\theta,\phi,\lambda)$ would be considered an obfuscation, and the extraction of one or more of the other radiance components of $L(\theta,\phi,\lambda)$ would become the objective of the remote sensing exercise.

In the following chapters we will deal with aquatic remote sensing from the perspective of a water researcher. While acknowledging the importance of the atmospheric processes to terrestrial and extraterrestrial geophysical science, and the importance of the boundary layer to terrestrial meteorology, we will regard the contributions from the atmosphere, the boundary layer, and the inland water bed as obfuscations that must be reluctantly accounted for or eliminated. We will thus concentrate on the extraction of the volumetric radiance from the totality of upwelling radiation $L(\theta,\phi,\lambda)$.

It would certainly be ideal if the volume of natural water under scrutiny could be brought up to the flight line of the aircraft or the orbit of the environmental satellite. This would allow remote measurements to enjoy the same privileges as *in situ* measurements. The absence of such an ideal situation, however, encourages the adoption of an experimental philosophy that mentally brings the aircraft or satellite to the water. Such a *Gedanken* experiment will consider the approach to interpreting the radiance spectrum recorded at a remote platform in terms of the composition of natural water bodies as being a three-step procedure:

1. Consider the availability or the development of models that would predict the responses of the remote sensor if it were actually flying or orbiting just beneath the surface of the water.

2. Consider the transference of these models through the air-water interface to predict the responses of the remote sensor if it were actually flying or orbiting just above the surface of the water.

3. Consider the application of atmospheric correction models to predict the responses of the remote sensor actually flying or orbiting at its altitude.

Once such suitable methods are in place, the radiance spectrum recorded at the remote platform may be converted into the corresponding volume reflectance spectrum that would be representative of the natural water body being remotely monitored. This subsurface volume reflectance must then be converted into the combination of organic and inorganic scattering and absorption centers that generated it.

1.8   THE BOUNDARY CONDITIONS OF THE CONSIDERED PROBLEM

The remaining chapters of this book will focus upon the optical properties of inland and coastal waters. These optical properties will be directed towards three objectives: (a) they will serve to describe the interactions of solar and sky radiation with natural water masses, (b) they will serve to provide a means of intercomparing different natural water bodies, and (c) they will serve to provide a linkage between the remotely sensed optical data and the biophysical activity that dictated that remotely sensed data. This concentration on inland and coastal waters, however, does not necessarily preclude the developed algorithms and methods from being applied to oceanographic waters since the algorithms and models developed for remotely monitoring optically complex waters can, in most instances, accommodate the remote monitoring of less optically complex waters as a special case. The converse, however, is rarely true, as most algorithms and models concerned with the remote sensing of mid-oceanographic waters are not readily expanded to accommodate inland and coastal waters.

Throughout this chapter we have discussed the propagation of radiation through aquatic media in terms of wavelength-dependent fluxes of photons which could be applicable over a wide range of the electromagnetic spectrum. In this book we will restrict our focus to that region of the electromagnetic spectrum encompassing the *photosynthetic available radiation (PAR)* (400 nm to 700 nm). The PAR also approximately encompasses the so-called *visible* spectral region, which is simply referred to as *light* (a λ range of about 390 nm to 740 nm). If essential to the development of algorithms, techniques, and models (as, for example, in the removal of non-aquatic contributions to the remotely observed upwelling radiance spectrum above the water surface), this spectral range will be exceeded. However, we will concentrate primarily upon the extraction of subsurface volume reflectance from radiance spectra recorded by remote sensors responsive to the PAR spectral range.

The early series of environmental satellites (such as the Landsat series) contained suites of *passive* remote sensing devices. Passive sensors are devices that faithfully respond to the optical field that emanates from the environmental target being monitored, whether those emanations be self-generated (such as thermal or phosphorescent aurae) or reflected from some natural source of downwelling radiation (such as solar and sky irradiances). Later generation environmental satellites (such as the Seasat series) also contained *active* remote sensing devices. Active sensors are devices that transmit a signal to the target being monitored and then measure the signal that is returned from the target. Active sensors, therefore, interact with the target, oftimes stimulating an optical transition within the target that results in a spectral signature that would not be observable if the target were not in an excited state. Passive sensors merely respond to the existing environmental situation, they do not actively influence the behavior of the target. The most familiar passive remote sensing devices are imaging camera systems and multispectral radiometer systems (including thermal and microwave frequencies). Active remote sensing devices include scatterometers, altimeters, synthetic aperture radar, and lasers. In this book, we will restrict our focus to data acquired by passive remote sensing devices.

As discussed earlier, the passive optical return from a natural water body is a direct consequence of the absorption and scattering centers encountered by photons propagating through that water body. The density and diversity of these absorption and scattering centers generally increase with increasing proximity to natural land masses. Such increased density and diversity are encountered in most inland and coastal aquatic regimes, and it is these areas that will be considered herein.

Consequently, the boundary conditions of the problem to be tackled in this book can be simply expressed by the words and phrases *visible wavelengths, passive remote sensing, aquatic scattering and absorption centers, inland and coastal waters, optical properties, algorithm development,* and *bio-optical models.*

We do not intend to discuss such related topics as satellite launchings, orbital dynamics, characteristics of sensor payloads, data gathering and transmission techniques, or computer techniques of data manipulation. Many excellent handbooks and user-guide documentation are readily available from agencies such as NASA, NOAA, the Canada Centre for Remote Sensing, and other government (and some non-government) agencies in a variety of countries. These documents list in great detail past, present, and planned satellite missions (both national and international), and in-depth descriptions of payloads and sensor characteristics, along with the acquired results from past and anticipated products of planned environmental satellite missions. The reader is encouraged to take advantage of this documentation to obtain such information.

Chapter 2

INCIDENT RADIATION

2.1 THE SUN AND EXTRATERRESTRIAL SOLAR RADIATION

Radiant energy emitted by the sun (solar radiation) supplies virtually all the energy for natural processes on the surface of the Earth and its atmosphere. The energy arriving at the Earth's surface as a consequence of the moon, the so-called "fixed" stars, the aurora borealis (northern lights), lightning, nocturnal sky luminescence, as well as cosmic rays and meteors, constitutes only about 10^{-5} to 10^{-8} parts of the incident solar radiation. Heat transmitted to the Earth's surface from its heated internal core is ~5,000 times less than heat energy incoming from the sun.[261]

The sun is a relatively small, faint, cool star about which the Earth rotates yearly at an average distance (center of gravity to center of gravity) of 1.497 $\times$ 10^{11} m (1 astronomical unit). The mass of the sun is ~333,400 times that of the Earth or 1.989×10^{30} kg. The rotation period of the sun about its own axis of rotation is a function of solar latitude, varying from about 26 days at the solar equator to about 34 days at the solar poles. An effective solar rotation period of 27 days is usually adequate for assessing and predicting recurring solar occurrences (such as sunspots and magnetic plage regions) that may affect the Earth's atmosphere.

Essentially, the sun is a sphere of gas heated by nuclear reactions occurring within its central core. Although the sun has no distinct phase boundaries, it possesses dynamically varying boundaries that represent differences in temperature, ionization levels, and magnetic field strengths. The solar atmosphere comprises three layers (photosphere, chromosphere, corona) that are in a state of constant visible change, manifesting as the appearances of prominences, solar flares, magnetic storms, and other dynamic features whose statistical descriptions, however, remain predictable and recurrent in a quasi-unchanged manner over time (a steady-state dynamics referred to as the *quiescent sun*).

The *photosphere* is the apparent solar surface and also includes the lowest layer of the solar atmosphere. The outside diameter of the photosphere is taken to be the diameter of the sun, namely 1.3914×10^9 m. The photosphere contains most of the solar mass and is, effectively, an optical boundary below which the solar gas is opaque. The *chromosphere* is a layer (~10,000 km thick) of transparent, glowing gas above the photosphere, and the *corona* or *solar crown* is the faint white halo in the region above the chromosphere. It is this halo that becomes visible during a total solar eclipse when the glare of the chromosphere is temporarily arrested. The "true" corona (there is also a component of the corona which is a consequence of sunlight scattered by dust particles that are at great distances from the sun and are not part of the solar atmosphere) is an irregular halo that surrounds the sun to a mean distance of approximately 1 solar radius, as well as local streamers that extend several solar diameters into space.

Energy leaves the sun either as electromagnetic radiation or as corpuscular radiation in the form of high-speed protons and other charged particles. Although both types of emitted radiation display temporal variations, the percentage change in corpuscular radiation is considerably larger than the percentage change in electromagnetic radiation. In this book we shall restrict our attention to electromagnetic solar radiation.

Electromagnetic radiation emitted by the sun encompasses a wide spectral region which may be divided into several wavelength intervals as follows (1 μm = 1000 nm):

gamma-rays ($\lambda < 10^{-5}$ μm),
X-rays (10^{-5} μm $< \lambda < 10^{-2}$ μm),
ultraviolet radiation (0.01 μm $< \lambda < 0.39$ μm),
visible radiation (0.39 μm $< \lambda < 0.74$ μm),
infrared radiation (0.74 μm $< \lambda < 3{,}000$ μm), and
microwave radiation ($\lambda > 3{,}000$ μm).

The visible radiation is subdivided into seven color intervals:

violet (0.39 μm $< \lambda < 0.455$ μm),
dark blue (0.455 μm $< \lambda < 0.488$ μm),
light blue (0.488 μm $< \lambda < 0.505$ μm),
green (0.505 μm $< \lambda < 0.575$ μm),
yellow (0.575 μm $< \lambda < 0.585$ μm),
orange (0.585 μm $< \lambda < 0.620$ μm), and
red (0.620 μm $< \lambda < 0.740$ μm).

The near-ultraviolet radiation lies in the spectral region (0.28 μm $< \lambda < 0.39$ μm); the near-infrared radiation in the spectral region (0.74 μm $< \lambda < 2.4$ μm). Together, the ultraviolet, visible, and near-infrared spectral intervals form the so-called *optical range*.

Over 95% of electromagnetic solar radiation is included within the spectral interval 0.29 μm $< \lambda < 2.4$ μm, with a maximum emissivity located at

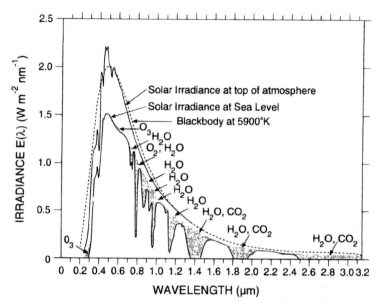

FIGURE 2.1
Extraterrestrial and sea-level solar irradiance spectra for an overhead sun. (Adapted from Valley, S. L., Ed., *Handbook of Geophysics and Space Environments,* Air Force Cambridge Research Laboratories, McGraw-Hill, New York, 1965). The spectral curve for a blackbody radiator at 5900°K is also sketched.

$\lambda = 0.4738$ μm [see Figure 2.1, adapted from Valley (1965), wherein are illustrated both extraterrestrial and sea-level solar irradiance spectra for an overhead sun]. Solar spectral radiation cannot be characterized by any single blackbody temperature distribution. At visible, near-infrared, and near-ultraviolet wavelengths, however, the solar emission is close to that of an absolute blackbody of temperature 5900°K (also illustrated in Figure 2.1). In the spectral region 0.30 μm < λ 0.40 μm, solar emission is less than that of the absolute blackbody (i.e., the sun emits less energy than the absolute blackbody). At lower values of λ (not shown in Figure 4.1), however, solar emission exceeds absolute blackbody emission, and at $\lambda \sim 0.1$ μm, the sun emits 2 to 3 times more energy than the absolute blackbody. At $\lambda < 0.05$ μm solar radiation departs dramatically from that of blackbody radiation. (This is due to a change in the mechanism of solar radiation generation. Below wavelengths of ~0.05 μm, photospheric radiation drops to a minimum, and the hot plasma of the upper chromosphere and lower corona becomes the prime source of solar radiation.)

There are relatively cold (temperatures ~4,600°K) regions in the photosphere called *sunspots.* The entire variety of these transient, non-stationary phenomena in the solar atmosphere is termed *solar activity* and is characterized by the *Wolf number* (also known as the *Zurich sunspot number* or the *mean sunspot number*), a number which is the sum of the total number of spots visible at a particular time plus ten times the number of sunspot groups. This solar activity affects the intensity of solar radiation at all wavelengths, the major impact, however, being to the X-ray and microwave emissions. However, since the contribution of these fluxes to the total solar radiation flux is small, sharp oscillations of solar activity do not significantly affect the magnitude of the total solar electromagnetic radiation flux. (This is not true of the corpuscular solar radiation wherein the large particulate bursts present in solar flares are directly related to solar activity and follow the 11-year sunspot cycle. Volumes of excellent work exist chronicling the formation and propagation of solar flare particles, the propagation of associated disruptions to the interplanetary magnetic fields, and the corresponding modulations to the galactic cosmic radiation impinging upon the Earth's atmosphere.)

The relative independence of the solar electromagnetic radiation with respect to the sunspot cycle enables introduction of the concept of a *solar constant*, E_{solar}. E_{solar} is defined as the total electromagnetic energy per unit time per unit area normal to the solar rays at one astronomical unit and outside the Earth's atmosphere. That is, the *extraterrestrial radiation* (the total electromagnetic irradiance impinging at the top of the Earth's atmosphere) can be considered constant. There is a seasonal variation in the extraterrestrial irradiance ($\sim$3.5%) resulting from the elliptical orbit of the Earth. Within this seasonal variation, however, the total extraterrestrial irradiance, E_{solar}, may be written:

$$E_{solar} = \sum_{\lambda = 0}^{\lambda = \infty} E(\lambda)d\lambda = \text{constant} \tag{2.1}$$

where $E(\lambda)$ is the irradiance per unit wavelength. The International Radiation Commission recommended a standard value of 1367 watts m^{-2} be adopted for E_{solar}.[395] Direct satellite observations, in agreement with model simulations, illustrate a variance of only 1 W m^{-2} over the 11-year solar activity cycle.

2.2   ATMOSPHERIC ATTENUATION AND GLOBAL RADIATION

Since atmospheric properties vary with height in a quasi-regular manner, it is convenient to describe the atmosphere in terms of a series of geocentric layers or shells, which, of course display both geographic and seasonal variabilities. Principal properties used to define these layers are temperature, pressure, and atmospheric composition. We will not reprise the excellent detailed documentation that exists on atmospheric physics and chemistry. We will only outline the atmospheric layers to assist in visualizing the optical impacts of the atmospheric components. The *troposphere* extends from the surface to a height of about 8 km to 10 km in the Arctic in winter to about 16 km to 18 km in equatorial regions. The upper limit of the troposphere is termed the *tropopause* and represents the height at which temperature decrease with height abruptly stops. Above the tropopause temperatures usually increase with height, very slowly at first remaining essentially isothermal to heights of 20 km to 25 km, then increasing quite substantially to a region of maximum temperature (just below 0°C) called the *stratopause* at an average height of about 50 km. The atmospheric shell between the tropopause and the stratopause is termed the *stratosphere* (and thus is the shell constrained by heights of about 15 km and 50 km). It is in the stratosphere that the maximum concentration of atmospheric ozone resides. Above the stratosphere is the *mesosphere* wherein the temperature again decreases with increasing height until a region of minimum atmospheric temperature (around −90°C) termed the *mesopause* is reached at heights around 90 km. Above the mesopause is an ill-defined layer termed the *thermosphere* extending perhaps as high as 150 km to 200 km, above which is the *exosphere* in which the temperature is totally under solar control.

Upon entry into the Earth's atmosphere, the constant extraterrestrial solar radiation undergoes absorption and scattering interactions with atmospheric components. The solar radiation is absorbed and scattered by air molecules and *aerosols* (a totality of suspended liquid and particulate matter). As a consequence of this scattering and absorption, two radiative components of differing natures reach the surface of the Earth: direct solar radiation, E_{sun}, attenuated by atmospheric absorption and scattering but still centred on the direction of the sun, and diffuse solar radiation (generally referred to as *skylight*) scattered by aerosols and air molecules, E_{sky}. Together, the direct solar radiation and the diffuse sky radiation incident upon the Earth's surface is termed the *global radiation, $E(\lambda)$*. That is, $E(\lambda) = E_{sun}(\lambda) + E_{sky}(\lambda)$.

A high-altitude aircraft or Earth-orbiting satellite viewing a natural water body is essentially viewing one optical medium through another optical medium. Understandably, therefore, the physics of light propagation through the atmosphere parallels the physics of light propagation through aquatic media in the manner we have been discussing in Chapter 1.

The *optical air mass, m,* of the atmosphere is defined as the ratio of the mass of an air column encountered by solar radiation emanating from the sun located at a zenith angle, θ, to the mass of an air column encountered by solar radiation emanating from the sun located directly overhead (zenith angle of 0°). The optical air mass, m, clearly increases with increasing solar zenith angle, and the closer the approach to an overhead sun, the more close becomes the fit to a secθ curve.

That is, for solar zenith angles $\theta < 60°$, $m \approx \sec\theta$.

For $\theta > 60°$, however, atmospheric refraction of solar rays becomes a consideration, and m may be approximated[343] by

$$m \approx 35/[1224 \cos^2\theta + 1]^{1/2}.$$

Just as the beam attenuation coefficient, $c(\lambda)$, describes the loss of radiant energy due to scattering and absorption in an aquatic medium, so also can an *atmospheric attenuation coefficient*, $c_s(\lambda)$, be introduced to describe the fraction of radiant energy removed per unit distance from a beam of solar radiation due to scattering and absorption as it traverses an infinitesimal distance within the atmosphere [$c(\lambda) \equiv a(\lambda) + b(\lambda)$]. The atmospheric attenuation coefficient $c_s(\lambda)$ may be expressed as the sum of the individual attenuation coefficients, $c_i(\lambda)$, of the principal atmospheric components. Atmospheric attenuation results from molecular scattering (Rayleigh), aerosols, and absorption (by gases) phenomena, namely:

$$c_s(\lambda) = \sum c_i(\lambda)$$
$$= c_R(\lambda) + c_a(\lambda) + c_g(\lambda) + c_w(\lambda) + c_{O3}(\lambda) \text{ m}^{-1} \qquad (2.2)$$

where the subscripts R, a, g, w, and O3 denote, respectively, Rayleigh scattering, aerosols, absorption by atmospheric gases (O_2, N_2, CO_2, CH_4, NO_2), water vapor, and ozone.

Equation (2.2) represents a *bulk optical property* of the atmosphere, which, as we shall discuss in Chapter 3, is a property of an attenuating medium that is considered as a composite entity with no focus on the specific contributions of the components of that medium to that optical property. Since the atmosphere is not a homogeneous attenuating medium, it would be of consequence here to accommodate the *specific optical properties* of the atmospheric components (i.e., the optical properties that can be attributed to individual scattering and absorption centers). In this manner a *mass attenuation coefficient*, $\beta_i(\lambda)$ the specific attenuation that can be attributed to a unit density of an atmospheric component as the solar beam propagates through the atmosphere can be defined. The attenuation coefficient contribution due to component i, $c_i(\lambda)$, would then be the product of its mass attenuation coefficient $\beta_i(\lambda)$ (units of $m^2 \text{ g}^{-1}$) and its density $\rho_i(z)$ (units of g m^{-3}), where z is the vertical height above sea level of the atmospheric point under consideration. The atmospheric attenuation coefficient $c_s(\lambda)$ may then be defined as the sum of the individual attenuation coefficient contributions:

$$c_s(\lambda) = \sum \beta_i(\lambda)\rho(z)$$
$$= \beta_R(\lambda)\rho_R(z) + \beta_a(\lambda)\rho_a(z) + \beta_g(\lambda)\rho_g(z)$$
$$+ \beta_w(\lambda)\rho_w(z) + \beta_{O3}(\lambda)\rho_{O3}(z) \qquad (2.3)$$

where the subscripts are as defined for equation (2.2).

Equation (2.3) serves as a prelude to the discussions in Chapter 3 wherein such considerations of the specific absorption and scattering properties pertinent to the aquatic medium provide a basis for the determination of the concentrations of co-existing organic and inorganic aquatic matter that collectively determine the upwelling radiance spectra recorded by remote sensing devices.

The *optical thickness*, $\tau(\lambda)$, of the atmosphere can then be defined as a dimensionless function of the mass attenuation coefficients $\beta_i(\lambda)$ and the densities of the principal atmospheric attenuating components:

$$\tau(\lambda) = m \int_0^{z'} \beta_i(\lambda)\rho_i(z)dz \qquad (2.4)$$

where m is the optical air mass, z is the vertical height above sea level of a point in the atmosphere, z' is the vertical distance to the top of the atmosphere, $\beta_i(\lambda)$ are the individual mass attenuation coefficients as defined above, and $\rho_i(z)$ are the height-dependent densities of each of the atmospheric constituents.

The inclusion of m in equation (2.4) ensures its application to any solar zenith angle θ. If the bulk optical property $c_s(\lambda)$ is either known or can be reasonably assumed, then the optical thickness of the atmosphere reduces to:

$$\tau(\lambda) = mc_s(\lambda)z'. \qquad (2.5)$$

The *atmospheric transparency coefficient*, $T(\lambda)$, a term comparable to the transmittance term, $T(\lambda)$, encountered in aquatic optics can then be defined as:

$$T(\lambda) = \exp[-\tau(\lambda)] \qquad (2.6)$$

In terms of the atmospheric transparency coefficient, the direct solar radiation, $E_{sun}(\lambda)$, reaching the Earth's surface may be expressed as:

$$\begin{aligned} E_{sun}(\lambda) &= E_{solar}(\lambda)\cos\theta(r_0/r)^2T(\lambda) \\ &= (r_0/r)^2E_{solar}(\lambda)\cos\theta[T_R(\lambda) + T_a(\lambda) \\ &\quad + T_g(\lambda) + T_w(\lambda) + T_{O3}(\lambda)] \end{aligned} \qquad (2.7)$$

where r_0 and r are, respectively, the average and current distances between the Earth and the sun. $T_R(\lambda)$, $T_a(\lambda)$, $T_g(\lambda)$, $T_w(\lambda)$, and $T_{O3}(\lambda)$ are the atmospheric transparency coefficients appropriate to Rayleigh scattering, aerosols, gases, water vapor, and ozone, respectively.

It is readily shown that:

$$\cos\theta = \sin\phi\sin\delta + \cos\phi\cos\delta\cos\Omega \qquad (2.8)$$

where ϕ is the geographic latitude of the observation location, δ is the solar declination, the angle reckoned from the normal to the celestial equator that a given hemisphere is tilted toward the sun (in the northern hemisphere δ varies from $+23°27'$ at the summer solstice to $-23°27'$ at the winter solstice with zero values at the vernal and autumnal equinoxes), Ω is the hour angle defined by $2\pi t/\Pi$ (Π is the duration of a solar day, namely 86,400 seconds, and t is true solar time reckoned from true solar noon).

The solar declination angle for a specific day of the year for either the northern or southern hemisphere may be found in published ephemerals. Alternatively, δ may be determined[383] from the equation:

$$\delta = 0.39637 - 22.9133\cos\psi + 4.02543\sin\psi - 0.3872\cos2\psi \qquad (2.9)$$
$$+ 0.052\sin2\psi$$

where ψ is the date expressed in angular degrees. Sequentially numbering the Julian days as $d = 0$ for January 1 to $d = 364$ for December 31, ψ is defined as $(360°)d/365$. Both ψ and δ are in degrees.

Substituting equations (2.8) and (2.9) into equation (2.7) will define the direct solar irradiance at the Earth's surface in terms of the geographic location, time of day and time of year, atmospheric attenuation, and extraterrestrial radiation.

Attenuation of direct solar radiation due to molecular scattering can be calculated from Rayleigh theory,[366] i.e.,

$$T_R(\lambda) = \exp(-0.33 \times 10^{-3} \times \lambda^{-4.09} \times m) \qquad (2.10)$$

from which it can be seen that the shorter wavelengths of direct solar radiation are the most susceptible to molecular scattering.

To estimate the selective absorption of $E_{sun}(\lambda)$ by atmospheric gases, the mixed gases may be individually considered in terms of their so-called *partial thicknesses* (a counterpart to the thermodynamic concept of partial pressures for an admixture of gases). The partial thickness of a designated atmospheric gas would be the height of the atmosphere (at normal temperature and pressure, namely 15°C and 1010 millibars) should the atmosphere be comprised solely of the designated gas. For atmospheric nitrogen this thickness is 6200 m; for atmospheric oxygen 1700 m; for atmospheric argon 74 m; for atmospheric neon 140 mm; for atmospheric helium 40 mm; for atmospheric krypton 8 mm. For non-elemental gases the heights of the homogeneous partial atmospheres can only be approximated. For atmospheric N_2O this height is $\sim$10 mm; for atmospheric CH_4 5 mm to 10 mm; for atmospheric CO_2 $\sim$2.5 mm, for atmospheric ozone $\sim$3 mm to 7 mm.[33] The total partial thickness of the atmosphere sums to about 8000 m, due almost exclusively to the nitrogen and oxygen components. In the spectral region (0.10–5) μm, representative principal absorption bands of atmospheric gases include the following:

O_2—an A-band centered at 0.76 μm and a very weak B-band centered at 0.69 μm. Between 0.76 μm and 0.80 μm, about 9% of the total absorption by a unit air mass occurs. O_2 absorbs all light at λ = 0.18 μm.

O_3—the Hartley-Huggins absorption bands in the region (0.20–0.35) μm and the Chappuis bands in the region (0.45–0.75) μm. The absorption between 0.50 μm and 0.60 μm accounts for about 1% of the total absorption by a unit air mass with a water vapor partial pressure of 2 mm. It is the Hartley-Huggins absorption bands that describe the protective ultraviolet (UV-B) screening function of the anthropogenically threatened stratospheric ozone layer. Ozone also absorbs solar radiation at λ < 0.19 μm, as well as between 0.24 μm and 0.29 μm (specifically, O_3 has unresolved absorption bands in the broad spectral interval from 0.20 μm to 0.33 μm).

CO_2—bands centered at 1.46, 1.60, 2.04, 2.75, and 4.27 μm. About 5% of the total absorption by a unit air mass is accounted for by CO_2 absorption between 2 μm and 3 μm.

N_2—relatively transparent to solar electromagnetic radiation down to 0.10 μm. Weak Lyman-Birge-Hopfield absorption bands exist in the 0.10 μm to 0.15 μm range but are not considered important to atmospheric absorption processes.

NO_2—absorbs all light at λ < 0.13 μm.

Quantitative estimates of the absorption by mixed gases can be obtained from a LOWTRAN-3 computer program technique.[362] For example, attenuation of direct solar radiation due to ozone absorption can be estimated from

$$T_{o3}(\lambda) = \exp(-Com) \tag{2.11}$$

where o is the total content of ozone in the air column, m is the optical air mass, and C is a coefficient estimated by Guzzi et al.[149]

The amount of water vapor in the atmosphere varies widely. In terms of partial thicknesses (at normal temperature and pressure), these variations lie in the range 6 cm to 75 cm. In the optical spectral range water has absorption bands in the UV region (0.01 μm to 0.11 μm; 0.12 μm to 0.15 μm; 0.15 μm to 0.19 μm) that play important roles in upper atmospheric energetics but are of little consequence to the troposphere since the bands fall in the region of strong light absorption by such atmospheric components as N_2O, N_2O^+, O_2^+, NO^+, O_2, and O_3 at altitudes of 50 km to 150 km. Water vapor also possesses very weak absorption bands in the visible wavelength region (the so-called rain bands in the interval 0.572 μm to 0.703 μm). However, water vapor is an effective absorber of solar infrared radiation, possessing several strong absorption bands in the interval 0.70 μm to 10.0 μm.

Attenuation of direct solar radiation due to water vapor absorption can be estimated from

$$T_w(\lambda) = \exp[-C_{1,2}(Wm)^y] \qquad (2.12)$$

where W is the precipitable water content in cm, $y = 1$ for weak absorption, $y = 1/2$ for strong absorption, m is the optical air mass, and $C_{1,2}$ is a coefficient estimated by Guzzi et al.[149]

Atmospheric aerosol is present as solid and liquid particles in a myriad of diverse forms (smoke, water and H_2SO_4 droplets, dust, ashes, pollen, spores, among others). Since the sizes of aerosol particulates are generally large in comparison to the wavelengths comprising the visible spectral range, aerosol scattering does not follow the Rayleigh $\lambda^{-4.09}$ law of scattering that describes the attenuation due to atmospheric molecular scattering [equation (2.9)]. Rather, for aerosols, the scattering becomes essentially spectrally neutral following a wavelength power law of exponent within the range -1 to 0.

To estimate the attenuation of solar radiation by atmospheric aerosol scattering, a simplified formula was presented by Ångström:[10]

$$T_a(\lambda) = \exp[-\beta\lambda^{-\alpha}m] \qquad (2.13)$$

where β and α are empirically determined coefficients.

Case studies [see Kondratyev et al.[237]] have shown that atmospheric aerosols can also provide absorption centers as well as scattering centers. During the Complex Atmospheric Energy Experiment (CAEnEx-70), tropospheric data were obtained (altitude range 0.3 km to 8.0 km) over desert terrain indicating that, in the wavelength interval 0.4 μm to 2.4 μm, aerosol absorption accounted for about 20% of the total aerosol attenuation and followed a λ^{-1} power law. The duality of aerosol absorption and scattering, the complexities of aerosol particulate distributions, the diverse morphology of atmospheric aerosols, as well as the spatial and temporal variations of their concentrations, have conspired to render the development of a standard approach to aerosol-induced attenuation of solar radiation an as yet unresolved issue. Standard atmospheres, however, do exist. Since aerosols are principal atmospheric attenuating agents, complexities that inhibit their predicability present major obstacles to extracting the impact of atmospheric intervention from remotely acquired data.

2.3 STRATOSPHERIC OZONE AND ULTRAVIOLET SOLAR RADIATION

Atmospheric ozone does not rival the domination of nitrogen, oxygen, and aerosols in atmospheric scattering and absorption in the visible wavelength region of the electromagnetic spectrum. It does, however, play a major role as a screening agent for the biologically-pernicious ultraviolet wavelength region of the solar spectrum. Atmospheric ozone is currently under seige from the catalytic action of man-made stable chlorine, bromine, and other industrial chemicals (chlorofluorocarbons or CFCs) that can remain chemically intact long enough to attain ozone altitudes. It would be of consequence, therefore, to include within this chapter a brief discussion of the atmospheric ozone layer.

Solar ultraviolet radiation (in the wavelength range 0.235 μm $< \lambda <$ 0.390 μm) is known to have a pathogenic effect on living organisms (sunburn and skin cancer are the two most familiar and dramatic effects). This UV spectral region is divided into three general wavelength intervals:

UV-A:—(0.320 μm $< \lambda <$ 0.390 μm) or (320 nm $< \lambda <$ 390 nm)
UV-B:—(0.280 μm $< \lambda <$ 0.320 μm) or (280 nm $< \lambda <$ 320 nm)
UV-C:—(0.235 μm $< \lambda <$ 0.280 μm) or (235 nm $< \lambda <$ 280 nm)

with the maximum pathogenic impact being associated with the shortest wavelengths (UV-C) and the minimum pathogenic impact being associated with the longest wavelengths (UV-A).

UV intensity observable at sea level rapidly decreases with decreasing wavelength, dropping many orders of magnitude as UV-A gives way to UV-B. The presence of atmospheric ozone (maximum concentrations observed in the stratosphere at altitudes of 25 km to 50 km) protects life on Earth by absorbing all of the UV-C and most of the UV-B radiation that would otherwise reach the Earth's surface. The UV-A passes virtually unimpeded through the ozone layer and reaches the surface of the Earth with irradiances 1 to 15 orders of magnitude higher than that of UV-B (the UV-A and UV-B irradiances are, of course, equal at 320 nm). Empirical models such as that of Green et al.[146] enable determinations of direct and diffuse downwelling UV irradiance as a function of latitude, time of day or season, and general atmospheric conditions. Figure 2.2 utilizes the Green et al. model to illustrate the UV-B irradiance spectra (in watts per m^2 per nm) that would be recorded under clear conditions at solar noon on the summer solstice at a latitude of 75°N. The rapid fall-off in UV irradiance as wavelength diminishes from 320 nm to 280 nm is largely a consequence of the absorptive properties of the stratospheric ozone layer.[78] The curves for five ozone layer thicknesses (varying from a maximum thickness of 0.350 atm-cm to a minimum thickness of

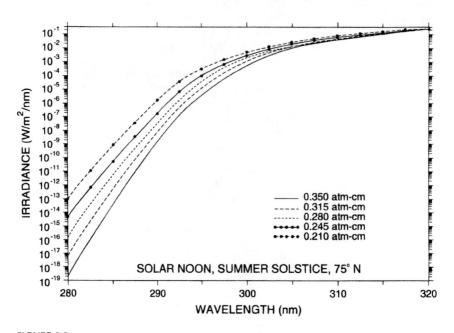

FIGURE 2.2
Ground-level clear-sky ultraviolet irradiance spectra that would be recorded under various stratospheric ozone concentrations at solar noon on the summer solstice at a latitude of 75°N.

0.210 atm-cm) illustrate that, while the UV-A may undergo less than an order of magnitude increase in sea-level irradiance as a consequence of such ozone depletion, the irradiance of the UV-B radiation may increase by as much as six orders of magnitude. Atmospheric ozone is considered a rare gas (accounting for only one of every 100,000 atmospheric gas molecules). Ground-level UV-B fluxes at $\lambda = 280$ nm under a "normal" ozone layer (0.350 atm-cm) are of the order of 10^{-19} watts per m^2 per nm, as compared to ground-level flux values > 1 watt per m^2 per nm at λ values in the middle of the visible spectral band. Despite the almost minuscule contribution of UV-B to the sea-level irradiance spectrum and the trace-gas nature of atmospheric ozone, these small amounts of ozone and UV-B radiation are major determinants of the quality of life on Earth.

The optical air mass, m, which is a function of the solar zenith angle θ, defines the relative length of the solar ray path in the atmosphere. As the sun approaches the horizon (i.e., $\theta \to 90°$), the degree of atmospheric attenuation increases and the solar radiation intensity rapidly decreases (the approximate $\sec\theta$ relationship). This decrease occurs in a spectrally selective manner, with short wavelength radiation generally decreasing more rapidly than longer wavelength radiation. Such spectral sensitivity of the optical air mass gives rise to short wavelength cut-off values (values of λ below which solar electromagnetic radiation will be almost totally attenuated by the atmosphere) that are functions of latitude and longitude as well as time-of-day, season, and atmospheric conditions at that latitude and longitude. This cut-off is particularly important at ultraviolet wavelengths. Even though glancing (i.e., high values of θ) solar UV-B radiation is absorbed more effectively than glancing solar UV-A radiation by the atmosphere, this reduced UV-B radiation may still be deleterious to living organisms.

Naturally produced atmospheric ozone (90% of which is stratospheric) resulting from photochemical reactions is subject to considerable spatial and temporal variations governed by the orbital motion of the Earth around the sun, the diurnal rotation of the Earth about its own axis, and the general circulation patterns present within the atmosphere. As a consequence of these dynamic constraints, the total ozone content in an anthropogenically undisturbed atmosphere is at minimum over the equator and at maxima over northern hemisphere mid-latitudes and the southern hemisphere Antarctic. Throughout the northern hemisphere, the stratospheric O_3 content is at maximum in the spring and at minimum in the fall. These natural and recurrent dynamical constraints dictate the geographic distribution of UV-B risk areas and the spectral cut-off wavelengths due to changes in the optical air mass m and the atmospheric transparency coefficient.

Direct measurements from Earth orbiting satellites have verified that the anthropogenic impact of CFCs has been the formation at stratospheric altitudes of regions of considerably reduced ozone content inappropriately termed *ozone holes.* In general, high latitude regions in both the southern and northern regions display the greatest reductions in ozone concentrations. The first conclusive evidence for this reduction in stratospheric ozone was presented by Farman *et al.*[103] wherein levels of ozone concentrations over Antarctica had declined ~40% over the decade 1975–1984. At the same time, however, mainly over northern hemisphere industrial regions, increases in ozone concentrations were observed. Data from 25 American stations suggested corresponding decreases in the global UV-B radiation of 0.1% to 1.1% per year. A 12% decrease in UV-B radiation between 1968 and 1993 was reported at Moscow.[409] Lack of universality of ozone trends in the air column, difficulties in monitoring changes in a minuscule and spectrally steep UV-B global spectrum, and the limited number of data sets have made non-controversial detection of long-term trends in global UV-B fluxes very difficult. Further, there is a small, but vocal, minority who remain skeptical about the seriousness of the ozone depletion scenario, arguing that clouds and the air pollution over cities can effectively combat any ozone thinning. Further, they trust that the Montreal Protocol (a landmark international agreement signed in 1987 that committed over 110 countries to reducing atmospheric CFC and halon emissions to pre-1986 levels; shortly thereafter, the London Protocol tightened the control measures and committed the countries to the phasing out of virtually all CFC and halon emissions by the year 2000) will solve the real or perceived UV-B problem. It is generally believed, however, that despite the legislative strategy of the most welcome Montreal Protocol, disruptive effects of enhanced UV-B radiation upon environmental ecosystems (while anticipated, such effects are far from being fully understood) will continue to manifest decades beyond compliance to the Montreal Protocol. The controversial nature of non-universal results, the knowledge gaps in impacts of enhanced UV-B radiation on terrestrial and aquatic ecosystems, and the difficulties associated with direct measurements of governing parameters testify to the complexity of the ozone depletion problem and underline the need for multidisciplinary ecosystem impact assessment studies.

Kerr and McElroy,[216] on the basis of direct spectral measurements of UV-B radiation over Toronto, Ontario, reported that the intensity of light in the vicinity of 300 nm increased by 35% in the spring of 1993, corresponding to the maximum reduction in total ozone also measured at Toronto during the same period.[215] These limited number of data points, however, represented the only data that could be confidently taken to be statistically distinct in four years of ozone/UV trend monitoring. This time period, unfortunately, also corresponded to an unusual combination of natural phenomena, including the eruption of the Mount Pinatubo volcano and a severe weather disturbance that swept North America.[264]

During the last decade, a 2.5% summer decrease in ozone concentration was recorded over Athens, Greece, resulting in a 5% increase in ground-level UV-B radiation.[410]

Analyses of the change in UV-B irradiance per unit change in atmospheric ozone show that the *absolute* change is highest at lower latitudes in summer, while the *percent* change is highest at higher latitudes in spring and fall. This raises the question of the importance of absolute versus relative change when assessing biological impacts of ozone depletion. (It must simultaneously be remembered, however, that the depletion of stratospheric ozone is currently non-uniform, being greater at the poles than at lower latitudes.)

UV-B radiation, like the photosynthetically available radiation (PAR) incident upon the air-water interface (as will be discussed in Chapter 3), can undergo reflection from or transmission through the interface. UV-B radiation, unlike the PAR, incident upon the air-water interface is highly diffuse in nature. This high percentage of diffuse-to-direct UV-B radiation results in an essentially constant reflection of 6% to 8% from the air-water interface (independent of solar zenith angle θ), as well as the transmission of UV-B radiation across the air-water interface being virtually unaffected by surface waves.

UV-B irradiance is highly attenuated by the presence of dissolved organic matter (yellow substance) within the water column. Figure 2.3 illustrates a plot of the depth of the 1% irradiance level for downwelling UV-B radiation $[z_{0.01} (\lambda_{UV})]$ as a function of dissolved organic carbon (DOC) concentration, a measure of dissolved organic matter. An average value of absorption

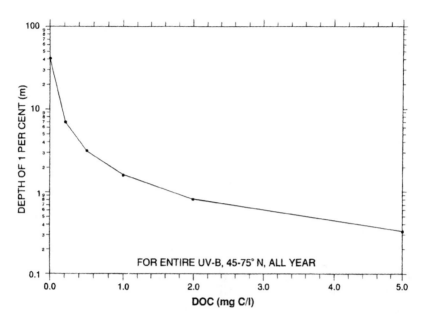

FIGURE 2.3
Depth of the 1% irradiance level for broadband UV-B radiation as a function of dissolved organic carbon concentration.

coefficient of DOC in the UV-B wavelengths from the limited number recorded in the scientific literature was used in generating Figure 2.3, wherein is considered the entire UV-B spectrum throughout the year. For a DOC concentration of 2 mg/l (average value for Lake Ontario waters), it is seen that 99% of the UV-B radiation is absorbed in the upper 1 meter (equivalently, 90% is absorbed in the upper 0.5 meter). A substantial number of inland and coastal waters display DOC concentrations higher than those found in Lake Ontario. Ladoga Lake, for example, with its DOC concentrations > 10 mg/l, loses 99% of the incident UV-B radiation in the upper 10 cm. The presence of suspended sediments will further and dramatically attenuate the UV-B radiation, thereby limiting the impacts of UV-B light to shallow water areas (wetlands, estuaries, littoral zones) and the flora and fauna that are dependent upon such areas for nutrients, spawning, food web, and/or habitat. For oceanic waters, however, the depth of the 1% UV-B irradiance level could be as large as 40 meters. We shall return to this topic in Chapter 10.

2.4 CLOUDS

Apart from gases and aerosols discussed in Section 2.2, a major determinant of the intensity of incident global radiation is the overhead cloud mosaic.

TABLE 2.1

Incident Direct Radiation (cal per cm² per min) under Clear and Cloudy
Russian Skies

	Solar Zenith Angle θ (degrees)						
Cloud Form	85	80	75	70	60	50	40
Cloudless sky	0.06	0.13	0.22	0.33	0.59	0.84	1.10
Cirrus	0.00	0.00	0.04	0.11	0.32	0.60	0.90
Altocumulus	0.00	0.00	0.00	0.00	0.00	0.12	0.31
Stratus	0.00	0.00	0.00	0.00	0.00	0.00	0.00

Table 2.1, taken from Kondratyev,[235] illustrates incident direct radiation (in calories per cm² per minute) measured under clear and cloudy skies over central Russia for a variety of solar zenith angles θ. The intensity variations displayed in Table 2.1 suggest it to be a virtual impossibility to track in a mathematically predictive manner the diurnal course of E_{sun} (λ) under the varying cloud covers routinely encountered in atmospheric monitoring. The global radiation under a sky comprised of a dynamically varying broken cloud cover will, understandably, vary in a highly discontinuous manner. Despite the perverse nature of cloud distributions and intrinsic features, however, empirical descriptions of cloud behavior have been developed and used to define monthly mean, seasonal mean, and annual mean values of global radiation E(λ) based upon *a priori* climatic and statistical knowledge and data regarding regional distributions of cloud type, cloud thickness, and cloud stability [see the discussions in Kondratyev[235] and many other existing texts dealing with atmospheric phenomena].

Provided it does not obscure the sun, scattered cloud cover does not affect the direct solar irradiance reaching the Earth's surface. This same cloud cover can, however, produce a significant increase in the diffuse sky irradiance reaching the Earth's surface, thereby increasing the total global radiation observable at a point under the cloud mosaic (5% to 10% increases beyond the global radiation associated with cloud-free atmospheres are routinely observed in mid-latitude stations as a consequence of non-storm-threatening cloud formations). Extensive cloud cover, however, (due to the high probability of solar obscurence) will almost always reduce the global radiation. Discussion of impacts of clouds on the global radiation will continue in Section 2.5.

In addition to impacting the global radiation, cloud cover produces a major impact on the ability of high altitude remote sensing devices to record visible light emanating from the terrestrial and aquatic ecosystems beneath them. Backscattered albedo from clouds overwhelm upwelling radiation from the non-atmospheric environment, and in almost all instances obscure the environmental target from the remote sensing device. Under a deep layer of cirrus clouds, as much as 20% of the downwelling E_{solar} is absorbed by the clouds, as much as 70% of E_{solar} is reflected back to space from the clouds, and as little as 10% of E_{solar} ends up as global radiation.[267] It is a small fraction of this reduced global radiation that constitutes upwelling radiation from natural waters that must then be redirected through the intervening cloud cover to the remote sensing platform.

2.5 DIFFUSE SKY RADIATION

To this point in our discussions we have been primarily concerned with the direct solar radiation component, $E_{sun}(\lambda)$, of the global radiation, $E(\lambda)$. The spectrum of the scattered diffuse component, $E_{sky}(\lambda)$, is comprised predominantly of short wavelengths, displaying a distinct maximum in the visible blue region between 0.425 μm and 0.450 μm. In the infrared region, about 96% of the scattered radiation is contained within the interval 0.7 μm to 1.7 μm, with less than 1% in the interval $\lambda > 2.27$ μm. Thus, while $E_{sun}(\lambda)$ covers a large spectral range from 0.250 μm to 3.0 μm (250 nm to 3,000 nm), $E_{sky}(\lambda)$ is essentially confined to visible and near-infrared wavelengths.

The direct solar radiance, $L_{sun}(\lambda,\theta)$ is defined by a relatively simple angular distribution, namely a very high radiance from the direction of the sun with a very rapid drop-off as the solar zenith angle is abandoned. The sky radiance, $L_{sky}(\lambda,\theta)$, is characterized by a complicated angular distribution determined by the atmospheric phase function (the atmospheric counterpart to the aquatic scattering phase function, $P(\theta,\phi)$, discussed in Section 1.4) and the optical thickness. As a consequence, under clear-sky conditions, a local maximum in $L_{sky}(\lambda,\theta)$ is invariably observed in the circumsolar zone of the sky with the total variation of $L_{sky}(\lambda)$ with zenith angle θ, being substantially greater in the sky quadrant containing the sun.

From computer simulations as well as direct observations performed in the transparent and cloudless atmosphere, a minimum sky radiance falls on the region in the solar vertical plane at an angular distance of about 90° from the position of the sun. This clear sky radiance distribution, following Moon and Spencer,[268] is illustrated in Figure 2.4. Herein are shown both the marked maximum near the sun and the marked minimum at about 90° from the sun. Under clear sky conditions, the spatial distribution of $L_{sky}(\lambda,\theta)$ may be considered symmetrical with respect to the solar zenith angle (and the 90° "anti-sun" zenith angle). Pokrovski,[320] as discussed in Walsh,[421] proposed a mathematical relationship to describe the clear sky radiance distribution of Figure 2.4:

$$L_{sky}(\theta, \zeta) = \eta\{(1 + \cos^2\zeta)/(1 - \cos\zeta) + t\}\{1 - \exp(-b\sec\theta)\} \quad (2.14)$$

where ζ = scattering angle, the angle that radiation is scattered from the solar incidence angle,

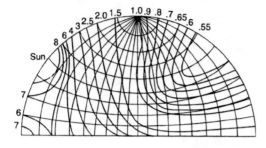

FIGURE 2.4
Clear-sky radiance distribution normalized to the zenith direction. (Adapted from Moon, P. and Spencer, D. E., *Illum. Eng.*, 37, 1942).

η = a scaling factor accounting for intensity of the incident solar beam upon encountering a scattering center within the atmosphere,

b = the atmospheric scattering coefficient,

t = an empirical constant allowing for the occurrence of multiple scattering, and

θ = solar zenith angle measured positively from the vertical.

The wavelength dependence of equation (2.14), while not specifically designated, is, nonetheless, implied. Steven[386] illustrated good agreement between direct observations and the use of equation (2.13) for sky regions beyond immediately circumsolar.

The ground level diffuse sky irradiance $E_{sky}(\lambda)$ for a cloudless sky was proposed by Makhotkin[255] as estimable from the mathematical expression:

$$E_{sky} = \frac{E_{solar}\cos\theta}{1 + \epsilon\tau\sec\theta} - E_{solar}\cos\theta \, \exp(-\tau\sec\theta) \tag{2.15}$$

where ϵ = the light backscattering probability in a cloudless atmosphere, which varies between about 0.14 and about 0.24,

θ = the solar zenith angle, and

τ = optical thickness of the atmosphere at λ = 0.55 μm.

Once again, the explicit wavelength dependencies of the parameters of equation (2.15) have been omitted for simplification. Equation (2.15), however, has been shown to consistently yield E_{sky} irradiance values exceeding those that are directly measured, due possibly to inappropriate atmospheric attenuation coefficients. Nonetheless, despite its quantitative shortcomings, equation (2.15) adequately describes qualitative aspects of variations within the diffusive component of the global radiation.

Uniform and thin (i.e., transparent to solar rays) cloud cover does not appreciably alter the general character of the spatial distribution of sky radiation. When clouds, particularly cirrocumulus and altocumulus, screen the circumsolar region of the skies, they can, however, promote increased diffusion resulting in sky radiation values as high as 15% of the direct radiation value. The presence of continuous thick non-circumsolar clouds can also dramatically alter the clear day sky radiance distribution shown in Figure 2.4. As cloud cover increases, the complicated azimuthal and zenith dependence of sky radiation reduces to a somewhat monotonic decrease in radiance from zenith to horizon[235] as θ increases from 0° to 90°. Under such conditions of extensive cloud cover, the zenith angular dependence of sky radiance can be described by a cardioidal distribution[268] of the form:

$$L_{sky}(\theta) = L_{sky}(90°)(1 + 2\cos\theta) \qquad (2.16)$$

Diffuse sky radiation, like direct solar radiation, is characterized by distinct diurnal variations. The maximum occurs at local noon, but the absolute value of this maximum depends upon the degree of atmospheric turbidity (the greater the atmospheric turbidity, the greater the absolute value).

Due to the differences in the spectral compositions of E_{sun} and E_{sky} (E_{sky} is far more abundant in short wavelength radiation than is E_{sun}), it is advantageous to consider global radiation in terms of its relative amount of diffuse radiation. For a cloudless sky, the sky radiation comprises about 10% of the global radiation. In high latitudes under considerable cloudiness the sky irradiance at high solar zenith angles is ~0.7 kw per m². Ground cover with high surface reflection coefficients can result in sky irradiance being reflected in large quantities (diffuse albedo). This diffuse albedo is retained near the surface due to secondary scattering and results in an overall increase in sky radiation over high surface reflectance regions of the terrestrial and aquatic environments. Thus, for middle and high latitudes, winter incident sky radiation is much stronger than summer incident sky radiation. Mathematical formulae have been suggested to describe the E_{sun} and E_{sky} irradiances under varying degrees of atmospheric composition and cloud cover. These empirical relationships require, of course, the use of coefficients that are spatially and temporally dependent.

Global radiation, $E(\lambda)$ [the sum of $E_{sun}(\lambda)$ and $E_{sky}(\lambda)$], in contrast to its component direct and scattered radiations, depends weakly upon the optical thickness τ (as τ increases, the global radiation slowly decreases). Thus, to calculate sea level $E(\lambda)$ for cloudless climes, Kondratyev[235] proposed the analytical formula:

$$E = \frac{2 - \sec\theta}{2 - \{\sec\theta \exp[\epsilon\tau(\sec\theta - 2)]\}} E_{solar}\cos\theta \qquad (2.17)$$

which, for typical atmospheric conditions ($\epsilon \approx 0.1$, $\tau \approx 0.3$) and $\sec\theta < 5$, can be reduced to:

$$E = E_{solar}\cos\theta/(1 + \epsilon\tau\sec\theta) \qquad (2.18)$$

where, as before, E_{solar} is the extraterrestrial solar radiation and θ is the solar zenith angle. Again, the λ dependency of equations (2.17) and (2.18) is implied.

From equations (2.18) and (2.8), it can be seen that the global radiation E will exhibit diurnal, seasonal, and latitudinal variability characterized by maximum values at local noon, in midsummer, and at the terrestrial equator, respectively.

The spectral distribution of the global radiation, $E(\lambda)$, in cloudless conditions is very close to that of E_{sky} (Kondratyev, 1969), characterized by a maximum in the interval 490–500 nm (weakly dependent upon solar elevation and essentially invariant in the interval 350–800 nm during daylight hours). As solar zenith angle θ increases, E_{sun} (λ) preferentially loses its blue and violet short wavelength radiations, resulting in a reddening of E_{sun}. Simultaneously, the percentage of E_{sky} (rich in blue and violet radiations) in $E(\lambda)$ increases. These compensating phenomena result in an essential independence of the spectral distribution of global radiation with solar location. However, for an overhead sun, a distinct blueing of the global radiation is observed.

As we have seen, the energy distribution in the global radiation spectrum can be highly variable under cloudy skies. In the presence of cloud cover, the global radiation could either increase or decrease. As a general rule, if the cloud cover is only partial and does not screen out the direct solar radiation, the global radiation will increase relative to that from a cloud-free atmosphere. Under continuous cloud cover, global radiation will always be less than that from a cloud-free atmosphere. This reduction in global radiation, observable for all cloud cover, is most pronounced in the presence of low altitude clouds (the most opaque to direct solar radiation). Although cloud cover and solar zenith angle are the major factors influencing global radiation, observations made under cloudless atmospheres have revealed marked dependencies of global radiation upon atmospheric transparency. As the atmospheric transparency decreases, so also does the global radiation noticeably decrease, particularly for near-vertical solar incidence (near-zero θ values). As mentioned earlier, the diffuse irradiance is known to be influenced by the underlying surface albedo (a consequence of the reflective properties of the terrestrial and aquatic surfaces). This albedo-dependence will also be observed in the global radiation.

2.6   WATER SURFACE ALBEDO

As we have defined in Section 1.5 and have used in Section 2.5, the term *albedo, A,* refers to the ratio of the spectral irradiance moving away from a surface to the spectral irradiance impinging upon that surface. Water surface albedo would, therefore, be defined as the ratio of the irradiance upwelling from the air-water interface to the irradiance incident upon the air-water interface. The interactions between the downwelling E_{sun} and E_{sky} irradiances and the air-water interface and underlying water column will be discussed throughout this book, beginning in Chapter 3. So also will be the nature of reflections and refractions across the interface, including Snell's law, the Fresnel equations, and the concept of internal reflection. However, it would be profitable here to discuss briefly the fate of light impinging upon an aquatic surface.

Natural radiation impinging upon a terrestrial surface in many (but not all) cases may be considered to be either absorbed by the terrestrial surface or reflected from it. The albedo from this terrestrial surface can then be defined in a straightforward manner as the upwelling (reflected) irradiance normalized to the downwelling (incident) irradiance. Since the incident irradiance is a composite of direct solar and diffuse sky irradiances, however, care must be exercised in treating them separately.

When considering the reflection of global radiation from the air-water interface, however, one must also be cognizant of the fact that the downwelling light penetrates the air-water interface. Photons undergo tortuous propagations within the water column, propagation paths that are determined by series of absorption and scattering events, the net outcomes of which are predominantly downward propagations until the flux of photons is totally attenuated. Some of the photons, however, are backscattered to the surface where a fraction is internally reflected back into the water column and a fraction is refracted back into the atmosphere. Thus, reflection at the air-water interface occurs both above the surface (due to downwelling direct and diffuse radiation) and below the surface (due to upwelling radiation). Surface reflection from the water is strictly defined as reflection of the downwelling global radiation. Albedo from the air-water interface, however, incorporates not only this upwelling surface radiation but some fraction of the upwelling subsurface radiation. Although global radiation consists of distinct direct and diffuse components, the upwelling radiation beneath the air-water interface may be approximated as being totally diffuse.

The albedo, A, of the water body would then be defined as the ratio E_{au}/E_{ad}, where E_{au} is the upwelling irradiance in air and E_{ad} is the downwelling irradiance in air. Following Jerlov,[196] the upwelling irradiance in air would be the sum of the reflection from the surface of the downwelling irradiance in air plus the upwelling subsurface irradiance that was not internally reflected at the air-water interface. Let E_{wu} be the upwelling irradiance in water. Further, let the reflectance of the downwelling irradiance in air, E_{ad}, be ρ_a and the reflectance of the upwelling irradiance in water at the interface, E_{wu}, be ρ_w. Then E_{au} may be written

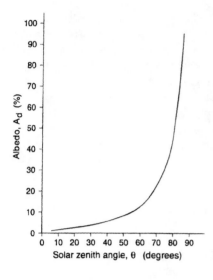

FIGURE 2.5
Dependence of direct solar albedo upon solar zenith angle (angle of incidence).

$$E_{au} = \rho_a E_{ad} + E_{wu} - \rho_w E_{wu} \qquad (2.19)$$

from which the water surface albedo A may be expressed as:

$$A = \rho_a + (1 - \rho_w)E_{wu}/E_{ad} \qquad (2.20)$$

The total global radiation reflectance, ρ_a, defined as the ratio of the incident and reflected global irradiances, may be expressed[198] in terms of the direct solar and diffuse sky radiation reflectances, ρ_d and ρ_s respectively, as:

$$\rho_a = \rho_d n + \rho_s(1 - n) \qquad (2.21)$$

where n is the fraction of global radiation comprised by solar radiation, i.e., n is the ratio $E_{sun}/(E_{sun} + E_{sky})$.

The direct solar reflection, ρ_d, is a function of solar zenith angle, viewing location, and condition of sea-state, and will be discussed in more detail in Section 3.3. The reflectance of polarized and unpolarized light from the air-water interface may be described by the *Fresnel reflectance formulae* which will be given in equations (3.2) and (3.3). Figure 2.5 sketches a composite curve illustrating the dependency of direct solar albedo, A_d, upon solar zenith angle θ. This composite follows results reported in Kondratyev[235] and is based upon equation (2.21) using Fresnel's reflectance formulae and direct observations by Ångström,[10] Kuzmin,[241] and others. It is seen that A_d monotonically increases with increasing θ from a value of 2.1% at near zenith angles to about 10% at about 65°. Beyond $\theta \sim 70°$, A_d increases sharply with increasing solar zenith angle approaching nearly 100% for grazing incidence.

Thus, direct solar radiation from a vertically overhead sun will almost totally (98%) penetrate a water body, while glancing direct solar radiation from a rising or setting sun will be almost totally reflected from that water body. The sky irradiance reflectance ρ_s is difficult to express quantitatively due to the complex radiance distributions that we have discussed in Section 2.5. Recall from equation (1.17) that the irradiance E and radiance L for the upper hemisphere are related by the polar coordinate equation

$$E = \int_0^{2\pi} \int_0^{\pi/2} L(\theta, \phi)\cos\theta\sin\theta d\theta d\phi. \tag{2.22}$$

For an isotropic (no preferred direction of arrival of $L_{sky}(\theta,\phi)$ at a point on the Earth's surface), equation (2.22) reduces to

$$E = 2\pi L \int_0^{\pi/2} \cos\theta\sin\theta d\theta. \tag{2.23}$$

If the E of equation (2.23) is taken to be the diffuse downwelling sky irradiance E_{sky} and θ is considered to be an incident angle i for which the Fresnel reflectance is $\rho(i)$, then the sky irradiance reflectance ρ_s for an isotropic sky radiance may be written as

$$\rho_s = \int_0^{\pi/2} \rho(i)\sin 2i di \tag{2.24}$$

Several workers have evaluated equation (2.24) and found a value of 6.6% for the sky irradiance reflectance, ρ_s, for the case of an isotropic sky radiance distribution.[52,123]

For a cardioidal sky radiance distribution [equation (2.16)], a value of 5.2% for ρ_s was calculated by Preisendorfer.[325]

Determinations of ρ_s using the Fresnel reflectance formulae in conjunction with isotropic or cardioidal sky radiance distributions display large discrepancies, undoubtedly due to the complexities involved in determining such essential parameters as the amount of radiation scattered by the water back into the atmosphere, the non-cardioidal distribution of cloudless sky radiation, the impact of clouds on the spatial distribution of sky radiation, and the actual state (departures from flatness) of the water surface.

Albedo of direct solar radiation E_{sun} is also governed by solar elevation, cloud cover, wind-roughening of the water surface, and water transparency (turbidity). From measurements in the Black Sea, Grishchenko[147] reported a non-monotonic dependence of the global radiation (sum of direct and diffuse) albedo upon solar zenith angle under relatively cloudy skies. Although sky albedo from the water surface rarely exceeds 6% for conditions of contin-

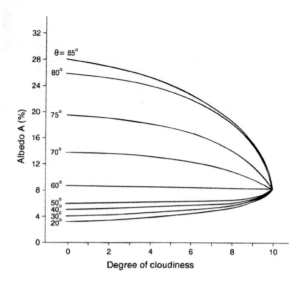

FIGURE 2.6
Dependence of water surface albedo on degree of cloudiness for various solar zenith angles.

uous low altitude cloudiness, values of ρ_s can be as high as 11% when clouds are observed near the horizon.[236] Studies by Ter-Markariants[394] of the effects of cloudiness on the water surface albedo of global radiation indicated that, with increasing cloudiness at $60° < \theta < 90°$, the albedo decreases, while at $0° < \theta < 60°$ the albedo increases. This is shown in Figure 2.6, wherein are plotted the surface water albedo, under varying degrees of atmospheric cloudiness, for a variety of solar zenith angles.

The effect of water surface roughening is very pronounced (see, for example, Cox and Munk[67]), especially at large solar zenith angles. Figure 2.7 is sketched in a manner similar to Figure 2.5 and illustrates the dependence of the water surface albedo upon solar zenith angle θ. Curve 1, comparable to the composite curve of Figure 2.5, is the calculated relationship from the Fresnel reflectance formulae (calculated for a flat water surface). Curve 2 represents a composite relationship, indicative of results obtained by such workers as Grischenko,[147] Hishida and Kishino,[168] Gordon,[142] Gushchin,[148] and others. Comparison of Curves 1 and 2 indicate that, for values of θ less than about 75°, the global radiation albedo as measured for natural waters substantially increases and, for values of θ greater than about 75°, the albedo substantially decreases in comparison to the albedo values for a flat water surface calculated from the Fresnel equations. These departures are a consequence of wind-induced inclinations of the air-water interface (surface roughness). The impact of roughness on the water albedo is determined not only by variations in the angle of incidence of the sun's rays, wave shadowing, and multiple scattering phenomena (at near grazing incidence), but also

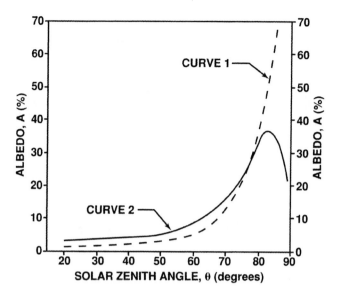

FIGURE 2.7
Effect of surface roughening (including foam and whitecaps) on water surface albedo.

by the formation of such surface features as white caps and foam streaks. Koepke[233] reports that the effective reflectance of white caps averages (22 ± 8)% and the effective reflectance of foam streaks averages (10 ± 4)%. In most instances foam patches and streaks are highly variable, requiring that a total reflectance be summed from histograms illustrating fractions of the aquatic surface definable by specific reflectance values. For near-surface wind speeds up to 8 m s^{-1}, the total reflectance due to foam is zero. Beyond this wind speed the foam reflectance gradually increases to values of 0.2% at a wind speed of 10 m s^{-1}, 0.9% at 15 m s^{-1}, and 5.1% at 25 m s^{-1}. The effective reflectance of foam streaks is essentially invariant with respect to λ in the visible spectral region. In the near-infrared region, however, the effective reflectance decreases rapidly with increasing λ due to absorption by liquid water. Superimposed upon this decrease are two intensive minima at 1.5 μm and 2.0 μm, respectively.

By comparison, Figure 2.8 illustrates the impact on surface reflectance of change in wave slope due to windspeed, but in the absence of foam and white caps. As the near-surface winds intensify in magnitude, the surface reflectance at large zenith angles dramatically decreases. The increase (as compared to the Fresnel equation calculations for a flat air-water interface) in reflectance at zenith angles < 75° illustrated in Figure 2.7 is seen to be due to the presence of foam and whitecaps.

For conditions of thick and intensive atmospheric cloud cover coupled with a rough water surface, Gushchin[148] reports decreases in global radiation

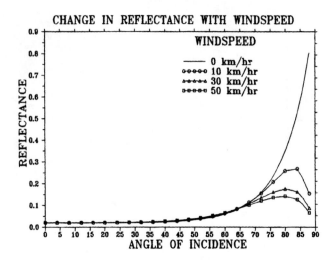

FIGURE 2.8
Effect of surface roughening (due solely to wind-induced changes in wave slope) on water
surface albedo.

albedo at zenith angles greater than ~55° and increases in global radiation
albedo at zenith angles less than ~55°. This can result in noticeable reductions
in the solar angle dependency of Curve 2 in Figure 2.7. For a nine-point
surface roughness and extensive overcast, the water surface albedo for global
radiation is approximately 7% at all solar angles, a not surprising result since
the isotropic diffuse sky irradiance reflectance ρ_s is consistently determined to
be 6.6%.

Particularly problematical to aquatic remote sensing is the mirror-image
reflection (sunglint) of the solar disk from the air-water interface under
relatively cloud-free (transparent) skies. Such high-intensity direct solar radiation albedo, when reflected into the field-of-view of a remote sensing device,
will invariably overwhelm the considerably smaller radiative return from
the water body being studied. For a flat air-water interface, the reflected
solar beam angle (as measured from the normal to the reflecting surface)
would be identical to the incident solar beam angle (i.e., θ). For a wave-
tilted air-water interface the angle of reflection will still equal the angle of
incidence. However, the normal to the reflecting surface now becomes a
variable that is controlled by wind-driven wave structures. The solution to
the problem of relating surface reflection to wave slopes falls within the
realm of analytical geometry (and to some degree statistics), and such was
the approach adopted by Cox and Monk[67]). Despite (or perhaps more appropriately because of) the elegant simplicity of their approach to wind-induced
wave slopes and the impacts of waves on global radiation albedo, the work

of Cox and Monk has become a staple in the literature dealing with the remote sensing of natural water.

We will return to the topic of surface reflection from natural waters and treat the issues of sunglint, wave slopes, and remote sensing viewing directions in considerably more detail in the next chapter.

Chapter 3

THE PROPAGATION OF ATMOSPHERIC PHOTON FLUXES INTO NATURAL WATER BODIES

3.1 OPTICAL CLASSIFICATION OF NATURAL WATERS

The degree of optical complexity of a natural water body is, in general, directly related to its proximity to land masses. Thus, the relative optical simplicity of mid-oceanic waters is usually denied the waters of lakes, rivers, and coastal regions. Optical *transmittance* $T(\lambda)$ is defined as the ratio of the radiant flux transmitted by an attenuating medium to the incident radiant flux impinging upon it. Transmittance is usually expressed for a specific thickness of the medium. Schemes for classifying the optical complexities of natural water bodies, as first proposed by Jerlov,[192, 193, 195] are based upon directly measured values of transmittance per meter of downwelling irradiance in the near-surface layer. Oceanic waters are classified[198] as Cases I, II, and III, with the irradiance transmittance of Case III waters being measurably lower than Case I waters, particularly at shorter wavelengths (300–550 nm). Coastal waters have added another nine classifications to this optical scheme, again with Case 9 coastal waters displaying substantially lower irradiance transmittance than Case 1 coastal waters, particularly at shorter wavelengths. Detailed discussions of these twelve optical classifications of oceanic and coastal waters, along with illustrative examples of irradiance transmittance spectra, can be found in Jerlov.[198]

Alternate schemes for optically classifying natural waters have been presented by Pelevin and Rutkovskaya,[306] who proposed that the waters be classified according to measured values of the irradiance attenuation coefficient $K(\lambda)$ at $\lambda = 500$ nm; by Smith and Baker,[373] who proposed that, due to chlorophyll being the principal ocean water colorant, oceanic optical classifications be based upon total aquatic content of chlorophyll and chloro-phyll-like pigments; by Kirk,[223] who proposed that freshwater optical classifi-cations be based upon the spectral absorption properties of particulate and soluble aquatic fractions; and by Prieur and Sathyendranath,[337] who pro-posed that natural waters be optically classified according to the relative proportions of the absorption due to the indigenous algal pigments, dis-solved yellow substances, and suspended materials. Despite the specific appropriateness of these classification schemes and despite the fact that the original transmittance measurements that led to the Jerlov classification scheme were performed utilizing broadband color filters (data from which are often at variance with the data obtained from the fine resolution submers-ible spectroradiometers currently in use), the Case I, II, and III ocean and the Case 1 to 9 coastal optical classification scheme remains in widespread use as a convenient means of at least quasi-objectively classifying optical complexity.

The Jerlov scheme optically classifies mid-oceanic waters as Case I due to their consistently high irradiance transmittance values (~85%–95% throughout the 300–550-nm range). Other than highly transparent lakes and some high-latitude river and estuary systems, most non-mid-oceanic global waters, including oceanic coastal zones, qualify as non-Case I. Most water bodies contained within or dominated by land masses will possess degrees of optical complexities that exclude them from being considered as Case I waters. These waters, however, could be characterized by degrees of turbidity that may even be substantially subordinate to some oceanic/coastal classifications.[397]

3.2 PHOTON INTERACTIONS WITH THE AIR-WATER INTERFACE

Before being recorded at a remote platform, the photon flux emanating from the volume of the water must interact with the air-water interface on two occasions, namely (a) subsequent to its downwelling propagation through the atmosphere and (b) subsequent to its downwelling and upwell-ing propagation through the aquatic medium. In each instance some of the impinging radiation will undergo reflection back into the original medium, and some will undergo refraction and transmission into the adjacent medium. For a calm surface the amount of reflection, refraction, and trans-mission are calculable from the indices of refraction of the respective media. These optical interactions between photon fluxes and the air-water interface are illustrated in Figure 3.1(a) for the above-water impinging photon flux and in Figure 3.1(b) for the subsurface upwelling photon flux. Figure 3.1 is an expanded version of Figure 1.6. $L_{da}(\lambda)$ represents a downwelling incident radiance of wavelength λ, the reflected portion of which is labelled $L_{ra}(\lambda)$

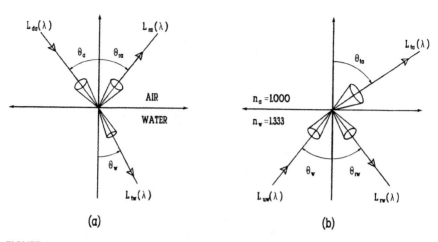

FIGURE 3.1
(a) Interaction of above-water downwelling incident radiance with the air-water interface. (b) Interaction of sub-surface upwelling radiance with the air-water interface.

and the refracted and transmitted portion of which is labelled $L_{tw}(\lambda)$. $L_{uw}(\lambda)$ represents an upwelling subsurface radiance of wavelength λ, the reflected portion of which is labelled $L_{rw}(\lambda)$, and the refracted and transmitted radiance is labelled $L_{ta}(\lambda)$. In both illustrations, θ_a represents the angle between the direction of photon flux propagation in air and the normal to the air-water interface, while θ_w represents the angle between the direction of photon flux propagation in water and the in-water normal. The index of refraction of air is labelled n_a and the index of water is labelled n_w. In Figure 3.1, the air-water interface is sketched as a straight line, indicative of the highly idealized situation of a calm or "flat" water body.

The refraction at such a calm surface is given by *Snell's Law*, namely:

$$n_a \sin\theta_a = n_w \sin\theta_w \qquad (3.1)$$

The index of refraction for air, n_a, is taken as unity, and the index of refraction of water, n_w, is generally taken as 1.333. The index of refraction of water varies slightly with temperature, salinity, turbidity, and wavelength. Tables of n_w for sea water may be found in Dorsey,[88] Lauscher,[242] Sager,[347] and others.

Two important consequences to remote sensing of refraction at the air-water interface are listed:

1. The entire above-water atmospheric hemisphere is condensed into an underwater cone subtending a half angle of about 48°36' (the angle whose sine is 1/1.333). Thus, all the radiance recorded by an above-water remote sensing device originates within this underwater cone.

2. The flux contained within a solid angle $d\Omega_w$ below the air-water interface will be spread into a larger solid angle $d\Omega_a$ (of magnitude $n_w^2 d\Omega_w$) above the air-water interface. Thus, the radiance emerging from the water surface will be decreased by a factor of $1/n_w^2$ (about 0.563).

In a homogeneous medium, light and other electromagnetic waves consist of sinusoidally oscillating electric and magnetic wave motions that propagate in accordance with *Maxwell's equations.* A stipulation of these equations is that the electric field density vector and the magnetic field density vector be aligned perpendicular to one another in the plane perpendicular to the direction of propagation. This *transverse* nature of travelling light waves results in light displaying a directional behavior known as *polarization.* It is sufficient to specify polarization solely in terms of the behavior of the electric field density. This is due to the fact that the corresponding magnetic field density can be readily determined from Maxwell's equations. Depending upon the direction of the electric field density vector in the plane perpendicular to the direction of propagation, the travelling light wave may be *plane, circularly,* or *elliptically polarized.* The travelling light wave could be *randomly polarized* (or *non-polarized*), that is, the electric field density vector has no preferred transverse directionality. When one considers reflection of light from a surface, it is convenient to regard the polarization of an incident light flux in terms of a polarization perpendicular to the plane of incidence or a polarization parallel to the plane of incidence.

The *Fresnel Reflectance formulae* readily compute the reflectance from the air-water interface. For polarized light:

$$\rho_\perp = \frac{\sin^2(\theta_i - \theta_r)}{\sin^2(\theta_i + \theta_r)} \tag{3.2}$$

and

$$\rho_\parallel = \frac{\tan^2(\theta_i - \theta_r)}{\tan^2(\theta_i + \theta_r)} \tag{3.3}$$

where $\rho_\perp$ and $\rho_\parallel$ are the reflectances for the perpendicular polarized and the parallel polarized components of the incident light, respectively. θi and θr are the angles of incidence and refraction, respectively. For non-polarized light:

$$\rho = 0.5 \frac{\sin^2(\theta_i - \theta_r)}{\sin^2(\theta_i + \theta_r)} + 0.5 \frac{\tan^2(\theta_i - \theta_r)}{\tan^2(\theta_i + \theta_r)} \tag{3.4}$$

It is seen from equation (3.3) that the reflectance $\rho_\parallel$ becomes zero when

$\tan^2(\theta_i + \theta_r)$ becomes infinite (i.e., for $\theta_i + \theta_r = 90°$). Thus, when the reflected and transmitted rays are at right angles to one another, it is impossible for light to be reflected, providing its polarizing electric field vector is parallel to the plane of incidence. The angle of incidence at which $\theta_i + \theta_r = 90°$ is termed *Brewster's angle* θ_B. While light with its electric field oscillating parallel to the plane of incidence will not be reflected, it will, however, be refracted and transmitted to the adjacent medium. Also, light with its electric field oscillating in a plane perpendicular to the plane of incidence will, at Brewster's angle, be both reflected and transmitted by the interface. It is readily seen that Brewster's angle in water, θ_{Bw}, can be determined from:

$$\tan\theta_{Bw} = n_a/n_w \tag{3.5}$$

and Brewster's angle in air, θ_{Ba}, can be determined from:

$$\tan\theta_{Ba} = n_w/n_a \tag{3.6}$$

For an aquatic index of refraction of 1.333, the in-air Brewster's angle is 53°08′ and the in-water Brewster's angle is 36°52′.

The amount of energy reflected from and transmitted through the boundary separating media of differing refractive indices is described in terms of *reflectivity r* and *transmissivity t*, which are comparable to the *reflectance* and *transmittance* terms we have previously encountered. Reflectivity and transmissivity are defined in terms of the indices of refraction of the adjacent media, the incident and refracted angles, and the *Poynting vector* **S**. The Poynting vector is defined as the vector product of the electric field density vector **E** and the magnetic field density vector **B** [$(c/4\pi)$ **E** X **B**], and gives the energy flow per unit volume in a direction **n** by means of the scalar product **S*n**. Using the concepts of reflectivity and transmissivity the case of oblique incidence (that is, for incident angles greater than Brewster's angle) can be shown to be as follows:

1. For light entering water (i.e., for light impinging upon a more optically dense medium, $n_w > n_a$), surface reflection of incident light polarized parallel to the plane of incidence (which was inhibited at Brewster's angle) is re-initiated and rapidly escalates to total reflection as the impinging flux approaches grazing incidence ($\theta = 90°$).

2. Incident light polarized perpendicular to the plane of incidence (which was not inhibited at Brewster's angle) also escalates to total flux reflection as grazing incidence is approached.

3. Unpolarized light entering the water behaves in a similar manner to light that is polarized perpendicular to the plane of incidence.

4. For light exiting the water (i.e., for light impinging upon a less optically dense medium, $n_a < n_w$), surface reflection of incident light polarized parallel to the plane of incidence (which was inhibited at Brewster's angle) is re-initiated with the reflectivity almost immediately rising to unity as the incident angle increases. Total internal reflection occurs, therefore, long before grazing incidence is reached.

5. Incident light polarized perpendicular to the plane of incidence and unpolarized light (which were not inhibited at Brewster's angle) also follow this almost instantaneous rise to unity reflectivity long before grazing incidence is attained.

6. For light exiting water, Snell's law defines the angle of refraction in air as the angle whose sine is the product of the ratio n_w/n_a and the sine of the in-water incident angle. A *critical angle of incidence* exists, therefore, at which the in-air exit angle is 90°. This critical incidence angle is seen to be the angle whose sine is n_a/n_w (for an n_w value of 1.333, this critical value of in-water incidence is 48°36' as compared to Brewster's angle of 36°52'). Beyond this critical incident angle in the denser of the two attenuating materials no transmission occurs. For angles beyond this critical angle *total internal reflection* occurs. That is, photons upwelling to the air-water interface with incident angles $\theta_w > 48°36'$ will not escape the aquatic medium.

Thus, light downwelling in the atmosphere and impinging upon the air-water interface can undergo refraction (thereby entering the water) or reflection (thereby remaining in the atmosphere). Light upwelling in the water and impinging upon the air-water interface can also undergo refraction (thereby re-entering the atmosphere) or internal reflection (thereby remaining in the water).

Tables and plots of calculated reflectivities and transmissivities as functions of in-media incident angles may be found in a number of physics textbooks. Values of the polarized reflectances $\rho_\perp$ and $\rho_\parallel$ from sea water may be found in Austin.[11] Figure 3.2(a) illustrates a plot of reflectances $\rho_\perp$, $\rho_\parallel$, and ρ versus the angle of incidence in air θ_a as calculated from the Fresnel reflectance formulae (3.2), (3.3), and (3.4), respectively, for $n_w = 1.333$. Some of the reflected/transmitted energy features listed above are evident from Figure 3.2(a). The ρ curve of Figure 3.2(a) is equivalent to Figure 2.5, which illustrated the zenith angle dependency of solar reflectance from the air-water interface.

Figure 3.2(b) illustrates a plot of reflectances $\rho_\perp$, $\rho_\parallel$, and ρ versus the angle of incidence in water (i.e., for upwelling subsurface radiation). The critical angle of incidence (beyond which total internal reflection occurs, i.e., reflectance = 1.0) for $n_w = 1.333$ is seen to be 48°36'.

The reflection of a direct beam of light from the sun (solar zenith angle θ) can be described in terms of the Fresnel reflectances ρ of equations (3.2), (3.3), and (3.4). As we have discussed in Chapter 2, the angular distribution

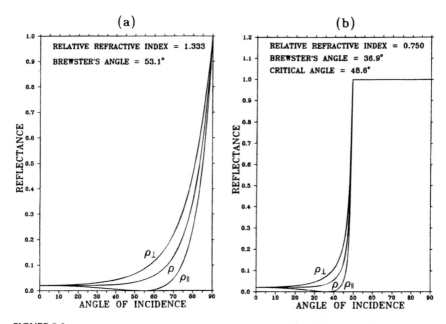

FIGURE 3.2
(a) Plots of $\rho\perp$, $\rho\|$, and ρ versus angle of incidence in air. (b) Plots of $\rho\perp$, $\rho\|$, and ρ versus angle of incidence in water.

of diffuse sky radiation can be a very complex function, varying from quasi-isotropic to a strong directional dependency. Consequently, the precise extent to which skylight is reflected by an air-water interface is difficult to determine. If the downwelling skylight is assumed to be defined by an isotropic radiance distribution,[198] a reflectance of 6.6% is obtained for a flat interface. If the downwelling skylight is assumed to be defined by a homogeneously overcast cardioidal radiance distribution,[325] this flat interface reflection is reduced to 5.2%. The presence of wave structures at the interface can impact the surface reflection for either atmospheric condition.

It would be of consequence, at this point, to recall the discussion of water surface albedo A [the ratio of the upwelling irradiance above the air-water interface $E_{au}(\lambda)$ to the downwelling irradiance above the air-water interface $E_{ad}(\lambda)$] presented in Section 2.6. Since both upwelling and downwelling light is reflected by the air-water interface, the water albedo A at any wavelength λ is [reiterating equation (2.20)]:

$$A = \rho_a + (1 - \rho_w)E_{wu} / E_{ad} \tag{3.7}$$

where ρ_a is the reflectance for the global radiation (i.e., the sum of the direct solar irradiance E_{sun} and diffuse sky irradiance E_{sky}), ρ_w is the reflectance of the upwelling irradiance E_{wu}. The global radiation reflectance ρ_a expressed in terms of the direct radiation reflectance ρ_d and the sky radiation reflectance ρ_s is [reiterating equation (2.21)]:

$$\rho_a = \rho_d n + \rho_s(1 - n) \tag{3.8}$$

where n is the fraction of global radiation that is direct (i.e., n is the ratio $E_{sun}/(E_{sun} + E_{sky})$.

Equations (3.7) and (3.8) define the total surface and volumetric energy return from the water normalized to the extant incident solar and sky radiation. In the discussions and sections immediately following, we will consider the reflection of direct solar light from the air-water interface, in particular the impacts of wave action and solar position on such reflection.

Two important consequences of surface reflection to remote sensing are the following:

1. A remote sensing device viewing the natural water body could be rendered unresponsive to portions of the reflected surface radiation if it were fitted with an appropriate polarizing filter and if it were observing the surface at Brewster's angle (as demonstrated by Borstad et al.[26] in their airborne spectrometer measurements of British Columbian coastal waters). However, at Brewster's angle the atmospheric air mass through which the device must view the aquatic target is increased to about 1.7 times the atmospheric air mass through which the water would be viewed from directly overhead (the secant of 53°08'). This added atmospheric intervention could substantially reduce the magnitude of the remotely sensed aquatic signal.

2. Surface reflection contains no volumetric information regarding the water column under scrutiny. Consequently, techniques must be devised to compensate for the surface reflection of both the solar (of which sunglint is one manifestation) and sky radiations.

3.3   THE IMPACT OF WAVE SLOPE ON REFLECTION FROM THE AIR-WATER INTERFACE

To this point in our discussion, the air-water interface has been idealized as a flat surface. Natural water, however, is a dynamic fluid that is seldom calm. The local wind field is a forcing function that inclines the surface, generating wave patterns that display spatial and temporal variations. Such patterns of air-water interface perturbations cause the observed reflectance of the surface to change. This change is particularly significant for downwelling radiation in the atmosphere at near-grazing incident angles and for upwelling radiation in the water body at near-internal reflection incident angles.

To consider this impact upon reflection from the aquatic surface, workers such as Duntley,[90] Cox and Munk,[67,69] Wu,[433] Gordon,[142] and Longuet-Higgins[249] have statistically related distributions of wave slopes with near-surface wind speeds. Their work has resulted in the generation of time-averaged sea-surface reflectances for windspeeds as high as ~20 m s^{-1}. At windspeeds substantially above 20 m s^{-1} correcting remote measurements for the impacts of wave activity becomes hazardous due to the presence of white caps. Tables and graphs of reflectance and transmittance at the air-water interface for various incident angles and wind speeds can be found in the above references as well as in Austin[11] and others. These graphs and tables illustrate that for most of the near-surface wind speeds and viewing geometries encountered in conventional remote sensing activities, the effect of the wind on the time-averaged reflectance does not pose a problem.

Wave action minimally impacts the time-averaging of the reflected sky irradiance. Surface roughening, however, has a major impact on the portion of the sky that is reflected by the air-water interface upward toward the remote sensing platform. In this regard the relative orientation of the platform, the sun, and the wave slopes are crucial to the solar glitter patterns that may or may not be readily observable by the remote sensing device. Thus, while wind speed is generally not a problem with regard to surface reflection of sky irradiance, it can be a serious problem with regard to surface reflectance of direct solar irradiance into the field-of-view of the remote sensing device. Therefore, the flight paths of aircraft and orbits of satellites are correlated with the field-of-view characteristics of sensor payloads in such a manner as to minimize the probability of the solar glitter pattern being recorded in the acquired data. Only a limited number of studies of sunglint patterns over water have been performed. Although the general consensus of most workers is to remove such patterns from data analyses schemes, Bukata and McColl[34] used aerial photography collected over Lake Ontario to illustrate that sun-glint can be a valuable tool in studying subsurface dynamical features such as episodes of large scale thermal upwelling.

Remote sensing of aquatic resources is performed under various combinations of viewing angles, time of day, and sea-state. It would, therefore, be instructive to illustrate the geometric relationships involved in the remote sensing of non-calm aquatic surfaces. The following account of the geometry of reflection, sunglint, the probability distribution of wave slopes, and horizon limitations follows the work of Cox and Munk,[67,68] Burt,[52] and Maul.[262]

Figure 3.3(a) illustrates a wind-direction-based coordinate system with the y-axis pointing upwind. The x-y plane defines a calm air-water interface with the positive z-axis directed vertically upward. Consider the sun to be located at a zenith angle ϕ_s measured from the z-axis and an azimuth angle θ_s measured from the y-axis toward the x-axis. (To be consistent with comparable discussions in the literature, we have elected to designate the zenith angle with the symbol ϕ as opposed to the symbol θ that we use throughout

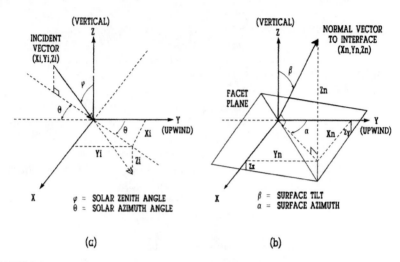

FIGURE 3.3
(a) Wind-direction-based coordinate system with the y-axis directed positively upwind, the z-axis directed positively upward, and the x-y axis defining an untitled air-water interface. (b) Angular coordinates of a wind-tilted wave facet in the wind-direction-based coordinate system.

this book. It is our intent that each topic, while intimately connected to the main theme of this book, should possess a quasi-self-contained character unto itself. We therefore trust that such a nomenclature departure will not prove to be problematical.)

Figure 3.3(b) illustrates the normal to a wave facet in this coordinate system, i.e., the unit vector (x_n, y_n, z_n) normal to a wind-induced departure from a flat air-water interface. Consider the wave facet to be defined by a surface tilt angle β (measured from the z-axis to the normal vector of the wavelet) and an azimuthal direction α (measured from the y-axis to the direction of steepest descent). This convention is different from that originally used by Cox and Munk, whose azimuthal angle was 180° from the azimuthal angle of this convention. However, the azimuth angles used herein correspond to the spherical coordinate angular values of the wave facet's normal vector, a correspondence that will simplify the upcoming analyses.

The wave slope m can then be defined as $m = \tan\beta$, with crosswind, z_x (projection of the wave slope on the x-z plane), and upwind, z_y (projection of the wave slope on the y-z plane), slope elevation components given by:

$$z_x = \delta z / \delta x = -m\sin\alpha$$

$$z_y = \delta z / \delta y = -m\cos\alpha.$$

The root mean square slope components σ_c^2 and σ_u^2 (crosswind and upwind, respectively) vary very nearly linearly with wind speed W (in m sec^{-1}) at 12.5 meters above sea level, and, as shown by Cox and Munk,[67] are given by:

$$\sigma_c^2 = 0.003 + 0.00192W$$

and

$$\sigma_u^2 = 0.000 + 0.00316W.$$

Wave slopes may then be standardized according to their mean square slope components as:

$$a = z_x/\sigma_c \quad \text{and} \quad b = z_y/\sigma_u.$$

The values of a and b using our angular conventions will always have the opposite sign to those of the Cox and Munk work.

For a given viewing geometry, the only wave facets that will reflect sunlight into the remote sensor are those whose tilt angles $\beta°$ lie in the range $\beta° \pm {}^1/_2\delta\tan\beta°$. This is the slope range required to reflect around the periphery of the sun. The sun's periphery, if plotted on a z_x versus z_y diagram, would trace out an ellipse. This *tolerance ellipse* is then defined as the range (i.e., tolerance) of possible (z_x,z_y) values for which a wave facet will reflect a highlight of the sun directly toward the remote sensing device.

As detailed in Cox and Munk,[69] the area of the tolerance ellipse, dA_t, is given by:

$$dA_t = (\Omega_{sun}\sec^3\beta\sec\omega)/4 \tag{3.9}$$

where Ω_{sun} is the solid angle subtended by the sun at the Earth's surface ($\sim6.8 \times 10^{-5}$ sr) and ω is the angle of incidence (i.e., the angle between the solar direction and the normal to the tilted wave facet).

If the probability of a particular wave slope occurring is $p(\beta,\alpha)$, then the probability $P(z_x,z_y)$ that waves will occur with slopes within the limits of $z_x \pm {}^1/_2\delta z_x$ and $z_y \pm {}^1/_2\delta z_y$ can be written:

$$P(z_x, z_y) = p(\beta, \alpha)\delta z_x\delta z_y. \tag{3.10}$$

In order to account for skewness in wavelet peaks, Cox and Munk used a Gram-Charlier series to define the probability density function $p(\beta,\alpha)$, which can be generalized in terms of σ, a, b, W, and the skewness parameters c_n as follows:

$$p(\beta, \alpha) = (2\pi\sigma_c\sigma_u)^{-1}\exp[-(1/2)(a^2 + b^2)] \times [1 - (1/2)c_{21}(a^2 - 1)b$$
$$- (1/6)c_{03}(b^3 - 3b) + (1/24)c_{40}(a^4 - 6a^2 + 3)$$
$$+ (1/4)c_{22}(a^2 - 1)(b^2 - 1) + (1/24)c_{04}(b^4 - 6b^2 + 3)$$
$$+ \cdots\cdots\cdots] \tag{3.11}$$

where $c_{21} = -(0.01 - 0.0086W)$
$ c_{03} = -(0.04 - 0.033W)$
$ c_{40} = 0.40$
$ c_{22} = 0.12$
$ c_{04} = 0.23.$

In equation (3.11) only the terms with coefficients c_{21} and c_{03} have an odd exponent of a or b. Thus, these are the only terms affected by our use of a different angular convention. To compensate, the signs of c_{21} and c_{03} have been changed relative to those of Cox and Munk. This allows the value of $p(\beta,\alpha)$ to maintain the current upwind/crosswind slope skewness. If the upwind/crosswind variations in slope distributions are ignored,

$$p(\beta, \alpha) = (\pi\sigma^2)^{-1}\exp[-(\tan\beta/\sigma)^2] \tag{3.11a}$$

where $\sigma = 0.003 + 0.00512W$.

The delta area $\delta z_x\delta z_y$ is equal to the area of the tolerance ellipse. Thus, the probability of occurrence of wave slopes that will directly reflect incident light toward the remote sensor is defined by:

$$P(z_x, z_y) = [p(\beta, \alpha)\Omega_{sun}\sec^3\beta\sec\omega]/4 \tag{3.12}$$

where ω is the angle of incidence measured from the normal to the tilted wavelet.

3.4 SATELLITE SENSOR, SUNGLINT, AND CURVATURE OF THE EARTH

To extend the discussion of sunglint to its impact on satellite-acquired data, we will illustrate the determination of surface slope required to produce a reflected solar image in the viewing direction of a satellite sensor. We will once again follow the approaches of Cox and Munk[67,68] and Maul.[262] However, we will include geometric impacts of the curvature of the Earth, as well as the geometric impacts of a sensor that can both tilt forward or backward (along the direction of the orbit) and scan from side to side from the satellite nadir point.

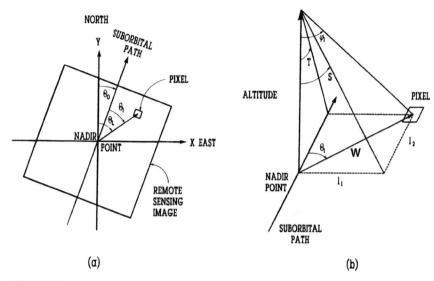

(a)   (b)

FIGURE 3.4

(a) Lab coordinate system in which the origin is at the satellite nadir point, the y-axis is directed positively northward, the x-axis is directed positively eastward, and the z-axis is directed positively upward. (b) Relationship among imager viewing angles, targetted pixel, and suborbital satellite path as measured in the Lab system.

To describe the conditions for which sunglint from a wind-roughened air-water interface will be recorded at a remote sensor (considered herein to be a satellite), we will first establish a reference co-ordinate system (which we shall refer to as the *Lab system*). The geometries of the sensor, the water surface, and the sun will be related to this Lab system.

Figure 3.4(a) illustrates the Lab co-ordinate system in which the origin is at the satellite's nadir point. The positive y-axis on the Earth's surface points north and the positive x-axis points east. The positive z-axis points vertically upward through the satellite. All azimuthal angles are measured from the y-axis toward the east. The satellite has an orbit azimuth θ_0 (relative to north) and an orbit altitude of a (in km). The satellite sensor is considered to be an imager that can tilt and scan with tilt T defined in the plane of the orbit (measured positively along the direction of the orbit) and scan S defined perpendicular to the orbital plane (positive to the right when looking along the direction of the orbit). The pixel (picture scene element) delineated in Figure 3.4(a) represents one location in the remotely sensed image. The sensor is taken to be downward viewing, so both T and S are considered within the limits $-\pi/2$ and $+\pi/2$. The suborbital path has an azimuth of θ_0 measured eastward from north. θ_I represents the viewing azimuth of the imager measured clockwise from the suborbital path. ϕ_I represents the nadir imager angle, and W [Figure 3.4(b)] represents the horizontal coverage of the satellite sensor (comprised of components l_1 and l_2 along the x and y axes, respectively).

The imager viewing geometry is defined by:

$$\tan S = l_1/a$$
$$\tan T = l_2/a$$
$$W = (l_1^2 + l_2^2)^{1/2} = a(\tan^2 S + \tan^2 T)^{1/2}$$
$$\tan\phi_I = W/a = (\tan^2 S + \tan^2 T)^{1/2}$$
$$\tan\theta_I = l_1/l_2 = \tan S/\tan T \text{ (or } \cos\theta_I = \tan T/\tan\phi_I\text{)}.$$

Defining the nadir angle as the polar angle measured from the nadir direction, the nadir and azimuthal look angles (ϕ_L and θ_L, respectively) defining the direction in which the sensor is directed in the Lab system to view the pixel are then obtained as

$$\phi_L = \phi_I$$

and

$$\theta_L = \theta_I + \theta_0.$$

Figure 3.5 schematically illustrates the geometric inter-relationships among the satellite orbit, sensor viewing directions, sea-state, and curvature of the Earth. The center of the Earth is at point C, and the angle β_c is the angular distance describing the great circle through the nadir point of the orbiting satellite and the geographic location of the pixel. The radius of the Earth is R (6368 km). The satellite imager is located at point I, at an altitude a along the z-axis which is directed positively from the center of the Earth through the satellite's nadir point. The imager geometry is as defined above via the look angles ϕ_L and θ_L.

The sensor will scan beyond the horizon when the nadir look angle $\phi_L \geq \sin^{-1}[R/(R + a)]$.

The curvature of the Earth results in a surface tilt being observed by the imager despite a flat air-water interface at the location of the pixel. This situation is sketched in Figure 3.6, wherein R, β_c, W, and a are as previously defined. The pixel containing the flat air-water interface is at location P. The nadir look angle required for the imager to view this pixel is ϕ_L. It is seen that at pixel P the curvature of the Earth results in a tilt of calm water equivalent to the angle β_c given by:

$$\sin\beta_c = y/R$$

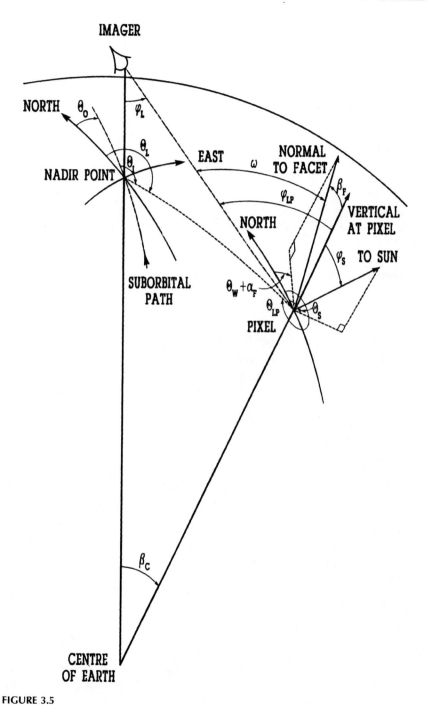

FIGURE 3.5
Geometric inter-relationships among satellite orbit, sensor viewing directions, sea-state, and curvature of the Earth as measured in the Lab system.

where y is determined from

$$y^2 + (R - h)^2 = R^2.$$

Since $y/(a + h) = W/a = \tan\phi_L$, h is determined from

$$h = \frac{-(2aW^2 - 2a^2R) + [(2aW^2 - 2a^2R)^2 - 4a^2W^2(a^2 + W^2)]^{1/2}}{2(a^2 + W^2)}$$

We shall now determine the latitude and longitude of the pixel location. The location of the pixel in the image of the sensor is seen (Figure 3.5) to lie along a great circle β_c° from the nadir point along the direction of θ_L.

Consider a spherical triangle on the Earth's surface with its apexes located at the North pole, the satellite's nadir point, and the pixel location. From spherical geometry, the latitude (LAT) of the pixel location in the Lab system is given by:

$$LAT = \sin^{-1}[\sin Lat_{Sat}\cos\beta_c + \cos Lat_{Sat}\sin\beta_c\cos\theta_L]$$

Similarly, the longitude (LONG) of the pixel location in the Lab system is given by:

$$LONG = Long_{Sat} - \tan^{-1}\left[\frac{\sin\theta_L}{\cos Lat_{Sat}\cot\beta_c - \sin Lat_{Sat}\cos\beta_c}\right]$$

The solar zenith angle ϕ_s and the solar azimuth angle θ_s at pixel P (located at LAT and LONG) are defined in terms of the pixel's LAT and LONG as well as the time of day that the pixel is being viewed. The time of day is described in terms of the solar declination angle ∂ and the solar Greenwich hour angle GHA.

The declination angle ∂ is defined as the solar latitude $\leq 23.5°$ north or south. The Greenwich hour angle GHA is defined as the angular equivalent of Greenwich mean time, i.e., if the local Greenwich mean time at the pixel is GMT, then the Greenwich hour angle is GHA $= 15° \times$ (GMT $- 12$). At noon in Greenwich, England, the GHA is $0°$. Tables of solar declinations ∂ may be obtained from ephemerides, nautical almanacs, or equation (2.7).

The meridian angle $m°$ is defined as the difference between the longitude of the pixel and the solar GHA, viz., $m \equiv$ LONG $-$ GHA.

The solar zenith angle ϕ_s and solar azimuth angle θ_s at pixel P, as expressed in the Lab system, can then be shown to be determined, from spherical triangle relationships, as:

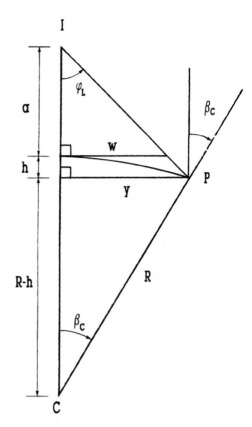

FIGURE 3.6
Sensor-observed tilt of a flat air-water
interface pixel due to curvature of the
Earth.

$$\cos\phi_s = \cos(90° - \text{LAT}°)\cos(90° - \partial°)$$
$$+ \sin(90° - \text{LAT}°)\sin(90° - \partial°)\cos m° \qquad (3.13)$$

and

$$\sin\theta_s = \sin(90° - \partial°)\sin m°/\sin\phi_s \qquad (3.14)$$

When the satellite sensor views a wind-roughened air-water interface, it is observing a surface tilted as a consequence of (a) the Earth's curvature and (b) the wave-slope. As evident from Figures 3.5 and 3.6, the component of the tilt due to the Earth's curvature (which, at the pixel, has a value equal to the angle β_c) in the imager's line of sight is always directed away from the imager. The component of the tilt in the imager's line of sight due to the wave-slope itself may be directed either away from or toward the imager. The total tilt angle in the line of sight of the sensor (as measured in the Lab system) will be designated as β_L.

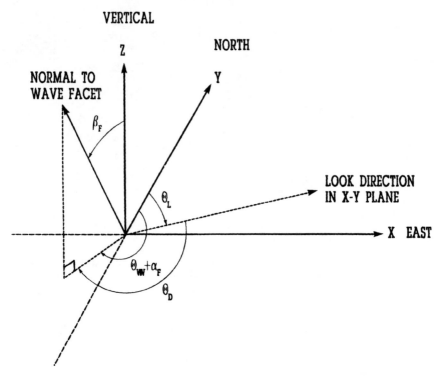

FIGURE 3.7
The look direction of the imaging sensor, the slope of the wave facet, and the component of tilt in the line of sight of the imager as measured in the Lab system.

To reiterate the purpose of this section, we seek to determine the surface tilts in the line of sight of a remote imager that would allow sunglint from that surface to be imaged. These surface tilts will be described in terms of the zenith and azimuthal angle components of β_L as measured in the Lab system.

Consider the Lab co-ordinate system of Figure 3.7 in which the facet and its normal is transposed to the origin. The z-axis and y-axis (North) are as previously defined. The sensor's azimuth look angle in the x-y plane θ_L (the sum of θ_I and θ_o) is illustrated. Let the azimuthal wind direction be θ_W (measured eastward from North to the upwind direction). Let the azimuthal angle of the slope of the wave facet be α_F (measured from the upwind direction to the line of steepest descent on the wave facet). In the Lab system, therefore, the azimuth of the wave facet is the angular value ($\theta_W + \alpha_F$). The zenith angle of tilt due to the wave facet in the Lab system is the angular value β_F (measured from the z-axis).

Let us define the difference between the azimuth of the look direction, θ_L, and the azimuth of the facet tilt, $(\theta_W + \alpha_F)$, as θ_D [Figure 3.7)], i.e., $\theta_D \equiv \theta_W + \alpha_F - \theta_L$.

The component of tilt in the line of sight of the imager is designated β_{LF} and is defined by:

$$\tan\beta_{LF} = \cos\theta_D \tan\beta_F$$

Therefore, the total tilt of the wave facet, β_L, in the line of sight of the imager is obtained as the sum of the tilt due to the curvature of the Earth and the tilt due to the wavelet itself, i.e.,

$$\beta_L = \beta_c + \beta_{LF}$$
$$= \beta_c + \tan^{-1}(\cos\theta_D \tan\beta_F) \qquad (3.15)$$

Consider a sensor possessing a nadir look angle ϕ_L. The minimum surface tilt in the line of sight of the sensor that will allow the sensor to focus upon a portion of the sky would be $(\beta_L)\text{MIN} = 0°$. Let $(\beta_L)\text{MAX}$ be the maximum surface tilt in the line of sight that would permit incident radiation from any region of the sky to enter the field-of-view of the sensor. If $\beta_L > 90° - \phi_L$, the surface is not visible from the location of the sensor. This maximum slope would be $\tan(90° - \phi_L)$. However, this slope could still result in the sensor viewing multiple surface reflections. To compensate for the possible recording of multiple reflections, Cox and Munk[68] set the maximum slope at half this value, i.e., set $\tan(\beta_L)\text{MAX} = 0.5\tan(90° - \phi_L)$. Thus, the range of surface tilts in the line of sight of the sensor that would allow sun-glint to be recorded by the sensor is considered to be given by:

$$0 \leq \beta_L \leq -\tan^{-1}(0.5\cot\phi_L).$$

For a flat air-water interface ($\beta = 0°$) close enough to a remote sensing device to be unimpacted by the Earth's curvature, reflected sun-glint will be observed at a nadir look angle equal to the solar zenith angle ϕ_s. As we have discussed, however, the angle of reflected sun-glint into a remote sensing device over waves will be a function of the wave tilt angles β and α, as well as the Earth's curvature tilt angle β_c. Since incidence and reflection angles are equal, the problem of surface solar glitter necessitates the definition of the incident and reflection angles in terms of the location of the normal to the air-water interface for varying combinations of β and α. Thus, the incident and reflected angles, while equal in magnitude, will not be defined by identical spherical angular coordinates.

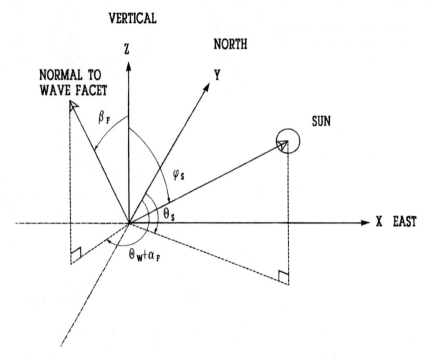

FIGURE 3.8
Pixel-centered coordinate system with x-axis directed positively eastward, y-axis directed positively northward, and z-axis directed positively upward from an untitled air-water interface at the pixel location. The solar zenith and azimuth angles and the normal to the wave facet are shown in the pixel-centered system.

To this point in our discussion we have used the Lab co-ordinate system based at the satellite nadir point with the z-axis vertically up to the satellite and the y-axis directed north. To solve the sunglint analyses, however, we must also determine the angle of surface tilt required to image the sun in the sensor's field-of-view as well as the angles of incidence and reflection for this imaging. To do so we will define another co-ordinate system (Figure 3.8) centered at the pixel. In this pixel-centered coordinate system, the y-axis will still point North and the x-axis will still point East. The z-axis, however, will point vertically upward from a calm air-water interface. Thus, the pixel-centered co-ordinate system has its z-axis rotated at an angle of β_c (along the sensor's look direction) from the z-axis of the Lab co-ordinate system.

In the pixel-centered co-ordinate system, the zenith and azimuth angles for the solar position are (ϕ_s, θ_s), for the normal to the surface are (β, α), and for the satellite position are (ϕ_{LP}, θ_{LP}). The azimuthal angle of the normal to

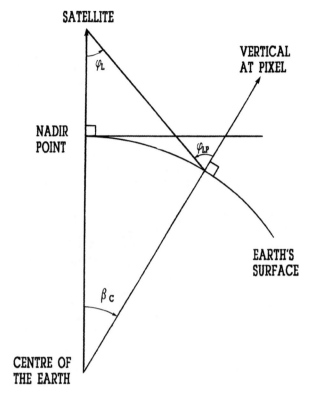

FIGURE 3.9
Location of the satellite in the pixel-centered coordinate system.

the flat water surface α may be expressed in terms of the azimuthal wind direction θ_W and the azimuthal wavelet slope direction α_F, i.e.,

$$\alpha = \theta_W + \alpha_F.$$

Figure 3.9 shows the relationships among the sensor look angle ϕ_L, the tilt of the vertical at the pixel due to the Earth's rotation β_c, and the zenith look angle from the pixel to the satellite ϕ_{LP}. The zenith look angle ϕ_{LP} from the pixel to the satellite is defined by

$$\phi_{LP} = \phi_L + \beta_c.$$

For sensor azimuth look angles θ_L in the range $0 \le \theta_L \le \pi$, the azimuth look angle from the pixel to the satellite, θ_{LP}, is defined by

$$\theta_{LP} = \pi + \theta_L.$$

For sensor azimuth look angles θ_L in the range $\pi \leq \theta_L \leq 2\pi$, the azimuth look angle from the pixel to the satellite, θ_{LP}, is defined by

$$\theta_{LP} = \theta_L - \pi.$$

Let the angle of solar incidence (i.e., the angle between the direction of the sun and the normal to the wavelet surface, measured in the pixel-centered co-ordinate system) be ω. This angle must be half the angular distance between the locations of the sun and sensor as measured in the pixel-centered co-ordinate system.

To solve for this angle of incidence ω, we will use the dot product of two unit vectors both anchored at the origin of the pixel-centered co-ordinate system. The first unit vector (incident) will point toward the sun, and the second vector (reflected) will point toward the sensor. Thus,

$$\cos2\omega = \sin\phi_s\sin\theta_s\sin\phi_{LP}\sin\theta_{LP} + \sin\phi_s\cos\theta_s\sin\phi_{LP}\cos\theta_{LP}$$

$$+ \cos\phi_s\cos\phi_{LP} \tag{3.16}$$

Equation (3.16) thus relates the incident solar beam angle ω to the angular co-ordinates of the sun and sensor in the pixel-centered co-ordinate system.

Let the incident solar beam direction be defined in the pixel-centered system by the vector (x_i, y_i, z_i). Let the reflected beam direction be defined by the vector (x_r, y_r, z_r), and let the direction of the normal to the wave facet be defined by the vector (x_n, y_n, z_n). We will now calculate the tilt (β, α) of the normal to the surface that will permit Fresnel reflection with incident and reflected angles equal to ω. To do so we shall invoke Alhazen's observational Law of Reflection, which states that the vector difference between incident and reflected rays must lie along the normal to the reflecting surface. (Alhazen's corresponding observational Law of Refraction states that the vector difference between the incident ray and the product of the refracted ray and the relative index of refraction $n(\lambda)$ must lie along the normal to the refracting surface). Thus:

$$x_r - x_i = 2x_n\cos\omega$$

$$y_r - y_i = 2y_n\cos\omega$$

$$z_r - z_i = 2z_n\cos\omega$$

where ω is the angle of incidence as measured from the normal to the tilted surface.

The incident vector components (x_i, y_i, z_i) are:

$$x_i = \sin\phi_s\sin\theta_s$$
$$y_i = \sin\phi_s\cos\theta_s$$

and

$$z_i = -\cos\phi_s.$$

The reflected vector components (x_r, y_r, z_r) are:

$$x_r = \sin\phi_{LP}\sin\theta_{LP}$$
$$y_r = \sin\phi_{LP}\cos\theta_{LP}$$

and

$$z_r = \cos\phi_{LP}.$$

The normal vector components (x_n, y_n, z_n) are:

$$x_n = \sin\beta\sin\alpha$$
$$y_n = \sin\beta\cos\alpha$$

and

$$z_n = \cos\beta.$$

Using the above equation sets to solve for (β, α) yields:

$$\cos\beta = [\cos\phi_{LP} + \cos\phi_s]/2\cos\omega \tag{3.17}$$

and

$$\cos\alpha = [\cos\phi_{LP}\sin\phi_{LP} - \cos\theta_s\sin\phi_s]/2\cos\omega\sin\beta \tag{3.18}$$

The tilt due to the Earth's curvature and the tilt due to the wind-roughened water surface must combine to produce the required total tilt value (β, α). Solving for the angle between the unit vector due to curvature and the required total tilt unit vector will yield β_F, the value of the zenith angle of tilt due to the wave facet that will allow sun-glint to be recorded by the sensor, viz.,

$$\cos\beta_F = \sin\beta\sin\alpha\sin\beta_c\sin\theta_L + \sin\beta\cos\alpha\sin\beta_c\cos\theta_L + \cos\beta\cos\beta_c. \tag{3.19}$$

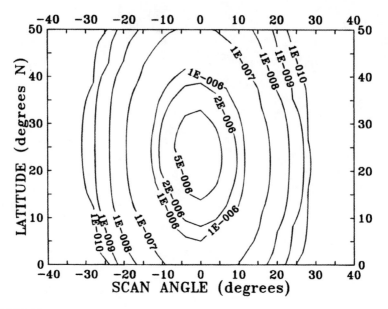

FIGURE 3.10
Sun-glint patterns for a suborbital satellite path crossing the equator at 12 o'clock local standard time at the summer solstice. The contours of the sun-glint are reflectivities plotted in terms of pixel LAT and sensor scan angle. The scanning sensor is at an altitude of 950 km with a tilt of $0°$.

This facet tilt must lie in an azimuthal direction α_F such as to ensure that its component in the x-y plane, when added to the component of the curvature tilt in the x-y plane, equals the component of the required total tilt in the x-y plane. Thus:

$$\cos\alpha_F = y_F / (x_F^2 + y_F^2)^{1/2} \qquad (3.20)$$

where

$$x_F = \sin\alpha\sin\beta - \sin\theta_L\sin\beta_c$$

and

$$y_F = \cos\alpha\sin\beta - \cos\theta_L\sin\beta_c.$$

Equations (3.19) and (3.20) determine, for a wave facet at the pixel location, the zenith and azimuth tilt angles (β_F, α_F) required to produce sunglint in a remote sensor of given look and scan angles for a given solar zenith and azimuth.

Figure 3.10 illustrates sun-glint patterns (plotted in terms of pixel LAT and sensor scan angle for a windspeed of 2 m s^{-1} from the North. The

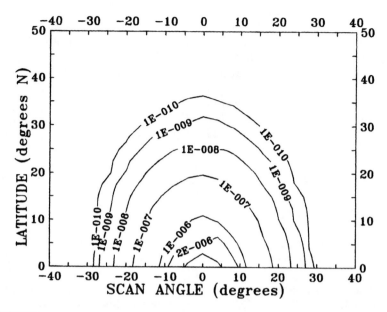

FIGURE 3.11
Sun-glint patterns as in Figure 3.10 but for a scanning sensor that is tilted 20° along the satellite suborbital path and away from the sun.

satellite suborbital path is taken to cross the equator at 12 o'clock local standard time at the summer solstice with an angle of inclination of 90°. The satellite is taken to be orbiting at an altitude of 950 km. Sun-glint patterns typically are elongated along the satellite suborbital path. This elongation is not exactly aligned with the satellite path in Figure 3.10 due to the sun not being located directly south of the satellite. The contours in Figure 3.10 are the reflectivities for solar radiance, and the reflected sun-glint values would then be determined as the product of these reflectivities and the incident solar radiance at the time of observation. Figure 3.11 illustrates sun-glint patterns in the manner of Figure 3.10 but for a sensor that is tilted 20° along the satellite suborbital path and away from the sun. Since the SeaStar satellite will have comparable orbital parameters and its payload sensor SeaWiFS can tilt to ±20° and scan to ±58.3°, figures such as Figures 3.10 and 3.11 can be used to determine appropriate SeaWiFS viewing angles to minimize sun-glint for other windspeeds and directions, other days of the year, and other suborbital points.

3.5   APPARENT AND INHERENT OPTICAL PROPERTIES OF NATURAL WATERS

Recall from Chapter 1 that, subsequent to the downwelling photon flux having penetrated the air-water interface, it will undergo attenuation in its subsurface propagation. The attenuation is described in terms of the total attenuation coefficient $c(\lambda)$, the scattering coefficient $b(\lambda)$, the absorption coefficient $a(\lambda)$, and the backscattering and forwardscattering probabilities $B(\lambda)$ and $F(\lambda)$, respectively. Subsurface reflectance at any level within the aquatic medium is usually described in terms of the volume reflectance $R(\lambda, z)$. The volume reflectance (also referred to as the irradiance reflectance) at a depth z beneath the air-water interface is the upwelling irradiance $E_u(\lambda, z)$ at that depth normalized to the downwelling irradiance $E_d(\lambda, z)$ at the same depth, i.e.:

$$R(\lambda, z) = \frac{E_u(\lambda, z)}{E_d(\lambda, z)} \tag{3.21}$$

In situ profiles of the volume reflectance are generally obtained from underwater spectroradiometers equipped with one upward-viewing and one downward-viewing sensor each possessing cosine responses.

Volume reflectance, being a local measurement of the upwelling and downwelling irradiance fields, is clearly dependent upon the radiance distribution of the underwater light regime at that location. Volume reflectance, therefore, qualifies as an *apparent optical property* of the water column (see Preisendorfer[326]). Similarly, the attenuation coefficients for downwelling irradiance and scalar irradiance are also apparent optical properties since, being spatial derivatives of the underwater light regime, they are dependent upon the distribution of that underwater light regime.

Baker and Smith,[14] however, have illustrated that irradiance attenuation coefficients in natural water bodies are not significantly altered *provided the composition of the water body remains essentially invariant.* This suggests that the optical behavior of natural waters is a consequence of qualities that are *inherent* to the water body itself. Such *inherent optical properties*[326] determine the impacts to the subsurface radiation field that manifest as the *apparent optical properties* that are readily measurable. As we have discussed earlier, the organic and inorganic materials comprising the water column contribute a specific set of absorption and scattering centers. It is these sets of organic and inorganic materials that attenuate the irradiance and modulate the propagative directions of the photon fluxes that are transmitted through the air-water interface. Inherent optical properties of a water body include the total attenuation coefficient $c(\lambda)$, the absorption coefficient $a(\lambda)$, the scattering coefficient $b(\lambda)$, the forwardscattering probability $F(\lambda)$, the backscattering probability $B(\lambda)$, the scattering albedo $\omega_o(\lambda)$, and the volume scattering function $\beta(\theta)$.

The volume reflectance was originally categorized as an apparent optical property of the water body by Preisendorfer.[326] Kirk[226] supports the premise that $R(\lambda, z)$ is more appropriately a property of the radiative field than a property of the water column and further notes that the volume reflectance is a dimensionless quantity, unlike the apparent and inherent optical properties that possess units. We are not uncomfortable with such concerns and arguments. However, we are also not uncomfortable with the Preisendorfer categorization that considers volume reflectance to be a property of the water column. In this book we will, therefore, adhere to regarding volume reflectance as an apparent optical property of the water column.

The only inherent optical property that is readily measured *in situ* is the total attenuation coefficient $c(\lambda)$. The inherent optical properties $b(\lambda)$ and $B(\lambda)$ are defined in terms of appropriate integrations of the volume scattering function $\beta(\theta)$. Thus, direct measurements of the inherent optical properties [other than $c(\lambda)$] require sophisticated instrumentation capable of determining the entire volume-scattering function to separate the scattering coefficient $b(\lambda)$ from the total attenuation coefficient $c(\lambda)$ $[c \equiv a + b]$. Therefore, apart from $c(\lambda)$, inherent optical properties must either be directly measured using very sophisticated instrumentation or be inferred from directly measured apparent optical properties.

A widely used approach to determine relationships among the measurable apparent and required inherent optical properties of natural water has been the development of models based on suitable simulations of the radiative transfer process. To arrive at a precise description of radiative transfer in an absorbing and scattering medium such as oceanic, inland, and coastal water regimes requires a solution of the classical *radiative transfer equation:*

$$\frac{dL(z, \theta, \phi)}{dr} = -cL(z, \theta, \phi) + L_*(z, \theta, \phi) \qquad (3.22)$$

where $L(z, \theta, \phi)$ is the radiance at depth z of a photon beam propagating in the direction (θ, ϕ) and dL/dr is the change in radiance along the direction r experienced by this photon beam due to the combined processes of absorption and scattering. c is the total attenuation coefficient appropriate to the medium. Thus, the first term on the right-hand side of equation (3.22), $[-cL(z, \theta, \phi)]$, represents loss due to attenuation. The second term on the right-hand side of equation (3.22), $[L_*(z, \theta, \phi)]$, represents gain due to scattering. $L_*(z, \theta, \phi)$, called the *path function*, is a consequence of the scattering occurring in every infinitesimal volume of the medium and is generalized in terms of the probability that a photon which is propagating along a direction other than (θ, ϕ) prior to a scattering event will propagate along the direction (θ, ϕ) subsequent to the scattering event. Extensive discussions on the form of $L_*(z, \theta, \phi)$ may be found in Gershun,[124] Preisendorfer,[326] Duntley,[92] and elsewhere.

We will not go into details of the *radiative transfer equation* here. Suffice to say that the radiative transfer equation as applicable to oceanographic and limnological remote sensing in the visible wavelength region (Chandrasekhar[59]) may be written:

$$\cos\theta \, \frac{dL(z, \theta, \phi)}{dz} = cL(z, \theta, \phi) - \beta(\theta, \phi; \theta_0, \phi_0)E_i\exp(czsec\theta_0)$$

$$- \int_{\theta'=0}^{\pi} \int_{\phi'=0}^{2\pi} \beta(\theta, \phi; \theta', \phi')L(z, \theta, \phi)\sin\theta'd\theta'd\phi' \quad (3.23)$$

where $L(z, \theta, \phi)$ = radiance of photons at depth z propagating in the direction (θ, ϕ),

dL/dz = the change in radiance with depth in the water due to scattering and absorption,

c = beam attenuation coefficient,

E_i = irradiance due to direct sunlight on a plane perpendicular to its propagation direction in the water (which is defined by zenith angle θ_0 and azimuth angle ϕ_0, both measured in-water),

$\beta(\theta, \phi; \theta_0, \phi_0)$ = volume scattering function that defines the probability that the in-water direct visible sunlight will scatter from its initial direction (θ_0, ϕ_0) to the direction (θ, ϕ), and

$\beta(\theta, \phi; \theta', \phi')$ = volume scattering function that defines the probability that the in-water diffuse visible radiance will scatter from its initial direction (θ', ϕ') to the direction (θ, ϕ).

Notice that in the spherical coordinate system representation of equation (3.23), the radiation is considered to be scattered *from* a direction (θ', ϕ'), and *from* a direction (θ_0, ϕ_0) *to* a direction (θ, ϕ). Equation (3.23) considers the underwater light field as both direct and diffuse. The first term on the right-hand side represents the attenuated radiance in the direction of propagation (θ, ϕ). The second term represents the direct solar irradiance scattered into the observed direction. The third term represents the diffuse radiance from outside the propagating direction beam scattered into the propagating direction. The second and third terms of equation (3.23) essentially define the *path function* L_* of equation (3.22). An excellent treatment of the radiative transfer equation, including detailed discussions of the boundary conditions required for its solution, may be found in Maul.[262]

3.6 MONTE CARLO SIMULATIONS OF PHOTON PROPAGATION

As we have seen, the inherent optical properties of an aquatic medium (in particular, the specific optical properties of the co-existing aquatic components), in conjunction with the incident light field and the sea-state, determine the nature of the subsurface radiation field. The radiative transfer process (i.e., the transfer of energy between an electromagnetic field and an attenuating medium) involves complex interactions among the photons and the scattering and absorption centers comprising the attenuating medium. These complexities have thus far prevented an exact solution of the radiative transfer equation and, therefore, prevented obtaining an explicit analytical relationship among the properties of the electromagnetic field and the inherent properties of the attenuating medium. Reviews of the attempts of workers to derive such mathematical relationships for natural waters may be found in Prieur and Morel[336] and Prieur.[335] These formulations will not be detailed here, other than to say that the more successful of the approaches generally consider the aquatic medium as a continuum of thin layers, concentrating upon the transmissive and reflective properties of a single layer, extending this behavior to the adjacent layer and so on, invoking the principle of invariance at each layer boundary [Beardsley and Zaneveld,[21] Plass et al.,[312,314] Sobolov,[381] Raschke,[339] and others].

Since explicit mathematical formulism describing the radiative transfer process is such a highly elusive commodity, semi-empirical relationships linking the properties of the radiation field to the inherent properties of the attenuating medium have been sought. The most widely used approach for obtaining such relationships has been Monte Carlo simulations of photon propagation. This approach has long been used in theoretical physics, one of its principle applications having been to describe the Brownian motion of atoms and molecules, another to describe the interactions of solar and galactic cosmic radiation with interplanetary magnetic field structures.

The application of Monte Carlo analyses to atmospheric and oceanic photon interactions was pioneered by Plass and Kattawar[309-311] and Kattawar and Plass.[211] Since then, a number of workers including Gordon and Brown,[134] Gordon et al.,[135] Kirk,[224,225,227] Jerome et al.,[204] Bannister,[17] among others, have developed and applied Monte Carlo simulations to relate empirically the inherent optical properties of natural waters to the radiation fluxes that are observed within them. Such an empirical approach is ideally suited to the stochastic nature of the propagation properties of individual photons. The geometric path followed by a single photon and the series of its lifetime encounters with scattering and/or absorption centers residing within its confining medium are best described in terms of a random process controlled by the nature of the individual photon and the nature of each individual scattering and absorption center. The lifetime of the photon in the attenuating medium is itself governed by the nature and densities of the scattering and absorption centers. The bulk inherent optical properties of the water column $[c, \omega_o, \beta(\theta)]$ provide probability distributions that can define the distance between successive interactions, whether that interaction is scattering or absorption, and the directionality of scattering interactions. Thus, these bulk properties can determine the probability that an individual photon will be absorbed or scattered in a specified direction within a given distance in the attenuating medium.

Monte Carlo simulations take advantage of the statistical behavior of photon interactions by utilizing a high-speed computer to follow the propagation of a large number of photons, considered individually, through an imaginary water column possessing a pre-selected set of optical properties. Random numbers are generated to select, in accordance with the pre-selected optical properties, such photon fate determinants as the pathlength between each successive photon encounter with the attenuating medium, whether that encounter be absorption or scattering, and, in the event of a scattering encounter, the direction in which the photon will be scattered. An incident radiation distribution defines the initial position and direction of propagation of the incident photon. The first encounter of the photon with the attenuating medium occurs at the first randomly-determined pathlength. Throughout the simulated propagation, statistical weighting techniques are utilized to determine the reduction in flux intensity due to absorption. A photon ceases to be tracked once its statistical weight falls beneath some pre-assigned value. The scattering angle is determined by the $\beta(\theta)$ function. After each encounter the photon trajectory is continued anew considering the modified photon (in either direction of propagation for scattering events or weighted intensity for absorption events) as the incident photon for the next encounter. The succession of encounters continues until the photon either achieves some minimum weighting value, exits the medium, or attains a pre-determined destiny such as arrival at a detector location. Snell's law and Fresnel's equation are used to accommodate surface reflection, refraction, and total internal reflection. For each scattering event the scattering angles are determined in a spherical coordinate system centered on the scattering event with the polar angle measured from the direction in which the photon is propagating. The polar angle is selected from the volume scattering distribution function through the use of a random number. The azimuthal angle (as measured in the plane perpendicular to the direction of propagation) is selected in the range 0° to 360° through the use of another random number. Pre-placed detectors record the total number of photons and their direction of arrival at a point or a depth in the water column in such a manner as to provide radiance, vector irradiance, or scalar irradiance values as required.

Clearly, the upwelling and downwelling irradiances and scalar irradiances, as well as the volume reflectance, may be determined in this manner at any depth within the water column. It is evident that the statistical accuracy of the Monte Carlo method is directly related to the number of individual photons tracked through the attenuating medium. Since many hundreds of thousands of photons (oftimes millions) are required to reliably simulate photon propagation through aquatic media, it is conceivable that, depending upon the number of adjacent attenuating media, the optical properties of those attenuating media, the number of measurement depths, and the number of values of each parameter required to consider a specific problem, a considerable amount of computer time could be required to obtain the desired accuracy in an optical property relationship. In this manner, however, a complete description of the underwater light regime may be readily, albeit painstakingly, obtained.

The accuracy of a Monte Carlo simulation involving N photons (of which n photons are detected) is given by a standard deviation σ of:

$$\sigma = \left[\frac{n(N-n)}{N}\right]^{1/2} \text{ photons} \qquad (3.24)$$

Selection of the initial conditions for Monte Carlo simulations include the directional distributions of the incident photon fluxes (e.g., direct solar radiation, diffuse skylight, angle of incidence, overcast atmospheric conditions). Other required inputs are the physical descriptions of scattering processes (e.g., the selection of Rayleigh scattering to describe molecular scattering within the aquatic medium and the selection of Mie scattering theory to describe particulate scattering). Suffice to say that *Rayleigh* or molecular scattering occurs when the wavelength of the impinging radiation is considerably larger than the diameter of the scattering center, and *Mie* scattering occurs when the wavelength of the impinging radiation is of the same order of magnitude or smaller than the diameter of the scattering center. Details and mathematical treatments of Rayleigh (molecular) and Mie (particulate) scattering phenomena may be found in Chandrasekhar,[59] Maul,[262] and numerous other optical texts.

Once Monte Carlo simulations of photon trajectories are completed, the desired relationships among inputted and outputted parameters involving apparent and inherent optical properties, bulk aquatic properties, directional and spectral character of the incident radiation fields, and radiance or irradiance levels are obtained from mathematical regressions of these data sets. Thus, curve-fitting is an integral component of Monte Carlo analyses.

In an epic work, Gordon *et al.*[135] utilized a Monte Carlo simulation of the radiative transfer process to relate the apparent optical properties of a water mass to its inherent optical properties. The following sections will discuss their and some other existing Monte Carlo simulation models.

3.6.1   The Gordon *et al.* Monte Carlo Simulation

Using volume scattering functions from Kullenberg,[240] Gordon *et al.*[135] used curve-fitting to Monte Carlo simulations to relate the apparent optical properties $K(z)$, $D_d(z)$, and $R(z)$ to the inherent optical properties c, ω_o, F, and B through the equations:

$$\frac{K(z)}{cD_d(z)} = \sum_{n=0}^{N} k_n(z)[\omega_0 F]^n \tag{3.25}$$

and

$$R(z) = \sum_{n=0}^{N} r_n(z)\left[\frac{\omega_0 B}{1 - \omega_0 F}\right]^n \tag{3.26}$$

where ω_0 = the scattering albedo (the fraction of the totality of scattering and absorption interactions that are due to scattering, i.e., b/c),

F = the forwardscattering probability,

B = the backscattering probability $(1 - F)$,

$K(z)$ = the irradiance attenuation coefficient for downwelling irradiance at depth z,

$R(z)$ = the volume reflectance at depth z,

$D_d(z)$ = the distribution function of downwelling irradiance at depth z (defined by the ratio of the downwelling scalar irradiance E_{0d} to the downwelling irradiance E_d at the depth z,

c = the total attenuation coefficient,

and $k_n(z)$ and $r_n(z)$ are sets of expansion coefficients. The wavelength λ dependencies of the terms in the above equations, while omitted for simplification, are, nonetheless, implicit.

Equations (3.25) and (3.26) can be inverted to allow the determination of inherent optical properties from measurements of $K(z)$, $R(z)$, and c (assuming that $D_d(z)$ can be estimated). That is:

$$\omega_0 F = \sum_{n=0}^{N} k_n'(z)\left[\frac{K(z)}{c(z)D_d(z)}\right]^n \tag{3.27}$$

and

$$\frac{\omega_0 B}{1 - \omega_0 F} = \sum_{n=0}^{N} r_n'(z)[R(z)]^n \tag{3.28}$$

Gordon et al. determined two sets of coefficients $r_n'(z)$, one set appropriate for solar angles $\leq \sim 20°$ (referred to as sun case) and one set appropriate for solar angles $\geq \sim 30°$ (referred to as sky case) as measured in the water column. The single set of $k_n'(z)$ coefficients is independent of sun angle.

The distribution function of downwelling irradiance at depth z, $D_d(z)$, was shown to be related to the inherent optical properties through the relationship:

TABLE 3.1

Expansion Coefficients for Relationships between Apparent and Inherent Optical Properties of Natural Waters[135]

Expansion Coefficient	Zeroth, First, Second, and Third Order Values				
	0	1	2	3	
k	1.0016	−0.9959	0.1089	−0.1527	
k'	0.9588	−0.7408	−0.3935	0.1776	
r	0.0001	0.3244	0.1425	0.1308	(sun)
	0.0003	0.3687	0.1802	0.0740	(sky)
r'	−0.0003	3.0070	−4.2158	3.5012	(sun)
	−0.0008	2.6987	−3.2310	2.8947	(sky)
d	0.0002	0.2991	0.1492	0.3115	(sun)
	−0.0037	0.3664	0.1689	0.0131	(sky)

$$D_d(z) - D_{d0}(z) = \sum_{n=0}^{N} d_n(z) \left[\frac{\omega_0 B}{1 - \omega_0 F} \right]^n \qquad (3.29)$$

where $d_n(z)$ are expansion coefficients and D_{d0} is the value of the downwelling distribution function at depth z for a water body defined by a scattering albedo $\omega_0 = 0.0$ (i.e., for a water body in which no scattering occurs). For direct solar irradiance corresponding to an in-water refracted angle of θ_0 just below the air-water interface (i.e., at depth $z = 0$), $D_{d0}(0)$ is given by $1/\cos\theta_0$ [as seen from equation (1.30)]. Since most natural light propagates in the direction of the direct solar beam during clear-sky days, $D_{d0}(0)$ can be reliably taken to be the inverse of the in-water refracted angle for ideal remote sensing conditions.

Equations (3.27), (3.28), and (3.29) then enable the apparent optical properties $K(z)$, $R(z)$, and the inherent optical property c (obtainable from *in situ* measurements) to be used in the determination of the inherent optical properties ω_0 and F. From ω_0, F, and c, we can then obtain a, b, and B.

In the Gordon *et al.* work, the water column was considered to be homogeneous and, therefore, the *inherent* optical properties were independent of depth. The *apparent* optical properties, however, display a depth dependency, a consequence of the change in the radiance distribution with depth. This depth dependency persists until, at large depth, an invariant *asymptotic radiance distribution* is attained. Once this asymptotic radiance distribution is reached, the apparent optical properties also become invariant and are independent of incident radiation distribution. In the Gordon *et al.* work, values of expansion coefficients are given in increments of attenuation lengths τ ($\tau = 1/c$). In remote sensing, the uppermost layer of natural water bodies is the dominant target. Consequently, in Table 3.1, we have listed the Gordon *et al.* coefficients k, k', r, r', and d applicable to just beneath the air-water interface (i.e., for $z = 0^-$). These values are those given for the upper-most attenuation length. Two entries are listed for each of the orders for the coefficients r, r', and d, the upper entry for the sun case and the lower entry for the sky case.

3.6.2 The Kirk Monte Carlo Simulation

Kirk[224] also used Monte Carlo simulations of photon propagation to determine relationships between inherent and apparent optical properties. Whereas Gordon *et al.* were concerned with Case I waters and restricted their analyses to $\omega_o \leq 0.9$, Kirk was concerned with turbid inland and coastal waters and determined relationships for $\omega_o \leq 0.968$ ($b/a = 30$). He used as a volume scattering function $\beta(\theta)$ measurements obtained in San Diego Harbor by Petzold[308] for which $B = 0.019$. For the case of vertical incidence, Kirk[224] obtained the relations:

$$K_d(z_m) = (a^2 + 0.256ab)^{1/2} \tag{3.30}$$

and

$$R(0) = 0.328Bb/a \tag{3.31}$$

where K_d is the attenuation coefficient for downwelling irradiance, z_m is the vertical depth of the mid-point of the photic zone (i.e., the mid-point between the 100% and 1% irradiance levels, which is the 10% irradiance level), and $R(0)$ is the volume reflectance just beneath the air-water interface.

In a later work, Kirk[227] incorporated the effect of non-vertical incidence and obtained the relations:

$$K_d(z_m) = (1/\mu_0)[a^2 + (0.473\mu_0 - 0.218)ab]^{1/2} \tag{3.32}$$

and

$$R(0) = (0.975 - 0.629\mu_0)Bb/a \tag{3.33}$$

where $\mu_o = \cos(\theta_o)$, θ_o being the in-water refracted angle.

For overcast conditions, equations (3.32) and (3.33) become, respectively:

$$K_d(z_m) = 1.168[a^2 + 0.168ab]^{1/2} \tag{3.34}$$

and

$$R(0) = 0.437Bb/a. \tag{3.35}$$

3.6.3   The Jerome *et al.* Monte Carlo Simulation

Through similar Monte Carlo simulations, Jerome *et al.*[204] obtained the relationships:

$$R(0) = (1/\mu_o)0.319Bb/a \qquad (3.36)$$

for $0 \leq Bb/a \leq 0.25$, and

$$R(0) = (1/\mu_o)[0.013 + 0.267Bb/a] \qquad (3.37)$$

for $0.25 \leq Bb/a \leq 0.50$
where $\mu_o = 0.858$ for overcast conditions.

At the time of their analyses, Jerome *et al.* noted that there appeared to be departures from the above relationships at large solar zenith angles, suggestive of a possible second order relationship between $R(0)$ and Bb/a, a relationship whose impact increased with increasing θ, indicating that an additional term should perhaps be added to equations (3.36) and (3.37).

Jerome *et al.* also noted that if the predictions of the Gordon *et al.*, the Kirk, and the Jerome *et al.* Monte Carlo simulation equations were intercompared for the same inherent optical properties, these predictions would be quite different.

With the appearance of discrepancy between the results, Gordon[133] further analyzed the relationship between volume reflectance and volume scattering function, attaining the equation:

$$R(D_d) = k(D_d - 1)R(1) + R(1) \qquad (3.38)$$

where D_d is the distribution function for downwelling light [which is equivalent to the $(1/\mu_o)$ term in equations (3.36) and (3.37) of the Jerome *et al.* simulation for a calm surface]. The parameter k, while dependent upon ω_o, is predominantly dependent upon the volume scattering function and must, therefore, be determined for each $\beta(\theta)$. Gordon determined that the values of k for the Gordon *et al.*,[135] the Kirk,[227] and the Jerome *et al.*[204] Monte Carlo simulations are 0.85, 1.0, and 1.15, respectively, a consequence of the $\beta(\theta)$ values selected for each Monte Carlo. These dissimilar k values account for their different predictions.

Kirk[228] also investigated the variation in volume reflectance with shape of the volume scattering function. He used six different volume scattering functions from Petzold,[308] but only a single ω_o value of 0.857. He analyzed the data considering the average cosine of scattering (μ_S) for each $\beta(\theta)$ as a variable. The value of μ_S was determined from:

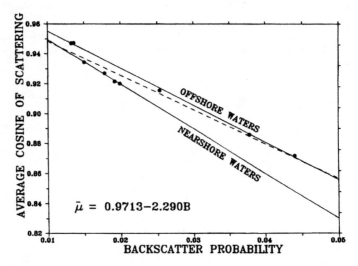

FIGURE 3.12
Average cosine of scattering plotted against backscatter probability for nine values of $\beta(\theta)$, five of which are appropriate for offshore waters and four of which are appropriate for near-shore waters.

$$\mu_s = \frac{\displaystyle\int_{4\pi} \beta(\theta)\cos\theta d\Omega}{\displaystyle\int_{4\pi} \beta(\theta) d\Omega} \tag{3.39}$$

Kirk then obtained the relationship:

$$R(0) = C(\mu_0)Bb/a \tag{3.40}$$

where $C(\mu_o) = M[(1/\mu_o) - 1] + 0.31$ and $M = 2.6\mu_s - 2.0$ [μ_S as defined from equation (3.39)].

Of the six $\beta(\theta)$ values employed by Kirk, one resulted in a significant deviation from equation (3.40), this deviation being a consequence of its associated M value. This $\beta(\theta)$ was for near-shore waters, whereas the other five $\beta(\theta)$ values were for offshore waters. This anomalous value becomes less of an anomaly if all nine of Petzold's volume-scattering functions (omitting the filtered water values) are considered.

Figure 3.12 illustrates the values of the average cosine of scattering (μ_S) plotted against the backscatter probability for these nine $\beta(\theta)$ values. Figure 3.12 strongly suggests the existence of two relationships between μ_S and B, as indicated by the two distinct curves we have sketched through the plotted points. The upper curve relates the five $\beta(\theta)$ values applicable to offshore waters, while the lower curve relates the four $\beta(\theta)$ values applicable to nearshore waters. These two relationships probably arise from the offshore scattering being dominated by organic particulates and the nearshore scattering being dominated by inorganic particulates. Treating all nine points of Figure 3.12 as a single data set will yield the relationship (segmented line):

$$\mu_s = 0.9713 - 2.290B.$$

The above relationship allows the term $C(\mu_o)$ of equation (3.40) to be written solely in terms of μ_o and B, eliminating the need for calculating the average cosine of scattering μ_S, i.e.,

$$C(\mu_o) = (0.9713 - 2.290B)[(1/\mu_o) - 1] + 0.31.$$

3.6.4   The DiToro Model

DiToro[85] assumed an exponential decay of both downwelling and upwelling irradiance with depth (i.e., compliance with Beer's Law). He further assumed that the volume scattering function $\beta(\theta)$ could be approximated as a delta function in the forward direction ($\theta = 0$), with the remaining directions displaying isotropy. The isotropic scattering fraction is given as $1 - \gamma$, where γ is the fractional anisotropy at $\theta = 0°$. The fractional anisotropy is related to the forwardscattering probability F through the relationship:

$$F = \gamma + [(1 - \gamma)/2] \tag{3.41}$$

To incorporate the effects of variations in solar incidence, DiToro utilized the sun-angle dependence described in the quasi-single scattering model of Gordon.[131] The governing equations of the DiToro exponential/quasi-single scattering model then become:

$$\gamma = \frac{m(1 - \kappa\mu_d)}{m(1 - \kappa\mu_d) + \kappa\mu_d} \tag{3.42}$$

$$\omega_0 = 1/[\gamma + m(1 - \gamma)] \tag{3.43}$$

where

$$\mu_d = \cos\theta_0 \approx 1/D_d$$

$$m = 1 - \frac{(1 - \ln2)}{2} + \frac{1}{2R(0)}\left[1 - \mu_d \ln\left(1 + \frac{1}{\mu_d}\right)\right]$$

$$\kappa = \frac{K(0)}{c}.$$

The terms $R(0)$, $K(0)$, and D_d are as previously defined, with numerical values appropriate to $\theta_0 = 0°$ and $z = 0$.

3.7   BULK AND SPECIFIC INHERENT OPTICAL PROPERTIES

Using the Gordon et al.[135] Monte Carlo model in concert with direct measurements of volume reflectance spectra and attenuation coefficients, Bukata et al.[36] determined spectral values of the inherent optical properties of coastal waters in Lake Ontario. Figure 3.13 illustrates these spectral determinations of c, b, a, F, ω_0, and B for data collected in 1977. F and b may be considered to be spectrally invariant. The variations in ω_0 and B are seen to be in opposition to the variations in a and c. The spectral structure observed in the backscattering probability B suggests that the volume scattering function $\beta(\theta)$ may be altered by the absorption characteristics of the water column. This is contrary to the premise that absorption and scattering are mutually independent processes and that in-water particulate backscattering is spectrally invariant. However, such scattering and absorption interdependence is consistent with the ocean color observations of Mueller.[277]

The non-Case I inherent optical properties depicted in Figure 3.13 represent *bulk inherent optical properties* of the water column. *Bulk* optical properties are those optical properties displayed when the water column is considered as a composite entity with no regard as to the specific component contributions to that property. Thus, bulk values of c, b, a, B, and F are consequences of the totality of scattering and absorption centers present within the aquatic medium. *Specific inherent optical properties* are those optical properties that can be attributed to the individual scattering and absorption centers comprising the water column under consideration. It is these *specific* inherent optical properties that must be determined if a water body is to be remotely sensed for estimates of the concentrations of its organic and/or inorganic components.

The bulk inherent optical properties of a natural water column, $a(\lambda)$, $b(\lambda)$, and $B(\lambda)b(\lambda)$ can be considered as the additive consequence of the individual specific inherent optical properties of the suspended and/or dissolved organic and/or inorganic materials present in the water column (in addition to the pure water itself). That is,

$$a(\lambda) = \sum_{i=1}^{n} x_i a_i(\lambda) \tag{3.44}$$

$$b(\lambda) = \sum_{i=1}^{n} x_i b_i(\lambda) \tag{3.45}$$

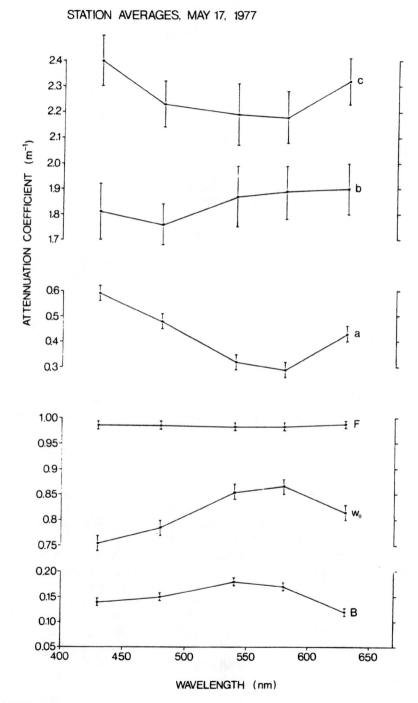

FIGURE 3.13

Spectral averages of the bulk inherent optical properties c, b, a, F, ω_o, and B for Lake Ontario coastal waters. (Adapted from Bukata, R. P., Jerome, J. H., Bruton, J. E., and Jain, S. C., *Appl. Opt.*, 18, 3926-3932, 1979.)

$$B(\lambda)b(\lambda) = \sum_{i=1}^{n} x_i B_i(\lambda)b_i(\lambda) \qquad (3.46)$$

where $B(\lambda)b(\lambda)$ = backscattering coefficient at wavelength λ, the product
of the backscattering probability $B(\lambda)$ and the scattering
coefficient $b(\lambda)$. Also written as $b_B(\lambda)$,

x_i = concentration of the ith component of the water column,

$a_i(\lambda)$ = absorption coefficient at wavelength λ for a unit concentration of aquatic component i,

$b_i(\lambda)$ = scattering coefficient at wavelength λ for a unit concentration of aquatic component i, and

$B_i(\lambda)b_i(\lambda)$ = backscattering coefficient at wavelength λ for a unit concentration of aquatic component i.

$a_i(\lambda)$, $b_i(\lambda)$, and $B_i(\lambda)b_i(\lambda)$ represent the amount of absorption, scattering, and backscattering that may be attributed to a unit concentration of an organic or inorganic component of a water column. They are widely referred to in the literature as the *specific absorption*, *specific scattering*, and *specific backscattering coefficients*, respectively, of the principal components of the natural water body in question. Since these parameters imply that a specific aquatic component will act as an effective target for a photonic interaction (bombardment resulting in absorption or scattering of the impinging photon), we have borrowed a term from atomic and nuclear theory and have come to refer to $a_i(\lambda)$, $b_i(\lambda)$, and $B_i(\lambda)b_i(\lambda)$ as the *optical cross sections* appropriate to the water column under consideration.[38] Consistent with this terminology, the units of optical cross sections are area per unit mass of aquatic component (m^2 mg^{-1}). Throughout the balance of this book we will consider the terms *absorption cross section* and *scattering cross section* to be interchangeable with the terms *specific absorption coefficient* and *specific scattering coefficient*, respectively.

The optical cross section spectra clearly provide the linkages between the bulk inherent optical properties of a water body and the concentrations of its organic and inorganic constituents. From the form of equations (3.44), (3.45), and (3.46), it is seen that, if values of the bulk aquatic optical properties a, b, and Bb can be obtained in concert with known values of the concentrations x_i of the co-existing aquatic constituents, it should be possible to determine the optical cross sections a_i, b_i, and B_ib_i of those aquatic constituents at any wavelength λ. Conversely, if the optical cross sections pertinent to a water body are known, it should be possible to estimate the co-existing concentrations of the aquatic constituents from directly measured (or remotely inferred) values of the bulk optical properties of that water body. *This is the governing principle behind the remote monitoring of the organic and inorganic constituents of natural waters, as well as the interpretation of remote measurements of aquatic color.*

To utilize equations (3.44), (3.45), and (3.46) requires a bio-optical model of the water body that realistically represents the organic and inorganic mix of suspended and dissolved indigenous matter that impacts water color. Obviously, a definitive optical model should consider the optical cross section spectrum of every species and sub-species of aquatic components present in natural water bodies (see the discussions in Chapter 4), as well as the possible temporal dependency of each component that undergoes an annual cycle. Such a model is clearly unattainable. To maintain a controllable number of variables in the development of an optical model, Bukata *et al.*[38,39,41] assumed that at any instant of time a natural water mass could be considered as a homogeneous combination of pure water, unique suspended organic material (represented by the chlorophyll *a* concentration corrected for phaeophytin contamination), unique suspended inorganic material (represented by the suspended mineral concentration SM), and dissolved organic material (represented by the dissolved organic carbon concentration DOC). An additional fifth component termed *non-living organics* (to account for detrital matter) was initially employed in their analyses. It was found for Great Lakes waters, however, that only a slight loss of generality occurs when the four-component optical representation that disregarded the contribution of non-living organics is considered. This is partly due to the fact that non-living suspended organics are rarely a dominant factor in the color of inland or coastal waters.

For a four-component optical model (SM, Chl, DOC, and pure water), the bulk optical property equations (3.44), (3.45), and (3.46) may be expanded as:

$$a(\lambda) = a_W(\lambda) + x_1 a_{Chl}(\lambda) + x_2 a_{SM} + x_3 a_{DOC}(\lambda) \tag{3.47}$$

$$b(\lambda) = b_W(\lambda) + x_1 a_{Chl}(\lambda) + x_2 a_{SM}(\lambda) \tag{3.48}$$

and

$$B(\lambda)b(\lambda) = B_W(\lambda)b_W(\lambda) + x_1 B_{Chl}(\lambda)b_{Chl}(\lambda) + x_2 B_{SM}(\lambda)b_{SM}(\lambda) \tag{3.49}$$

where the subscripts *W*, Chl, SM, and DOC refer to the pure water, unique organic, unique inorganic, and dissolved organic components of the water column, respectively; x_1, x_2, and x_3 are the concentrations of chlorophyll *a*, suspended minerals, and dissolved organic carbon, respectively. To partially compensate for the presence of non-living organics, the concentration x_1 in this case would usually refer to the concentration of chlorophyll uncorrected for phaeophytin contamination (unless it can be safely assumed that the non-living organic concentration was low enough to render its contribution to the optical processes insignificant. In that case x_1 would refer to the corrected chlorophyll concentration). The components of *a* (a_{Chl}, a_{SM}, and

a_{DOC}), of b (b_{CHL} and b_{SM}), and of Bb ($B_{Chl}b_{Chl}$ and $B_{SM}b_{SM}$) are the specific absorption coefficients, the specific scattering coefficients, and the specific backscattering coefficients, as a function of wavelength λ (i.e., the *optical cross section spectra*) of the indigenous organic and inorganic components of the water column.

Note that in equations (3.48) and (3.49) the scattering and backscattering cross sections b and Bb for DOC are taken as zero for all values of λ. As we will discuss in Section 4.4, it is reasonable to ascribe significant absorptive but insignificant scattering properties to dissolved organic matter. Numerical values of the optical cross sections for pure water may be found in Hulbert,[177] Smith and Baker,[375] and Reference 280.

Once the optical cross section spectra are obtained, the bulk optical properties $a(\lambda)$, $b(\lambda)$, $B(\lambda)b(\lambda)$, $c(\lambda)$, and $\omega_o(\lambda)$ may be readily generated for water columns made up of any pre-selected amounts of chlorophyll a, suspended minerals, and dissolved organic carbon. If these bulk optical properties are then used to estimate the corresponding sub-surface volume reflectance spectrum $R(\lambda, z)$ [for example, through the use of equations (3.36) and (3.37) from the Jerome *et al.* Monte Carlo simulation model or through the other Monte Carlo simulation models described in this chapter], then, in principle, *in situ* or remote measurements of volume reflectance spectra could be compared with an existing catalog of simulated volume reflectance spectra for water columns of known component concentrations. This can then be used as a means of estimating the component concentrations from volume reflectance measurements. Such comparison, however, is subject to stringent stipulations:

1. The natural waters must be appropriately defined within the limitations of the restricted number of aquatic components that can be considered. We have considered a four-component water column with no regard as to sub-species of chlorophyll-bearing biota, distinctiveness of mineral types, non-living organics, or type of dissolved organic matter. We will discuss aquatic composition in greater detail in Chapter 4. However, it is clear that the more optically complex a water column becomes (i.e., the greater the number of components that affect the optical behavior of the water column), the less reliable will be the four-component optical model approach described herein. Although an extension to the ideal number of components may be unrealistic, an extension to accommodate a reasonable number of additional components would not seem impossible.

2. The optical cross section spectra must be known. Since these cross sections reflect the regional diversities of geography and biology as well as the adaptations of chlorophyll-bearing biota to their surroundings, temporal and spatial variations of the cross section spectra must either be directly determined, theoretically predicted, or be ignored by restricting focus to temporal and/or spatial dimensions where the assumption of constancy can be justified.

3. Since this approach has assumed the water column to be homogeneous, there is no provision incorporated to account for possible sub-surface layering effects. These layering effects can become particularly problematical to attempts at remotely monitoring *aquatic primary production* and *aquatic carbon budgets*, topics to which we will return in Chapter 9.

4. No provision is incorporated for the presence of chemical impurities. Small concentrations of most chemicals (by comparison with the concentrations of chlorophyll, suspended minerals, and dissolved organic carbon) produce negligible absorption and scattering impacts, and their detection is beyond such *in situ* optical measurement techniques. If remote sensing of chemical impurities is attempted, such attempts would best be oriented around relating impacts of chemical presence to variations in organic and/ or inorganic materials that do generate measurable optical signals.

3.8 DEPENDENCE OF THE UNDERWATER LIGHT FIELD ON SOLAR ZENITH ANGLE AND SKY IRRADIANCE

As we have seen, the nature of the underwater light field is (apart from its critical dependence upon the optical cross section spectra pertinent to the local aquatic organic and/or inorganic components) dependent upon the spectral composition and directional character of the photon flux impinging upon the air-water interface. These spectral and directional properties are determined in part by the solar zenith angle (i.e., local time) and in part by the local atmospheric conditions. The diurnal variation in the solar zenith angle results in a closely cosinal downwelling solar irradiance distribution centered on the time of day at which the sun is most nearly directly overhead. In addition, increasing atmospheric path lengths are encountered by the solar photons as their zenith angles approach grazing incidence. As evident by dramatic dawns and sunsets, the preferential scattering properties of the atmosphere results in spectral changes to the solar light observed at the Earth's surface. Consequently, the incident sunlight associated with low solar elevations (i.e., solar zenith angles approaching 90°) is heavily weighted toward the red wavelength region of the visible spectrum. At these grazing angles, however, surface reflection dominates over air-water interface transmission. This results in the skylight (which is spectrally weighted toward the blue wavelength region of the visible spectrum) becoming dominant in the downwelling radiation immediately below the air-water interface. For a fixed aquatic composition, therefore, one would expect a colour change [i.e., a change in the sub-surface volume reflectance spectrum $R(\lambda)$] in the water column which systematically tracked the changes in solar zenith angle. Jerlov,[197] using *in situ* measurements of *color index* [defined in this case as the ratio of the subsurface radiances $L(\theta, z, 450 \text{ nm})$ and $L(\theta, z, 520 \text{ nm})$ for each value of solar incidence angle θ at several depths z] obtained at stations in the Mediterranean Sea, reported that for values of solar zenith angles θ < 85° this solar angle dependence was not important, particularly at small depths. For values of θ > 85° the maximum changes in the color index occurred for clear as opposed to turbid waters.

The effect of solar zenith angle on the depths of the sub-surface irradiance levels had been approached along both theoretical and experimental avenues for some time, and the general consensus [as discussed in Jerlov and Steeman Nielsen[199] and Jerlov[198]] was that such an effect, if indeed existent, was certainly very minor. In many instances variations in the depths of subsurface irradiance levels with solar zenith angles as measured at a fixed station were indiscernible from the experimental scatter of the measurements themselves. Workers such as Jerlov and Nygård[200] reported variations with solar zenith angles, while others such as Bethoux and Ivanoff[23] reported an essential invariance in the depths of subsurface irradiance levels with solar zenith angles. Still others [e.g., Højerslev[170] who reported ~12% variation in the depth the 1% irradiance level for 20° ≤ θ ≤ 80°] reported a small dependence upon solar zenith angle. However, most of these early measurements failed to consider the actual optical properties of the waters in which the measurements were performed. This was particularly evident insofar as scattering phenomena were concerned. In some clever early work Whitney[426] combined a simple cosine law correction factor for the direct solar beam and an isotropic radiance distribution for the diffuse component of the incident radiation to determine the effects of the solar angle and the direct-to-diffuse character of the downwelling radiation on the in-water pathlength. But, here too, the model ignored the scattering occurring within the water column.

The proliferation of Monte Carlo approaches to the general problem of photon propagation through attenuating media enabled an investigation of the relationships among solar zenith angle, water type, and variations in the depth of irradiance levels. Recall that the scattering albedo ω_0 is a convenient means of classifying water types in terms of the degree of scattering and absorption occurring in the water. While ω_0 provides no information regarding the *absolute* numbers of scattering and absorption interactions occurring within a water column, it does indicate the *relative* number of scattering and absorption interactions that are occurring. Further, the apparent optical properties of a water column are affected by the angular distribution of the propagating photon flux (in addition, of course, to the inherent optical properties of the water column). It is logical, therefore, to anticipate that the relationships among the apparent and inherent optical properties of the aquatic system would also display a dependence upon the angular distribution of the propagating photon flux, i.e., a dependence upon solar zenith angle. Thus, the Monte Carlo analyses work of Plass and Kattawar,[310] Gordon and Brown,[134] Gordon et al.,[135] Kattawar and Humphreys,[212] Kirk,[224,227,228] Jerome et al.,[202,204] Gordon,[133] and others quite appropriately dealt with the

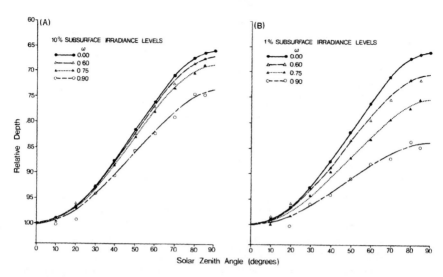

FIGURE 3.14

Relative depths of the (A) 10% and (B) 1% PAR irradiance levels (Monte Carlo determinations) as a function of solar zenith angle θ for ω_o = 0, 0.60, 0.75, and 0.90. (From Jerome, J. H., Bruton, J. E., and Bukata, R. P., *Appl. Opt.*, 21, 642–647, 1982. With permission.) Relative depths have been normalized to a unit depth for θ = 0°.

relationships among the optical properties with a systematic exploration of these relationships at a variety of solar zenith angles and a variety of direct-to-diffuse conditions of incident radiation.

Figure 3.14 is taken from Jerome et al.[202] (their Figure 2) and illustrates the anticipated solar zenith angle impacts on the relative depths of the 10% and 1% PAR irradiance levels (as determined from Monte Carlo simulations) for hypothetical water columns defined by ω_o values of 0.60, 0.75, and 0.90. Also included is a curve for ω_o = 0 (a water column containing no scattering interactions) using the simple cosine correction model of Whitney.[426] Figure 3.14 considers the special case of the downwelling radiation flux being comprised solely of a direct beam incident at the solar zenith angle θ. No diffuse component is considered. The relative depths have been normalized to a unit depth for a solar zenith angle θ = 0° (i.e., the sun directly overhead). Depth variations in the 10% irradiance level are seen to be 26% for ω_o = 0.90, 31% for ω_o = 0.75, 33% for ω_o = 0.60, and 34% for ω_o = 0. The corresponding variations for the depths of the 1% irradiance level are 16%, 25%, 29%, and 34%, respectively.

Clearly, the solar angle dependence of the depth of an irradiance level is significantly altered by the optical properties of the water column. This would certainly explain the contradictory nature of early investigations wherein solar angle dependence was observed in some instances and not in others. Measurements were probably made in waters displaying significantly different optical properties. Studies performed in waters characterized by low values of ω_0 would detect a solar angle dependence, while studies performed in waters characterized by high values of ω_0 would not.

For the special case in which the incident radiation is considered to be completely diffuse, there is, of course, no direct solar incidence. Therefore, no solar zenith angle dependence is possible, and only one depth for each irradiance level would be observed throughout the day.

The general case of incident radiation, however, is a composite of both direct and diffuse radiation. Figure 3.15 is also taken from Jerome et al.[202] (their Figure 3) and illustrates anticipated solar zenith angle impacts on the relative depth of the 1% irradiance level (as determined from Monte Carlo simulations) for a hypothetical water column defined by $\omega_0 = 0.60$ under several varying conditions of direct-to-diffuse incident radiation fields. The relative depths of the 1% irradiance level have been normalized to the depth of the 1% irradiance level for a solar zenith angle of 0° (vertical incidence). Curve A represents the anticipated solar zenith angle impact for a direct solar beam (no diffuse component at any zenith angle). Curve B represents this relationship for a clear day in which the percentage of diffuse radiation is 10% at small solar zenith angles and gradually increases to ~25% at large solar zenith angles. Curves C and D steadily increase the percentage of diffuse radiation at both small and large solar zenith angles, and Curve E illustrates the relationship for a near-isotropic (90% diffuse component at all solar zenith angles) incident radiance distribution. It is evident that the variation in recorded depths of the 1% irradiance level is larger for clear days (highly collimated incident radiation) than for overcast days (highly diffuse incident radiation). Quite prominent in Figure 3.15 is the inversion point at larger solar zenith angles which appears, in general, as the percentage of diffuse incident radiation increases. For clear days this inversion occurs at $\theta > \sim 75°$. As the percentage of diffuse incident radiation increases, however, this inversion moves toward smaller values of θ. Data collected at the National Water Research Institute in Ontario, Canada, over the past two decades have borne out the above Monte Carlo analyses. The anticipated irradiance level/solar zenith angle dependencies upon the scattering properties of natural waters have been observed in the optically complex waters found in Lakes Erie and Ontario, as well as in the less optically complex waters found in regions of Lakes Superior and Huron. However, as seen from Figure 3.15, the variances in incident radiation encountered during clear day measurements do not greatly affect the irradiance level/solar zenith angle dependence for $0° \leq \theta \leq 60°$. Daylight determinations of PAR irradiance depth profiles centered on the local time of near-zenith sun for clear-sky conditions should, therefore, minimize the impact of solar zenith angle variations on those determinations.

Direct subsurface measurements or determinations of subsurface phenomena which rely upon precise knowledge of PAR (for example, aquatic

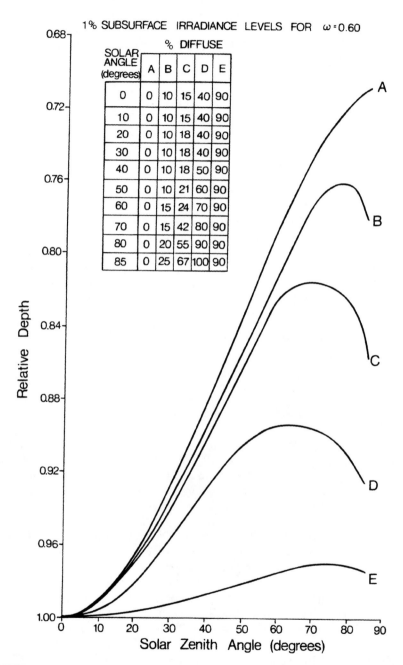

FIGURE 3.15
Solar zenith angle impacts on the relative depth of the 1% PAR irradiance level for $\omega_o = 0.60$ and incident radiation fields comprised of a variety of direct and diffuse components. (From Jerome, J. H., Bruton, J. E., and Bukata, R. P., *Appl. Opt.*, 21, 642–647, 1982. With permission.) Relative depths have been normalized to a unit depth for a solar zenith angle of 0°.

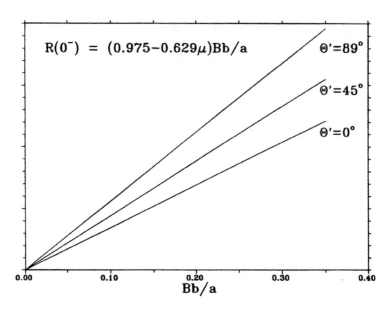

FIGURE 3.16
Linear relationships between volume reflectance just beneath the air-water interface and Bb/a for three different in-water incidence angles θ'. (Adapted from Kirk, J. T. O., *Limnol. Oceanogr.*, 29, 350–356, 1984.)

photosynthesis and primary production) must slavishly adhere to minimizing or otherwise adjusting for the dependencies of irradiance level depths upon variations in both solar zenith angle and the direct-to-diffuse character of the incident flux above the air-water interface.

The volume reflectance spectrum $R(\lambda, z)$ [the ratio of the upwelling irradiance $E_u(\lambda, z)$ to the concurrent downwelling irradiance $E_d(\lambda, z)$], as we have seen, is strongly dependent upon the solar zenith angle and the percentages of direct and diffuse radiation comprising the incident flux. In a similar manner, the apparent optical properties of the water column are also strongly dependent upon the variations in these parameters.

Figure 3.16 is taken from Kirk[227] (his Figure 3) and illustrates the linear relationships between the volume reflectance just beneath the air-water interface $R(0)$ and Bb/a for three different incident angles θ.

Figure 3.17 is taken from Bukata *et al.*[45] (their Figure 3) and illustrates the solar zenith angle dependence of the *irradiance attenuation coefficient for PAR*, K_{PAR} (the irradiance attenuation coefficient for the downwelling irradiance in the wavelength interval 400 nm to 700 nm. Although such a broadband value of attenuation coefficient ignores its wavelength infrastructure, it is consistent with the general practice of using broadband irradiance values in the determination of irradiation and primary production) for three different water types defined by $\omega_0 = 0.60$, 0.75, and 0.90, respectively. Since the

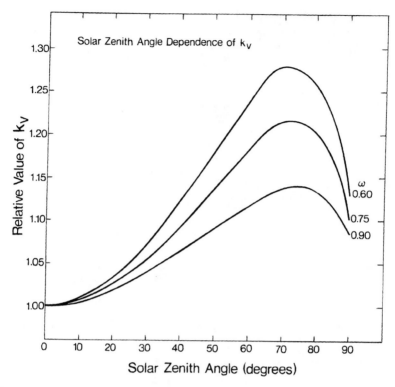

FIGURE 3.17

Solar zenith angle dependence of the vertical attenuation coefficient for PAR (labelled K_y) for water masses defined by $\omega_o = 0.60$, 0.75, and 0.90. (From Bukata, R. P., Jerome, J. H., and Bruton, J. E., *J. Great Lakes Res.*, 15, 327–338, 1989. With permission.)

irradiance attenuation coefficient of PAR is an obvious function of solar time, *in situ* determinations of irradiation and primary production will suffer from inaccuracies based on the usage of a constant rather than a time-dependent irradiance attenuation coefficient. Such inaccuracies may be minimized, however, by a judicious selection of solar zenith angles at which to determine a single "constant" value of K_{PAR}. For the latitudes containing the Laurentian Great Lakes, Bukata et al.[45] show this solar zenith angle range to be 36° to 70°, depending upon location and time of year.

The Monte Carlo results of Kirk[227] provide a convenient means of calculating the dependence of the irradiance attenuation coefficient $K(\lambda)$ appropriate to the euphotic zone on (a) solar zenith angle and (b) the diffuse/total incident irradiance ratio. If the bulk absorption coefficient $a(\lambda)$ and the bulk scattering coefficient $b(\lambda)$ are known, the euphotic zone irradiance attenuation coefficient is calculable from the following set of equations:

For direct incident solar irradiance:

$$K_{\text{direct}}(\lambda) = \frac{a(\lambda)}{\mu_0} [1 + (0.473\mu_0 - 0.218)b(\lambda)/a(\lambda)]^{1/2} \qquad (3.50)$$

For diffuse incident sky irradiance:

$$K_{\text{sky}}(\lambda) = \frac{a(\lambda)}{0.856} [1 + 0.168b(\lambda)/a(\lambda)]^{1/2} \qquad (3.51)$$

For incident irradiance comprised of a diffuse sky fraction F:

$$K(\lambda) = FK_{\text{sky}}(\lambda) + (1 - F)K_{\text{direct}}(\lambda) \qquad (3.52)$$

All the terms in equations (3.50), (3.51), and (3.52) are as previously defined.

Chapter 4

COMPOSITION OF
NATURAL WATERS

4.1 THE AQUATIC FOOD CHAIN (TROPHIC LEVELS)

Natural waters are complex physical-chemical-biological media comprising living, non-living, and once-living material that may be present in aqueous solution or in aqueous suspension. Together with air bubbles and inhomogeneities resulting from small-scale water eddies, these components determine the bulk optical properties of natural water bodies.

The principal living organisms present in water columns are *plankton*, a collective term encompassing all vegetable and animal organisms suspended in water (either hovering or floating), unable to resist the current, and not rigidly connected to the confining basin. Plankton include animal organisms (*zooplankton*), algal plant organisms (*phytoplankton*), bacteria (*bacterioplankton*), and lower plant forms such as algal fungi. These organisms represent the lowest level of feeding in a system of alimental food chain relationships involving not only the plankton (and their essential nutrients), but also higher forms of aquatic life. This alimental food chain is schematically shown in Figure 4.1 (as will be discussed in Section 7.9, interplay among aquatic nutrients and aquatic life defines the *trophic status* of the water column, and thus the food chain dynamics illustrated in Figure 4.1 represents a procession of *trophic levels*), which illustrates the nutrient-life-death-decomposition cycle of natural waters. Phytoplankton consume nutrients (biogenes) from their confining waters and in the presence of subsurface sunlight synthesize these nutrients into organic matter through the process of *primary production*, a topic to which we shall return in Chapter 9. Zooplankton, represented in Figure 4.1 by the categories protozoa rotifers, cladocera, large obligatory predators, small obligatory predators, and small non-obligatory predators, graze on phytoplankton. As consequences of their vital functions and of their mortality, zooplankton generate secondary organic matter. Bacterioplankton decompose this organic matter, and as a consequence of these low level food chain dynamics, a dissolved organic matter (DOM) component

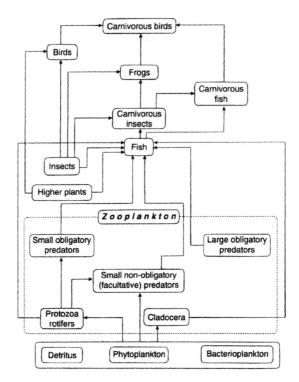

FIGURE 4.1
Generalized flow diagram of the aquatic food chain (trophic levels).

(termed the *autochthonous DOM* component) is indigenous to natural waters. Lakes and coastal waters are also receptacles of DOM inputs from surficial run-off and river discharges which contribute an *allochthonic DOM* component. These run-off and river discharge inputs also introduce a suspended inorganic matter (SIM) component to non-Case I waters.

The higher order life forms involved in the food chain dynamics of the aquatic ecosystem are also sketched onto Figure 4.1. The zooplankton, as well as insects and higher order plants, serve as food for fish that, in turn, extend the food chain to aquatic and terrestrial carnivores and herbivores.

Following Drabkova,[89] Figure 4.2 illustrates a generalized flow diagram of the formation of dissolved organic matter in natural waters. The autochthonous DOM component is shown as a result of the interlinkages among the plant organisms (algae and macrophytes), herbivorous and carnivorous animal organisms (zooplankton), bacteria, and fish. Metabolic releases of living organisms and secondary products of organisms permanently removed from the food chain are included within the flow diagram. The term *detritus* in both Figures 4.1 and 4.2 refers to small particles of organic

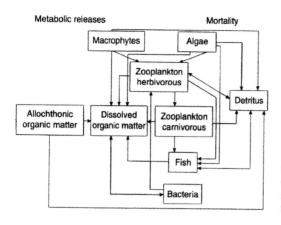

FIGURE 4.2
Generalized flow diagram of the formation of dissolved organic matter in natural waters.

or partially mineralized matter formed from decayed plants and animals as well as their excretions. The allochthonic DOM component contributes to both the DOM and the detrital components.

Both the DOM and the SIM (along with the phytoplankton, zooplankton, and bacterioplankton) dictate the manner in which the subsurface light field is distributed in the natural water body and, therefore, should be considered in terms of their specific absorption and scattering optical cross section spectra. However, the extreme diversity of the organic and inorganic components of inland and coastal waters, coupled with the temporal and spatial variability of the composition of these waters, presents a severe obstacle to such a consideration. Thus, it becomes incumbent to select a manageable number of these components (or surrogates to these components) if an adequate physical model is to be utilized to effectively describe the optical properties of a natural water body under scrutiny. The remainder of Chapter 4 will briefly discuss the variety of organic and inorganic matter encountered in non-Case I waters.

4.2 PURE WATER

In the consideration of natural water, *pure water* is taken to imply water that is free from the optical effects of terrestrially and/or atmospherically derived organic and inorganic matter. Thus, *"pure water"* would be defined as a *chemically* pure substance comprised of a mixture of several water isotopes, each of different molecular mass. The absorption coefficient $a_w(\lambda)$, the scattering coefficient $b_w(\lambda)$, and the total attenuation coefficient $c_w(\lambda)$ used in the additive optical properties equations of Chapter 3 refer to such water wherein the only absorption and scattering is a consequence of the water molecules. Numerical values of the absorption and scattering properties of pure water have been presented in Hulbert,[177] Smith and Baker,[375] and others.

A detailed overview of experimental data on the optical characteristics of pure water may be found in Reference,[279] and Table 4.1 lists the values of $a_w(\lambda)$, $b_w(\lambda)$, and $c_w(\lambda)$ therein as well as those reported in Smith and Baker.[375] The spectral range of Table 4.1 encompasses the UV-B, UV-A, visible, and near-infrared wavelengths.

The absorption and scattering coefficients listed in Table 4.1 refer to a water temperature T of 20°C. In the temperature range 0.5°C ≤ T ≤ 26°C, the variation in $c_w(\lambda)$ across the spectrum is <15%, with the largest of this variation being observed in the region of maximum attenuation near 750 nm, along with a slight shift in the location of the maximum itself. In addition to this temperature dependence, $a_w(\lambda)$, $b_w(\lambda)$, and $c_w(\lambda)$ also exhibit a pressure dependence which, however, can be neglected for most cases of aquatic remote sensing.

Recall from equation (1.40) that $c(\lambda) \equiv a(\lambda) + b(\lambda)$. From Table 4.1 it is readily seen that for $\lambda > \sim580$ nm, scattering by water molecules becomes essentially insignificant when compared to absorption by water molecules. Thus, the attenuation of light of wavelengths greater than ~580 nm becomes essentially a consequence of molecular absorption, and $c_w(\lambda) \sim a_w(\lambda)$. For values of λ in the range 400–520 nm, however, scattering by water molecules is dominant to absorption by water molecules.[279] Thus, in the blue region of the spectrum, $c_w(\lambda)$ is primarily dictated by $b_w(\lambda)$.

Figure 4.3 is a composite of $c_w(\lambda)$ spectra taken from a variety of sources (including the values listed in Table 4.1). Note that the vertical scale for $c_w(\lambda)$ increases by a factor of 10 at $\lambda = 580$ nm. It is seen that pure water strongly absorbs in the red wavelength region of the visible spectrum, an absorption band that is peripheral to stronger near-infrared absorption at $\lambda > 700$ nm. Two points of inflection are observed at $\lambda = 604$ nm and $\lambda = 514$ nm, corresponding, respectively, to the 5th and 6th harmonics of the water molecule valence vibration at 3 μm.[393] The absorption in the ultraviolet spectral region is inherent to electronic transitions occurring within the water molecule.

Thus, due to the dominant role of molecular scattering at small values of visible λ and the dominant role of molecular absorption at large values of visible λ, pure water (which is generally regarded as a colorless fluid) displays a blue hue. This blue color is readily apparent in mid-oceanic waters as well as pristine inland and coastal waters that are infertile and are not subject to loadings dominated by terrestrial matter.

According to Rayleigh scattering theory, the spectral dependence of molecular scattering follows a $\lambda^{-4.09}$ relationship. The directional scattering at visible wavelengths by water molecules (described by volume scattering functions) has been modelled by Morel[270] and given as:

$$\beta_w(\theta) = \beta_w(90°) (1 + 0.835) \cos^2\theta \qquad (4.1)$$

TABLE 4.1

Optical Properties of Pure Water: (Multi-author)[MA] and (Smith and Baker)[SB]

λ (nm)	$n(\lambda)$ (m^{-1})	$a(\lambda)^{MA}$ (m^{-1})	$b(\lambda)^{MA}$ (m^{-1})	$c(\lambda)^{MA}$ (m^{-1})	$a(\lambda)^{SB}$ (m^{-1})	$b(\lambda)^{SB}$ (m^{-1})
250	1.377	0.190	0.032	0.220	0.559	0.0443
300	1.359	0.040	0.015	0.055	0.141	0.0201
320	1.354	0.020	0.012	0.032	0.0844	0.0153
350	1.349	0.012	0.0082	0.0202	0.0463	0.0103
400	1.343	0.006	0.0048	0.0108	0.0171	0.0058
410					0.0162	0.0052
420	1.342	0.005	0.0040	0.0090	0.0153	0.0047
430					0.0144	0.0042
440	1.340	0.004	0.0032	0.0072	0.0145	0.0038
450					0.0145	0.0035
460	1.339	0.002	0.0027	0.0047	0.0156	0.0031
470					0.0156	0.0029
480	1.337	0.003	0.0022	0.0052	0.0176	0.0026
490					0.0196	0.0024
500	1.336	0.006	0.0019	0.0079	0.0257	0.0022
510					0.0357	0.0020
520	1.335	0.014	0.0016	0.0156	0.0477	0.0019
530	1.335	0.022	0.0015	0.0235	0.0507	0.0017
540	1.335	0.029	0.0014	0.0304	0.0558	0.0016
550	1.334	0.035	0.0013	0.0363	0.0638	0.0015
560	1.334	0.039	0.0012	0.0402	0.0708	0.0014
570					0.0799	0.0013
580	1.333	0.074	0.0011	0.0751	0.108	0.0012
590					0.157	0.0011
600	1.333	0.200	0.00093	0.2009	0.244	0.0011
610					0.289	0.0010
620	1.332	0.240	0.00082	0.2408	0.309	0.0009
630					0.319	0.0009
640	1.332	0.270	0.00072	0.2707	0.329	0.0008
650					0.349	0.0007
660	1.331	0.310	0.00064	0.3106	0.400	0.0007
670					0.430	0.0007
680	1.331	0.380	0.00056	0.3806	0.450	0.0006
690					0.500	0.0006
700	1.330	0.600	0.00050	0.6005	0.650	0.0005
710					0.839	0.0005
720					1.169	0.0005
730					1.799	0.0005
740	1.329	2.250	0.00040	2.2504	2.38	0.0004
750	1.329	2.620	0.00039	2.6204	2.47	0.0004
760	1.329	2.560	0.00035	2.5604	2.55	0.0004
770					2.51	0.0004
780					2.36	0.0003
790					2.16	0.0003
800	1.328	2.020	0.00029	2.0203	2.07	0.0003

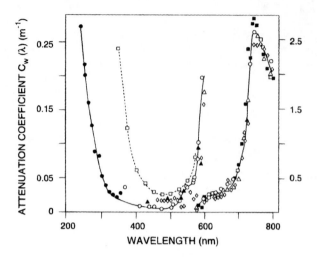

FIGURE 4.3
Composite of $c_w(\lambda)$ spectra (including values listed in Table 4.1). Note that the vertical scale is increased by an order of magnitude at $\lambda = 580$ nm.

where $\beta_w(\theta)$ =   volume scattering function at wavelength λ and scattering angle θ and

$\beta_w(90°)$ =   volume scattering function at wavelength λ and scattering angle 90°. The value of $\beta_w(90°)$ has been estimated (at $\lambda = 550$ nm) as 0.93×10^{-4} m^{-1} sr^{-1}.

From equation (4.1), it is seen that $\beta(180°) = \beta(0°)$. That is, in a pure water medium, scattering in the backward direction is numerically equal to scattering in the forward direction. Such is distinctly not the case for natural waters containing absorption and scattering centers other than water molecules.

4.3   DISSOLVED SALTS AND GASES

The salinity of mid-oceanic waters can be considered as being constant, varying within the narrow range (32–37.5) kg of dissolved mineral per m^3. By comparison, inland and coastal waters display a salinity range of (0–42) kg per m^3. Enigmatically, mid-oceanic waters display a globally stable ion composition. Considering the principal ions in descending order of their percentage contribution to the totality of aquatic metallic ions in solution (namely, Cl$^-$, Na$^+$, Mg^{++}, SO$_4^{--}$, Ca^{++}, CO$_3^{--}$), there is virtually no significant variance in the absolute concentration values to within 99% of the totality of dissolved minerals.[5]

Inorganic salts dissolved in natural waters affect the absorption and scattering occurring within the water column. However, the most significant absorption due to dissolved salts is observable[367] at ultraviolet, rather than visible, wavelengths since the electronic absorption bands of dissolved inorganic salts are at values of $\lambda < 300$ nm.

Visible light scattering by dissolved inorganic salts, which displays the same spectral dependence as visible light scattering by pure water, accounts for $\sim(20-30)\%$ of the total scattering in oceanic waters of average salinity 35 kg m^{-3}. Molecular scattering of light by mid-oceanic pure water plus dissolved organic salts is generally insignificant when considered in relation to the total light attenuation (absorption plus scattering) appropriate to these waters. Molecular scattering, however, from the pure water/dissolved salt combination assumes greater significance when specific directional scattering is considered. The ratio of $\beta_{mol}(\theta)$ (i.e., scattering from water and dissolved salt molecules) to $\beta(\theta)$ for the water column (i.e., the scattering from pure water, the dissolved inorganic salts, and all other suspended and/or dissolved matter residing in the mid-oceanic water), $\eta = \beta_{mol}(\theta)/\beta(\theta)$, can reach substantial magnitudes for moderate to large scattering angles. Shifrin[367] has assembled data from a number of sources to illustrate that, while a value for η of 0.10 is appropriate for mid-oceanic waters for scattering into the forward hemisphere, values of $\eta > 0.85$ are encountered for scattering into the backward hemisphere.

Due to the concentrations of terrestrially derived matter and the concomitant enhanced fertility that contrast Case I and non-Case I waters, the η values appropriate to inland and coastal waters (except for those rare locations such as the Sargasso Sea or infertile lakes wherein the water body is afforded a set of environmental conditions enabling it to optically resemble an ocean) are considerably smaller, and the role of dissolved inorganic salts in impacting a directional nature to $\beta(\theta)$ is dramatically reduced.

Natural waters are an integral component of the carbon, nitrogen, oxygen, and other elemental cycles that link the atmospheric, terrestrial, and aquatic ecosystems. Thus, these elements are invariably found in various chemical combinations throughout global water systems. The relatively low temperatures that characterize large, deep temperate lakes result in high solubility of oxygen, nitrogen, carbon dioxide, and other generally less abundant gases in these waters. For example, the dissolved oxygen content of the open waters of Ladoga Lake varies from 9.2 to 14.8 mg per litre. The dissolved nitrogen content in the same waters is about half the oxygen content.

In general, the concentrations of CO_2 in the open areas of inland waters are very small compared to the oxygen and nitrogen concentrations, and usually do not exceed 1 mg per liter. Spectral analyses of dissolved aquatic gases reveal that only oxygen absorbs light of visible wavelengths. However, even in oxygen-rich waters, the concentrations of dissolved oxygen are insufficient to produce consequential changes to the bulk absorption properties of natural waters.

4.4 DISSOLVED ORGANIC MATTER

Dissolved organic matter (DOM) concentrations in natural waters, as discussed in Section 4.1, are consequences of either photosynthetic activity of phytoplankton (autochthonic) or direct inputs of terrestrially derived matter (allochthonic). The transformation of phytoplankton into DOM is a consequence of photolysis, hydrolysis, and bacterial decomposition of the phytoplanktonic cellular structures. Of the total organic matter resulting from phytoplankton photosynthesis, up to 20% can be released to the aquatic environment through metabolic egestion.[144] These egesta include a) intermediate metabolic products of small molecular mass (e.g. glycolic acid) resulting from phytoplankton photosynthesis involving a deficiency of nutrients and/or an excess or deficit of incident light and b) final metabolic products of larger molecular mass (hydrocarbons, peptides, aldehydes, ketones, enzymes, as well as inhibitors and/or stimulators of growth). Obviously, the metabolic release of dissolved organic matter by phytoplankton cells is greater in fertile waters than in less fertile waters. These egesta are, in general, assimilated by bacterioplankton. However, in nutrient-limited waters, the phytoplankton themselves may compete with the bacterioplankton for such egesta assimilation.

When phytoplankton cells decompose in the water column (or in the bottom sediments), the organic matter is chemically transformed, through microbial action on time-scales of days to weeks, to carbon dioxide and inorganic nitrogen, sulfur, and phosphorous compounds. Over 80% of the principal remains of the decomposed phytoplankton cells (under aerobic conditions) result in such substances as CO_2, CH_4, NH_3, and H_2O, among others. In addition, the decomposition process results in the creation of a variety of complex polymers generally referred to as the so-called *water humus*, or *humic substances*. These humic substances comprise both water-soluble and water-insoluble fractions, the insoluble fraction, referred to as *humin* being present as suspended particulates of a variety of particle sizes and molecular weights. The water-soluble fraction comprises dissolved organic carbon in the forms of *humic acid* and *fulvic acid*. The soluble and insoluble fractions of humic substances are quite similar in chemical composition, albeit different in molecular weight.

The presence of water-soluble humic substances in natural waters is readily apparent in inland and coastal waters (when their presence is not otherwise obfuscated by suspended particulates) by varying degrees of yellowness that are observable in such waters. This yellowness is a consequence of the temperature-dependent Maillard reaction[210] involving amino acids, carbonates, and phenol-bearing matter. In the presence of an alkaline medium the Maillard reaction leads to the formation of yellow and brown melanoids. These melanoids, however, comprise only a small fraction $\sim$(10–40)% of the aquatic DOM. As a consequence of this yellow hue, the dissolved aquatic humus is generally referred to as *yellow substance (YS)*, although such humic matter has been historically referred to in the literature by a variety of terms such as *gelbstoff, aquatic humic matter, yellow organic acids, humolimnic acid, gilvin,* and others.

Since allochthonic DOM is absent from mid-oceanic (Case I) waters, the yellow substance matter resident therein is totally autochthonic. Further, the nutrient levels in Case I waters are consistently lower than those found in non-Case I waters. Thus, oceanic waters generally contain lower concentration levels of autochthonic DOM than do inland and coastal waters. The presence of this considerably reduced level of DOM concentration results in yellow substances being less readily apparent to the human eye in oceanic waters. Since autochthonic DOM is a direct consequence of the presence of phytoplankton, it would be reasonable to expect that the aquatic humus would covary with the chlorophyll content of the water body. This characteristic of Case I waters provides an algorithm-development opportunity for oceanic remote sensing that is not available for remote sensing of inland and coastal waters.

DOM concentrations in oceanic waters are generally in the range (0.001–0.005) g Carbon per m^3 (it is generally considered that carbon constitutes approximately half the DOM by weight, and thus the concentration of DOM is usually given in units of grams of carbon per unit volume), which is about four orders of magnitude below the concentration of dissolved salts. However, the optical impact of DOM is greater than that of dissolved salts. Although DOM does not significantly impact scattering within the water column, DOM considerably increases absorption. From Romankevich,[345] the depth/concentration profile of oceanic DOM (C_{DOM}) generally displays a maximum in the upper 100 m, remains relatively constant to depths of about 250 to 300 metres, beyond which it continuously decreases with depth. In inland and coastal waters, however, DOM concentrations can be considerably higher than in oceans. From various measurements reported in the literature, values of C_{DOM} range from as low as (1 or 2) g C per m^3 to as high as (20–25) g C per m^3 (depending upon the trophic status of the water body), with average concentrations near 9 or 10 g C per m^3. For example, low intermediate values of C_{DOM} (average value of $\sim$2 g C per m^3) are observed in Lake Ontario,[37] and higher intermediate values of C_{DOM} (average value of 8 g C per m^3) are observed in Ladoga Lake.[280]

The light absorption properties of dissolved aquatic humus have been readily determined from laboratory measurements performed upon filtered samples. Results of such studies (Yentsch;[434] Zepp and Schlotzhauer;[437] Bricaud *et al.*;[31] Davies-Colley and Vant;[80] Carder *et al.*;[55] Højerslev;[172] and others) show that, as a consequence of electronic transitions, absorption by dissolved organic matter dramatically decreases with increasing values of λ. Unfortunately, spectral absorption signatures in the visible region of the electromagnetic spectrum do not provide a method of distinguishing autochthonic DOM from allochthonic DOM. There are non-spectral techniques based upon ratios of permanganate to bichromate oxidizability that may provide such information (a ratio > 0.40 being indicative of waters containing predominantly allochthonic DOM and a ratio < 0.40 being indicative of autochthonic DOM), but such techniques[369] will not be detailed here. Due to the concern regarding the deleterious impacts to the aquatic environment of enhanced ground-level UV-B fluxes associated with stratospheric ozone depletion (see Section 2.3), the absorption of ultraviolet radiation by aquatic humus has become a research focal point. It is possible, therefore, that absorption signatures in the ultraviolet region of the electromagnetic spectrum may provide some insight into the optical delineation of sources of the humic matter.

The spectral dependence of the absorption coefficient for yellow substance, a_{YS}, can be reasonably described[31,437] by the exponential formula:

$$a_{YS}(\lambda) = a_{YS}(\lambda_0) \exp[-s(\lambda - \lambda_0)] \qquad (4.2)$$

where s is a slope parameter that is assumed to be independent of wavelength λ. The relative constancy of the shape of the $a_{YS}(\lambda)$ versus λ curve of equation (4.2) enables the selection of a reference wavelength, λ_0, upon which to compare the yellow substance concentrations of different waters. The selection of $\lambda_0 = 400$ nm is certainly appropriate since it represents the maximum a_{YS} in the visible spectrum. A λ_0 value of 440 nm, however, is also frequently encountered in the published literature.

Although equation (4.2) is based upon the premise of a single slope s being appropriate for the water sample being investigated, this slope is a function of the basin in which the water body resides as well as the location and depth at which the absorption is measured. Values of s have been observed to vary within the limits 0.011 and 0.021. Carder et al.[55] observed $s = 0.0194$ for the predominantly fulvic acids in the Mississippi plume and $s = 0.0110$ for the predominantly humic acids in the Gulf of Mexico, suggesting that fulvic acids (FA) display a steeper absorption curve slope with increasing wavelength than do humic acids (HA). Carder et al. also report that FA absorption is much lower than HA absorption at the same wavelength. The $a_{YS}(\lambda)$ is, therefore, dependent upon the ratio of FA to HA in the water body under consideration. Freshwater humus is largely comprised of fulvic acids (Ghassemi and Christman,[125] Zepp and Schlotzhauer,[437] Visser[413]). The observed $a_{YS}(\lambda)$ spectra for natural waters would, therefore, be influenced by the FA/HA ratio appropriate to the water body and thus be integral to the remote sensing of that water body.

For oceanic waters Plass et al.[313] have reported values of a_{YS} that vary from 0.0240 m^{-1} at 400 nm to 0.0125 m^{-1} at 440 nm to 0.0003 m^{-1} at 700 nm. Values of a_{YS} observed in non-Case I waters display a very large geographic variance, dependent on the trophic status of the natural water body. For the encapsulated waters of the Baltic Sea, a_{YS} at λ = 400 nm has been reported[198] in the range (0.60–0.68) m^{-1}. For lakes in the Netherlands a_{YS} at λ = 440 nm have been reported[81] in the range (0.91–22.6) m^{-1}, close to values of a_{YS} at λ = 440 nm found for a variety of Australian lakes.[221]

Routine laboratory protocols for analyses of collected water samples do not normally include direct determinations of a_{YS}. An approximation of a_{YS} may be obtained by a spectrophotometer scan of a filtered sample (to remove the optical impact of suspended matter) throughout the visible spectrum. The value of $a(\lambda)$ thus obtained, once the absorption coefficient for pure water, $a_w(\lambda)$ is subtracted, becomes an estimate of $a_{YS}(\lambda)$.

High correlations between $a_{YS}(\lambda_o)$ and dissolved organic carbon concentration (DOC) have been observed for some water bodies (correlation coefficients between 0.80 and 0.96). For these conditions, an accurate determination of DOC concentration can yield a reliable estimate of a_{YS}, and *vice versa*.

To date very little attention has been directed toward the possible effects of DOM on the scattering properties of natural waters. (We, too, in our analyses of the volume reflectance/water quality relationships in non-Case I waters find it convenient to set $b_{DOM}(\lambda)$ equal to zero.) Although such neglect of the scattering from organic matter that resides in the water column in true molecular aqueous solution is justified, there is a fraction of the "dissolved" organic matter that is not in true molecular solution. This fraction of what is conveniently (but erroneously) included within the umbrella term of DOM contains organic matter in colloidal form which can impact the bulk scattering coefficient characterizing the water column. Further, since laboratory samples of DOM are obtained by filtration removal of matter in suspension, the definition of "dissolved" is a direct function of the filter pore size and the sampled particle size distribution. It is highly unlikely, therefore, that $a_{DOM}(\lambda)$ is determined for a 100% true solution. Particulates with diameters less than the filter pore [generally (0.2–0.4) μm] are, therefore, included within the filtered sample, contributing not only to a neglected b_{DOM} term, but also to an overestimated a_{DOM} term.

4.5 SUSPENDED MATTER

All natural water bodies inescapably contain a suspended matter component comprised of organic and inorganic material under the collective term *seston*. Seston is extremely diverse in origin and in composition, and includes mineral particles of terrigenous origin, plankton, detritus (largely residual products of the decomposition of phytoplankton and zooplankton cells as well as macrophytic plants), volcanic ash particles, particulates resulting from *in situ* chemical reactions, and particles of anthropogenic origin.

The presence of the terrigenous suspended particles is a consequence of river discharge, shore erosion, long and short range transport of atmospheric particulates followed by dry deposition. These particulates are diverse in shape and size. According to Adamenko *et al.*,[1] fine clay particulates rarely exceed (3–4) μm in diameter, silt particles are in the range (5–40) μm, very fine grain sands are in the range (40–130) μm, and coarser grain sands are in the range (130–250) μm.

About 90% of the coarse fraction of the terrigenous matter loading of ocean waters is found within the nearshore zone, with concentrations of terrigenous matter of particulate diameter greater than 1 μm being virtually absent from Case I waters. The impact of terrigenous suspended matter on the optical properties of Case I waters, therefore, is inconsequential. The morphology and hydrology of inland and coastal waters, however, can result in coarse terrigenous particulate concentrations that range from optically insignificant to optically overwhelming. The elemental composition of such suspended matter includes silicon, aluminum, and iron, generally in the form of oxides. From Jerlov,[198] suspended particle (sediment) concentrations in the surface waters of oceans are in the range (0.02–0.17) g per m^3. Ivanoff,[185] however, reports this range to be (0.8–2.5) g per m^3, with values up to 18 g per m^3 being present in bays. Both authors report the suspended minerals comprise between 40% and 80% of the suspended matter. The disturbing discrepancy between the concentrations of oceanic sediments reported by the two authors may be inherent to lack of standardization or other shortcomings in the laboratory analyses techniques. Our above statement that the impact of suspended terrigenous matter has a negligible optical impact on oceanic water color attests to our bias toward the range given by Jerlov. Concentrations of suspended sediments of the magnitude reported by Ivanoff (based upon the optical cross section spectra for suspended matter discussed in this book) would contradict the distinct blueness displayed by the global oceans. In Ladoga Lake seston concentrations display a range (0.1–12.0) g per m^3, with suspended minerals comprising 50% to 97% of this concentration.[280] Suspended mineral concentrations in Lake Ontario have been reported[38] in the range (0.2–8.9) g per m^3.

Another form of suspended aquatic colorant may arise from extended regions of precipitates such as iron and manganese hydroxides and calcium carbonate. Substantial concentrations of these insoluble precipitates can produce significant selective spectral absorption and scattering within a water body. The color of $Fe(OH)_2$ is pale blue-green, the color of $Fe(OH)_3$ is reddish brown, and the color of $Mn(OH)_2$ is dark grey to black. $CaCO_3$ precipitate, however, is milky-white and generally referred to as *whitings* (named after the powdered chalk used in making paints and not after the global food fish of the same name). Precipitates are a consequence of chemical activity (in the case of hydroxides) or separation of solute and solvent in supersaturated solutions (in the case of whitings). Consequently, the occurrence of lake-wide precipitate patterns is dependent upon not only the right combination of local aquatic components, but also appropriate pH (the negative logarithm of the aquatic hydroxyl ion concentration, a measure of the lake acidity/alkalinity), local presence of crystallization centers (for whiting events), and a triggering mechanism to disrupt the precarious stability of the supersaturation. Clearly, such conditions are not met at every global location, and thus the occurrence of aquatic browning, blackening, and whitening are observed as local phenomena. For example, iron and manganese hydroxide precipitates have been observed in lakes in the Isthmus of Kola in northwestern Russia.[388] Whitings have been observed as short-lived and seasonal phenomena in some of the Laurentian Great Lakes[391] and Lake Sevan in Armenia.[403]

Natural suspended organic matter is restricted to plankton and detritus, plankton comprising zooplankton, algae, bacteria, and algal fungi, and detritus comprising fragments of decayed plants and animals along with their excretions.

Zooplankton, being grazers of algae, detritus, and bacteria, are an integral component of the trophic status of natural waters, and, as such, are in dynamic equilibrium with the other components of the aquatic food chain. Thus, provided there is no untoward disruption in the food chain, highly productive waters would be expected to display higher zooplankton concentrations than less productive waters. Depending upon the trophic status of inland or coastal waters, the concentrations of zooplankton can easily exceed several hundreds of thousands of plankters per m^3, varying in sizes from 30 μm to > 2 mm.

The spatial distribution of zooplankton is controlled by current patterns, thermal structure, specific features in reproductive cycles (species, health, and gender), the presence and concentrations of nutrients, and the presence and populations of planktivores. This multi-dependence results in a zooplankton population mosaic (patchiness) being a common feature of inland and coastal waters, a patchiness which becomes more pronounced, due to nutrient limitation, as the trophic status of the water diminishes.

Zooplankton are generally considered to be ineffectual aquatic colourants, and consequently are ignored in most water color models. This ineffectiveness is a result of their small concentrations (compared to the many orders of magnitude higher concentrations of phytoplankton and bacterioplankton). However, due to the ubiquitous patchiness of their spatial distributions, it is conceivable that there are locations at which such populations are substantial enough to be worthy of colorant status (e.g., in alpine lakes or saline lakes that appear reddish). Further, since zooplankton are grazers of phytoplankton and phytoplankton are chlorophyll-bearing biota, chlorophyll is invariably present within the digestive tracts of zooplankters. This presence of chlorophyll and its derivatives would suggest that absorption and scattering cross sections of zooplankton might display comparable features (albeit not comparable spectral values) to the cross sections of phytoplankton, although we are not aware of existing data that would support or refute this hypothesis.

The coloration of phytoplankton cells, be they unicells or colonies of cells, is dependent upon their pigment composition (see Chapter 9). This pigment composition results in green algae or Chlorophyta (which are dark green), red algae or Rhodophyta (which are dark red), blue-green algae or Cyanophyta (which may be olive-green, yellow-green, pink, violet, or brown), dinoflagellates or Pyrrhophyta (which are reddish), diatoms or Bacillariophyceae (which are brownish), among others. Phytoplankton cells and colonies exist in a variety of shapes (filaments, ribbons, stars, etc.) and sizes. Reynolds[341] and Petrova[307] present statistical data on particle size distributions of phytoplankton which show that unicellular algae possess nominal mean maximum dimensions in the range (5–40) μm. The corresponding range for filaments would be (80–150) μm, for ribbon colonies (60–several hundred) μm, and for mucilaginous colonies (50–200) μm.

Non-turbid, oligotrophic (see Section 7.9) lakes in the global temperate zones (the northern and southern regions of the Earth between the tropical and the polar zones) display great diversities in indigenous species and shapes of phytoplankton. For example, over 120 species and varieties of algae have been documented in Lake Michigan[387]), over 150 species and varieties in Lake Erie,[281] and over 380 species and varieties in Ladoga Lake.[307] The majority of algae indigenous to inland waters are diatoms, green algae, and blue-green algae. Red algae, which are prominent in oceanic waters, are absent from inland waters.

It is widely observed that the relative populations of the phytoplankton classifications comprising natural water bodies display a strong and (in the absence of abnormal external forcing functions) predictable seasonal cycling, with each of the seasons displaying specific phytoplankton population compositions.

Apart from the seasonal species succession and spatial patchiness of phytoplankton distribution, such factors as thermal density stratification, light and nutrient availability, hydrooptical properties of the water column, life-cycle dynamics, and meteorological parameters, produce fluctuations in the vertical profiles of phytoplankton concentration. As a result, the depth location of the phytoplankton concentration maximum in inland and coastal waters varies in both space and time.[275]

Phytoplankton pigments serve as the collectors and suppliers of energy for the process of *photosynthesis* (see Section 9.1), the basic mechanism in plant growth. There are three basic types of photosynthesizing agents: *chlorophylls, carotenoids,* and *phycobilins.* Chlorophylls and carotenoids are present in all algal species. Phycobilins are additionally present in blue-green algae and dinoflagellates. There are several types of chlorophyllous pigments, differing chemically and designated as chlorophylls *a, b, c,* and *d.* There are also several sub-species[422] of chlorophyll *c.* All green algae contain chlorophyll *a,* and many contain chlorophylls *b* and/or *c.* It is still uncertain as to whether or not chlorophyll *d* plays a role in photosynthesis.

In freshwater systems, the green algae molar ratio of chlorophyll *a* to chlorophyll *b* is ~3:1. This is comparable to the ratio generally observed for oceanic green algae. Phytoplankton containing the pigment chlorophyll *c* are characterized by a chlorophyll *a* to chlorophyll *c* ratio which is also ~3:1, illustrating the predominance of chlorophyll *a* in algae.

Carotenoids are pigments generally dissolved in fatty tissue. Even though there are considerably more known carotenoids than known chlorophylls, the predominant carotenoid present in aquatic algae is β-carotin (the β-carotin molecule is a long unsaturated carbon chain with specific end groups). The molar ratio of concentrations of carotenoids to chlorophylls may be less than or greater than unity, and varies from plankton species to plankton species. A ratio of 1:3 is observed for green algae; a ratio of 2:1 for diatoms; a ratio 3:2 for dinoflagellates; a ratio of 1:2 to 2:1 for brown algae; a ratio of 1:1 to 2:5 for red algae.

Phycobilins are protein-bound pigments comprised of open chains of conjugated double bonds. Together with the chlorophylls and the carotenoids, the phycobilins account for the phytoplankton contribution to water color.

From laboratory absorption spectra performed on dissolved pigments (such spectra are dependent upon the particular solvent employed), the principal chlorophyll absorption occurs in the short wavelength blue and the long wavelength red regions of the visible spectrum. Absorption in the blue region is more intensive than in the red for chlorophylls *a* and *b,* and considerably more intensive for chlorophyll *c.* Carotenoid absorption peaks in the blue spectral region. Phycobilin absorption (depending upon species) peaks at green, yellow, or short red wavelengths. As a result, chlorophylls appear green, carotenoids appear orange, and phycobilins appear either red or blue.

In addition to the optical impacts of chlorophyllous algal pigments, there is also a related optical impact due to the products of decomposed chlorophyll (*phaeophytin* and *phaeoforbid*) resulting from oxygen hydrolysis (the removal of a magnesium atom from the chlorophyll molecule). With such a decomposition of chlorophyll, the absorption in the blue displays a peak shift toward shorter λ values and the absorption in the red displays a peak shift toward longer λ values. The intensities of the absorption bands of both these chlorophyll derivatives, however, are considerably weaker than the intensities of the absorption bands of chlorophyll. Nevertheless, a correct estimate of the concentration of chlorophyll derivatives is essential to indicate the portion of non-living phytoplankton in the water column. (The partitioning of chlorophyll *a* and phaeophytin concentrations has direct implications to the estimation of aquatic primary production, a topic to which we shall return in Chapter 9.)

Bacterioplankton may be unicell, colonial, or filamentary organisms representing lower lifeforms of various shapes and sizes (cells or colonies varying in sizes from ~1 μm to ~100 μm). Physiological groups of bacterioplankton include ammonificators, nitrificators, thiobacteria, ironbacteria, carbon-decomposing bacteria, and others. Although most bacterioplankton are colorless, some species do display coloration.

Bacterioplankton produce organic matter either through chemical reactions (chemosynthesizing) or through light-induced reactions (photosynthesizing). Photosynthesizing bacteria, which may be more widespread in freshwaters than has been previously believed,[179] include green and purple thiobacteria (unicellular or colonial strictly anaerobic bacteria) and nonsulfur purple bacteria (unicellular or colonial non-obligatory aerobes capable of growth on inorganic nutrients in anaerobic conditions).

Number counts of bacterioplankton are of the order of 10^5 cells per cm^3 in oligotrophic waters, a figure that becomes considerably larger as the water becomes more eutrophic, reaching number counts of the order of 10^9 cells per cm^3. The surface waters of Ladoga and Onega lakes, for example display bacterioplankton in the number count range (180–540) and (200–270) $\times$ 10^5 cells per cm^3, respectively.[89]

In general, the vertical profile of bacterioplankton concentration follows the vertical temperature profile since cold waters suppress the rate of bacterial reproduction. It is not surprising, therefore, that bacterial number counts display sharp increases in the summer and autumn months, and a sharp decrease in the winter months.

The colorless bacterioplankton, of course, do not produce an optical impact on the aquatic color, although they likely contribute to scattering. The colored bacterioplankton, however, possess pigments that are comparable to those of phytoplankton (and, indeed are referred to as bacterial chlorophylls a, b, c, and d). The absorption spectra of the bacterial chlorophylls are similar to those of algal chlorophylls, except the peaks are shifted toward longer λ values. The purple coloring that is observed in some species of bacterioplankton, however, suggests the presence of non-chlorophyllous pigments.

Algal fungi are colorless, chlorophyll-free lower plant forms consisting of assemblages of cellular filaments or *hyphae* containing spores. The characteristic sizes of spores are in the range (3–15) μm, whereas the lengths of hyphae fragments range from several mm to several cm. In Ladoga Lake thirty-two species, representing five classes, of algal fungi have been observed. Algal fungi, along with bacterioplankton, are the sole transformers of organic matter in natural waters into compounds digestible by most aquatic microorganisms.

The number counts of algal fungi within a water body are highly variable throughout all seasons except the summer. Although the algal fungi are virtually independent of abiotic factors, in the summer there is a distinct dependence of fungi counts on water temperature and concentration of organic matter. The general distribution of freshwater algal fungi is character-ized by enhanced concentrations in the coastal zones and estuaries.

The heterotrophic nature of algal fungi suggests that they, along with all forms of indigenous plankton, are good indicators of the ecological state of aquatic basins. The spore concentration in the water column is one indica-tion of trophic status, and an intensive proliferation of microflora with corres-ponding successions of dominant algal fungi can be indicative of anthropogenic eutrophication. For example, the spore count in the offshore (pelagic) waters of Ladoga Lake in 1986 increased from previously recorded values of $(6–7) \times 10^3$ diaspores per cm^3 to an all-time high value of 32.5×10^3 diaspores per cm^3.

The lack of fungal coloration, coupled with their low concentrations, results in algal fungi producing no optical impact on natural waters.

4.6 AIR BUBBLES

In the upper layers of natural waters exist hydrosols comprised of trapped air bubbles, each with a relative refractive index of 0.75. The sizes of air bubbles vary[185] from 0.01 μm to 1000 μm. The histogram of the size distribution of this type of hydrosol depends upon meteorological conditions and has no stable parameters.

The presence of a large number of purely scattering centers (bubbles) in near-surface waters is capable of dramatically altering the radiation distri-bution in the water column. These scattering centers result in (a) an attenua-tion of downwelling radiation and (b) an increase in the backscattered upwelling radiation. This, in turn, results in a modified observable volume reflectance which may be erroneously attributed to the presence of sus-pended organic or inorganic particulates.

4.7 CUMULATIVE OPTICAL IMPACT OF THE AQUATIC COMPONENTS

As discussed in this chapter, there is a wide variety of species and sub-species of organic and inorganic materials present in freshwater bodies in suspension and/or solution. The radiance spectrum upwelling from the freshwater body is a direct consequence of the cumulative spectrally selective absorption and scattering that can be attributed to each of these materials. To convolve each and every one of these materials into a unified bio-optical model with its own specific absorption and scattering cross section spectra, as illustrated in equations (3.15) to (3.17), is clearly an unattainable goal. A more attainable goal, however, would be reducing the variety of aquatic materials to a smaller, more manageable number while minimizing the compromise to the optical model. The pigments residing in phytoplankton (chlorophylls, carotenoids, and phycobilins), the scattering and absorption of suspended inorganics, and the absorption properties of yellow substances must be prominently considered in any such optical model. These colorants have been included within the bio-optical model we have presented in Chapter 3, although we have utilized the predominance of chlorophyll a as justification to ignore the other chlorophyllous pigments. Second generation non-Case I optical models might add Chl$b$ and Chl$c$ to the equation. Carotenoids and phycobilins might also be considered. In this regard, however, due to the essentially inconsequential optical contribution of suspended inorganic matter to Case I water color, the compartmentalization of phytoplanktonic pigmentation is proceeding with greater success in ocean waters than it is in fresh waters.

We feel, at this time, that the exclusion of zooplankton, algal fungi, dissolved gases and salts, and air bubbles does not significantly detract from the applicability of the four component model of the previous chapter. We do, however, feel that the inclusion of individual phytoplankton and bacterioplankton pigments could greatly improve the predictive capabilities of a unified bio-optical model. Such improvement should, of course, be considered. However, the harsh realities of the optical complexities of inland and coastal waters and the need to match the number of solvable independent equations with the number of unknowns restricts the realistic options accordingly.

The optical complexities of natural waters in general (and inland and coastal waters in particular) briefly presented in this chapter dramatically illustrate the need to extract the optical signature of each component from the cumulative optical signature of the totality of components. To simply correlate the concentrations of a single component with the optical return from the water body (as we will see comprises a large number of remote sensing concentration extraction algorithms reported in the literature) ignores not only the specific optical properties of the multitude of organic and inorganic materials comprising the water body, but also their cumulative impact on the recorded signal at each λ throughout the visible spectrum. This signal is strongly dependent upon both the time that the signal was recorded and the composition of the water column at that particular time. Such single component algorithms, therefore, describe a local condition that only exists at the time the correlation was performed (such correlations are currently being referred to in the literature as *local algorithms*). The value of such local algorithms to the remote sensing of a water body of more than one-component, is, of course, open to debate. What is less debatable is an algorithm or methodology that yields multi-component as opposed to single component concentrations (i.e., one that will reliably extract each component concentration from the cumulative optical signal of many components). The specific optical properties (cross section spectra) of the aquatic components provides such a methodology, and follows the general philosophy that in order to remotely extract the concentration of a desired aquatic component concentration in a non-local manner, it is essential to simultaneously extract the concentrations of its co-existing organic and inorganic optical competitors.

Thus, despite the obvious limitations of the multi-component concentration extraction model discussed in this book (as well as comparable models based upon specific optical properties of indigenous aquatic matter), we feel that such an approach far exceeds the capabilities of the "local algorithms" in widespread use.

Chapter 5

THE EFFECTS OF CHLOROPHYLL, SUSPENDED MINERALS, AND DISSOLVED ORGANIC CARBON ON VOLUME REFLECTANCE

5.1 OPTICAL CROSS SECTIONS

We have seen that the combined processes of absorption and scattering control the manner in which impinging radiation propagates through a natural water body. The nature and magnitude of these absorption and scattering processes are controlled by the bulk optical properties of the aquatic medium. The bulk optical properties are, in turn, direct consequences of the amounts of scattering and absorption that may be attributable to each optically significant organic and inorganic component comprising the natural water body. These *absorption* and *scattering cross sections* (also referred to as *specific absorption* and *specific scattering coefficients*), therefore, provide the direct linkages between the optical properties of a natural water body and its composition.

The optical cross section spectra of indigenous organic and inorganic matter are directly responsible for the radiance distribution emerging from a water body in the visible region of the electromagnetic spectrum (and thus directly responsible for the observed color of natural waters). The remote estimation of aquatic component concentrations is, therefore, contingent upon precise knowlege of the cross section spectra appropriate to the water body being remotely monitored. Since chlorophyll pigments are among the

135

principal ocean colorants, it logically follows that numerous workers have directed considerable effort toward obtaining the optical cross sections of chlorophyll-bearing biota. The results of such determinations, however, have been far from universally applicable. Certainly there has been general agreement concerning rather broad ranges of the numerical values of such cross sections, as well as the realization that the optical cross sections are algal species-dependent (and, therefore, temporally and spatially dependent). Thus, a major obstacle to the remote estimation of chlorophyll concentrations from space has been a lack of such precision information concerning the cross section spectra of chlorophyll-bearing biota, particularly for those indigenous to major non-Case I waters (i.e., for inland, estuarine, and coastal regimes).

Morel and Bricaud[274] have shown that the absorption cross sections of chlorophyll a vary not only with species of chlorophyll-bearing biota, but also with variations in cell age, cell structure, and previous optical history within the same species. Theoretical considerations presented by Kirk[219,220] coupled with the directly observed variations in absorption spectra reported by such workers as Smith and Baker,[374] Kirk and Tyler,[229] Morel,[272] Mitchell and Kiefer,[266] Bricaud and Stramski,[32] Bukata et al.,[46] Gallie and Murtha,[120] and others have illustrated these departures from cross section constancy. Figure 5.1, taken from Bukata et al.[47] (their Figure 6), depicts an intercomparison of a number of independent determinations of the absorption cross section spectrum of chlorophyll a (uncorrected for phaeophytins). Apart from Curve 3, which was obtained from laboratory cultures (and is the average absorption spectrum for fourteen different algal species), the data in Figure 5.1 were determined from field investigations in waters from such diverse sources as the North Central Pacific, the California coast, an upwelling region off the coast of Peru, the Sargasso Sea, Ladoga Lake in northern Europe, and Lake Ontario in central North America. Six of the seven curves in Figure 5.1 display similar spectral shapes, the data from the Sargasso Sea being the only obvious exception. The limited data of Figure 5.1 notwithstanding, it is noted that the two curves representing inland waters (Curves 1 and 2) display absorption cross section spectra values that are consistently larger at the longer wavelengths than those determined for oceanic and coastal waters.

Figure 5.2 illustrates the backscattering cross section spectra for chlorophyll a determined by Bukata et al.[46] for Lake Ontario in 1979 and Ladoga Lake in 1988/1989. A distinct double peak at ~550 nm and ~620 nm is clearly seen. Even though they were obtained a decade apart, the near-identical Ladoga Lake and Lake Ontario chlorophyll absorption cross section spectra of Figure 5.1 and the very similar chlorophyll backscattering cross section spectra of Figure 5.2 (which would be even more sensitive to variations in the physical characteristics of the cell than would the absorption

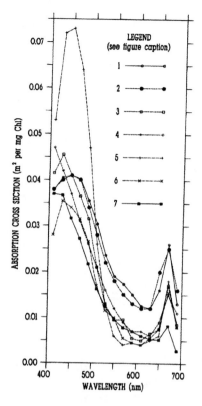

FIGURE 5.1
An intercomparison of a number of independent determinations of the absorption cross section spectrum for chlorophyll *a* uncorrected for phaeophytins. (From Bukata, R. P., Jerome, J. H., Kondratyev, K. Ya., and Pozdnyakov, D. V., *J. Great Lakes Res.*, 17, 470–478, 1991. With permission.) Curves 1 and 2 are data from Bukata *et al.*, 1991; Curve 3 is data from Morel, 1988; Curve 4 is data from Smith and Baker, 1978; Curves 5 and 6 are data from Bricaud and Stramski, 1990; Curve 7 is data from Mitchell and Kiefer, 1988.

cross section spectra) suggest that Ladoga Lake and Lake Ontario are characterized by optically comparable populations of chlorophyll-bearing biota.

In general, due to their proximity to land masses, non-Case I waters display a larger number of principal colorants as well as greater ranges in the concentrations of indigenous colorants than do Case I waters. Therefore, in addition to the necessity of contending with the temporal and spatial variabilities of the optical cross section spectra of the indigenous chlorophyll-bearing biota, remote sensing of these waters must contend with the temporal and spatial variabilities of additional indigenous materials of terrestrial origin. Inorganic cross section spectra are dependent upon the particle shape, particle size distribution, and refractive index of the terrigenous materials characteristic of a particular basin, as evidenced by the dissimilar cross section spectra reported by Morel and Prieur,[273] Whitlock *et al.*,[425] Prieur and Sathyendranath,[337] Bukata *et al.*,[46] Gallie and Murtha,[120] and others.

There are few published values for the absorption and backscattering cross section spectra of suspended minerals [$a(\lambda)_{SM}$ and $b_B(\lambda)_{SM}$, respectively]. Figure 5.3 is an intercomparison of absorption cross section spectra obtained for non-chlorophyll matter by several workers. The absorption cross section

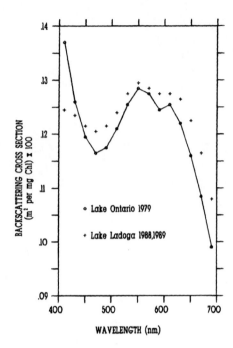

FIGURE 5.2
Backscattering cross section spectra for indigenous chlorophyll-bearing biota for Lake Ontario and Ladoga Lake. (From Bukata, R. P., Jerome, J. H., Kondratyev, K. Ya., and Pozdnyakov, D. V., *J. Great Lakes Res.*, 17, 461–469, 1991. With permission.)

spectra for Lake Ontario,[42] Ladoga Lake,[46] and Chilko Lake in British Columbia, Canada,[120] were determined for concentrations of suspended minerals, and, as such, are directly comparable. The suspended inorganic matter in these three inland water bodies display a distinct minimum in absorption cross section spectra in the 590 nm to 630 nm interval and a distinct maximum at lower wavelengths. These results are comparable (although technically not *directly* comparable) to the spectrum taken from Morel and Prieur[273] and the spectrum taken from Prieur and Sathyendranath.[337] The former is a specific absorption spectrum for "suspended and dissolved material apart from algae" and the latter the specific absorption spectrum for "nonchlorophyllous particles." These two cross section spectra, therefore, include effects of matter other than suspended minerals. Workers such as Roesler *et al.*[344] and Witte *et al.*[428] illustrate absorption cross section spectra which monotonically decrease with increasing wavelength (i.e., no inversion as shown in the spectra of Figure 5.3). These, and the majority of other published works, however, pertain to detrital material of organic origin, as opposed to the curves of Figure 5.3, which pertain to suspended particulates of non-biological origin.

Figure 5.4 illustrates an intercomparison of the limited number of existing backscattering cross section spectra for suspended minerals. Once again Lake

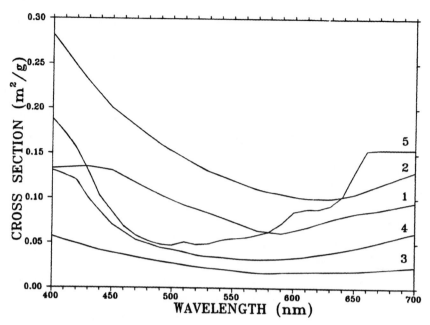

FIGURE 5.3

Absorption cross section spectra for non-chlorophyll matter. Data shown are from Bukata *et al.*, 1985 (Curve 1); Bukata *et al.*, 1991 (Curve 2); Gallie and Murtha, 1992 (Curve 3), from Morel and Prieur, 1977 (Curve 4); Prieur and Sathyendranath, 1981 (Curve 5).

Ontario, Ladoga Lake, and Chilko Lake data from Bukata *et al.*[46] and Gallie and Murtha,[120] respectively, are shown, along with spectra derived from Whitlock *et al.*[425] for three rivers in Virginia, U.S.A. [The spectra were derived on the assumption that backscattering from suspended minerals in rivers can be confidently considered to overwhelm backscattering from algae and that the relationship between $b_B(\lambda)_{SM}$ and the concentration of suspended mineral is linear at all values of λ. $b_B(\lambda)_{SM}$ could then be approximated as $b_B(\lambda)$ divided by the SM concentration]. Apart from the cross section spectrum from Ladoga Lake, the backscattering cross section spectra depicted in Figure 5.4 all show a relatively uneventful increase toward values of shorter λ. This would appear to be in reasonable agreement with Morel,[269] who suggests the particulate scattering within water column be expressed as a power law distribution in λ. Some mid-frequency flattening is evident in the Lake Ontario data, but, in general, with the exception of the Ladoga Lake data, a drop of ~35% in the backscattering cross section of suspended minerals over the visible wavelengths is a feature of Figure 5.4. We are not aware, however, of any other natural water body displaying decreased values of

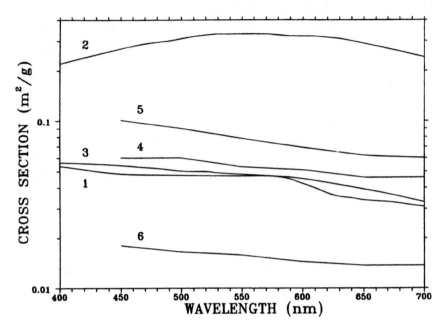

FIGURE 5.4
Backscattering cross section spectra for suspended minerals. Data shown are from Bukata *et al.*, 1991 (Curves 1 and 2); Gallie and Murtha, 1992 (Curve 3); and Whitlock *et al.*, 1981 (Curves 4, 5, and 6).

$b_B(\lambda)_{SM}$ such as observed in the waters of Ladoga Lake. In order to develop remote sensing monitoring capabilities, it is mandatory that this shortcoming in the number of reported optical cross section spectra be overcome.

There is considerably better agreement in spectral shape (albeit with limited data sets) among the measured cross sections of dissolved organic materials (Unoki *et al.*,[405] Bricaud *et al.*,[31] Bukata *et al.*,[38] Gallegos *et al.*,[118]) suggesting that the absorption for dissolved organics can be considered to be less variable than absorption and scattering for particulates. The optical properties of pure water may, with greater confidence, be considered invariant.

5.2 MULTI-COMPONENT OPTICAL MODEL FOR NATURAL WATERS

In Section 3.7 we indicated that once the optical cross section spectra for the indigenous organic and inorganic components were known, the bulk optical properties $a(\lambda)$, $b(\lambda)$, and $b_B(\lambda)$ could be generated for water columns made up of any pre-selected combinations of co-existing concentrations of those components from the additive equations:

$$a(\lambda) = \sum_{i=1}^{n} C_i a_i(\lambda) \tag{5.1}$$

and

$$b_B(\lambda) = \sum_{i=1}^{n} C_i (b_B)_i(\lambda) \tag{5.2}$$

where $a_i(\lambda)$ and $(b_B)_i(\lambda)$ are the absorption and backscattering cross section spectra of the aquatic component i, and C_i is its concentration.

As discussed in Section 3.6, Monte Carlo simulations of the radiative transfer process could then be used to relate the subsurface volume reflectance spectra just beneath the air-water interface, $R(\theta,0,\lambda)$, to these bulk inherent properties. Recall that, for a vertically incident photon flux $(\theta = 0°)$[204]:

$$R(0°, 0, \lambda) = 0.319 b_B(\lambda)/a(\lambda) \tag{5.3}$$

for $0 \leq b_B(\lambda) \leq 0.25$, and

$$R(0°, 0, \lambda) = 0.013 + 0.267 b_B(\lambda)/a(\lambda) \tag{5.4}$$

for $0.25 \leq b_B(\lambda) \leq 0.50$.

The ideal bio-optical water model incorporates the cross section spectra pertinent to each "optically active" organic and inorganic aquatic component, whether that component be in suspension or in solution. This would then determine the optimal number of cross sections and concentrations [numerical value of i in equations (5.1) and (5.2)] that must be determined or estimated. Although a variety of chlorophyllous pigments, suspended inorganic particulates, and dissolved organic material (in addition to living organisms such as zooplankton) are colorants of non-Case I water bodies, we will consider the approach outlined in Bukata et al.[41] and Kondratyev and Pozdnyakov[238] and restrict the optical model to four components, viz. pure water, chlorophyll a (Chl), suspended minerals (SM), and dissolved organic carbon (DOC). It is this four-component optical model that we will use to illustrate the impact on the subsurface volume reflectance of varying the concentrations of chlorophyll, suspended mineral, and dissolved organic carbon, either independently or in concert.

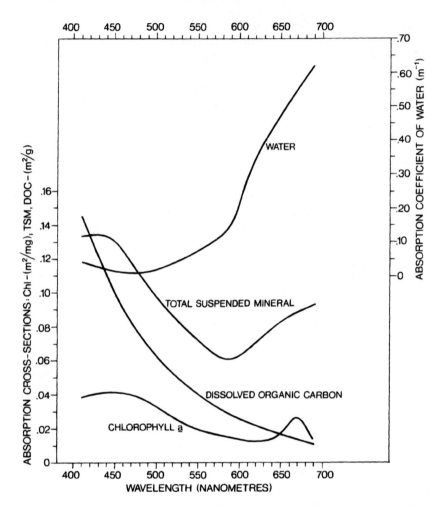

FIGURE 5.5
Absorption cross section spectra for chlorophyll, total suspended mineral, and dissolved organic carbon concentrations indigenous to Lake Ontario. (From Bukata, R. P., Bruton, J. E., and Jerome, J. H., *Rem. Sens. Environ.*, 13, 161–177, 1983. With permission.) Also shown are the absorption coefficients for pure water.

Figures 5.5 and 5.6, taken from Bukata *et al.*[40] (their Figures 1 and 2), illustrate absorption $a_i(\lambda)$ and scattering $b_i(\lambda)$ cross section spectra, respectively, calculated for the chlorophyll a (Chl), suspended mineral (SM), and dissolved organic carbon (DOC) indigenous to Lake Ontario. Also shown on the figures are the optical cross section spectra for pure water taken from Hulburt.[177] Values of the backscattering probability $B(\lambda)$ were determined by computer-fitting directly measured volume reflectance spectra to predicted volume reflectance spectra using the calculated values of $a(\lambda)$ and $b(\lambda)$. To

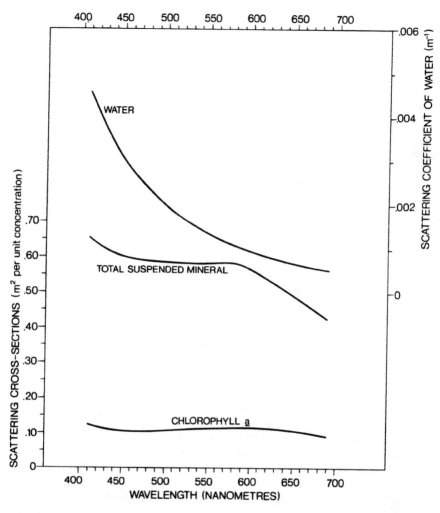

FIGURE 5.6

Scattering cross section spectra for chlorophyll and total suspended mineral concentrations indigenous to Lake Ontario. (From Bukata, R. P., Bruton, J. E., and Jerome, J. H., *Rem. Sens. Environ.*, 13, 161–177, 1983. With permission.) Also shown are the scattering coefficients for pure water.

a good approximation, $B(\lambda)$ as determined by Bukata *et al.*[40] for Lake Ontario waters, were represented as the spectrally invariant values:

$$B_{Chl} = 0.011$$
$$B_{SM} = 0.080$$
$$B_{w} = 0.500$$

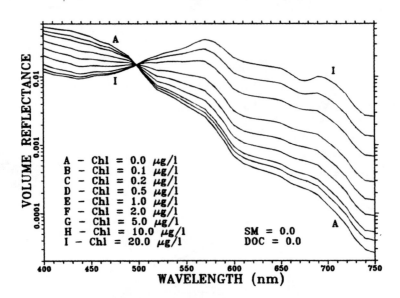

FIGURE 5.7
Volume reflectance spectra for various chlorophyll concentrations in a water column for which
the SM and DOC concentrations are kept fixed at zero.

Values of $b_B(\lambda)$ can then be readily considered as the products of $b_i(\lambda)$
with the appropriate value of B [e.g., $(b_B)_{Chl}(\lambda) = 0.011b_{Chl}(\lambda)$].

Elsewhere,[42] we have used the Lake Ontario optical cross section spectra
to illustrate the generation of volume reflectance spectra from pre-selected
co-existing concentrations of Chl, SM, and DOC. The Lake Ontario cross
section spectra were determined at 20-nm intervals across the visible spectrum. However, to herein illustrate impacts on volume reflectance of varying
Chl, SM, and DOC, more finely resolved cross section spectra will be culled
from the literature, and it will be those values of cross section spectra that will
be used. Hoepffner and Sathyendranath[169] have presented spectral values at
2-nm intervals of the absorption coefficients of phytoplankton indigenous
to the North Atlantic normalized to the wavelength 440 nm. These values
along with the wavelength value for 440 nm taken from Mitchell and Kiefer[266]
will be used for $a_{Chl}(\lambda)$.

Curves of $b_B(\lambda)$ for nine species of phytoplankters are reported in Ahn
et al.[4] The upper limit for the species depicted in their Figure 7 can be
reasonably represented by a constant value of $(b_B)_{Chl} = 0.0010 \text{ m}^2 \text{ mg}^{-1}$ (a
value not too dissimilar from that resulting from the $b_{Chl}(\lambda)$ curve of Figure
5.6 and the B_{Chl} value of 0.011 determined for Lake Ontario). This wavelength
invariant value of 0.0010 for $(b_B)_{Chl}(\lambda)$ will be used in the upcoming analyses.

To obtain cross section spectra appropriate to suspended minerals, we
curve fit data reported in Witte et al.[428] for Calvert clay and arrived at
the relationships

$$a_{SM}(\lambda) = 0.201 - 0.245(10^{-3})\lambda \tag{5.5}$$

and

$$(b_B)(\lambda) = 0.0316 - 0.844(10^{-5})\lambda. \tag{5.6}$$

Consistent with unpublished Lake Ontario data, we will define the absorption cross section spectrum for DOC to be

$$a_{DOC}(\lambda) = 0.173\exp[-0.0157(\lambda - 400)]. \tag{5.7}$$

Spectral values for $a_w(\lambda)$ and $b_w(\lambda)$ were taken from Smith and Baker[375] [See Table 4.1]. A spectrally invariant value of $B = 0.500$ was taken for pure water.

Then, using these optical cross section spectra in conjunction with equations (5.1) and (5.2), the specific inherent optical properties of the water column may be converted into the bulk optical properties $a(\lambda)$ and $b_B(\lambda)$. Using these bulk optical properties of the water column, equations (5.3) and (5.4) will readily yield the volume reflectance spectrum that would be recorded just beneath the air-water interface if a photon flux were vertically impinging upon a water column possessing those particular bulk properties.

Since the bulk optical properties of a water column are the integrated consequences of the optical cross section spectra and the concentrations of co-existing aquatic components, it is evident that volume reflectance spectra may be readily generated (once the optical cross section spectra are either directly measured or can be reliably inferred) for any hypothetical water body containing pre-selected concentrations of chlorophyll, suspended minerals, and dissolved organic carbon. In such a manner, the impacts of varying the concentrations of these organic and inorganic components, either individually or in concert, can be dramatically illustrated. We will illustrate these impacts in the following sections of this chapter.

5.3   IMPACT OF CHLOROPHYLL ON VOLUME REFLECTANCE SPECTRA

Utilizing the selected optical cross section spectra, Figure 5.7 illustrates the family of volume reflectance spectra just beneath the air-water interface, $R(0^-,\lambda)$ (at 5-nm increments from 400 nm to 740 nm), for a hypothetical water column in which the SM and DOC concentrations are kept fixed at zero while the Chl$a$ concentrations C_{Chl} are allowed to possess concentrations between 0 and 20.0 μg/l. It is seen that water columns with little or no chlorophyll concentration (nearly pure water) display a pronounced volume reflectance in the blue region (short wavelength) and a minimal volume reflectance in the red region (long wavelength) of the visible spectrum. The volume reflectance for nearly pure water dramatically reduces beyond $\lambda =$ ~700 nm. Increasing the chlorophyll concentration in such waters (i.e., waters characterized by an absence of SM and DOC concentrations) tends to decrease the volume reflectance at blue wavelengths, while simultaneously increasing the volume reflectance at green and red wavelengths. For chlorophyll concentrations in excess of ~$(1 - 2)$ μg/l the volume reflectance spectra are already characterized by well-defined minima in the blue and maxima in the green. As chlorophyll concentrations continue to increase, a well-defined maximum in the red also develops.

An obvious feature of Figure 5.7 is the common spectral pivot point at ~497 nm indicating that at λ ~497 nm subsurface volume reflectance is independent of the chlorophyll concentration [i.e., $(\partial R / \partial C_{Chl}) = 0$ at $\lambda = $ 497 nm for the chlorophyll cross section spectra chosen]. Since Figure 5.7 was generated for a two-component water column (pure water and chlorophyll), the pivot point represents an equilibrium between the optical properties of these two components. By integrating equation (5.3) with respect to chlorophyll concentration, it can be shown that the volume reflectance will be independent of chlorophyll concentration at that value of λ for which

$$\frac{(b_B)_{Chl}}{a_{Chl}} = \frac{(b_B)_w}{a_w} \tag{5.8}$$

i.e., at that value of λ for which the backscatter to absorption ratio of chlorophyll is equal to the backscatter to absorption ratio of pure water.

The effect of inserting a small fixed concentration of suspended minerals (0.10 mg/l) into the water column (which now becomes a three-component water column) is illustrated in the family of volume reflectance spectra of Figure 5.8. Concentrations of chlorophyll are varied within a water column for which the DOC concentration is kept fixed at zero and the SM concentration C_{SM} is kept fixed at 0.10 mg/l. In this case the addition of chlorophyll to the water column results in a volume reflectance decrease in the blue region of the visible spectrum that proceeds at a rate more nearly equal to the rate of volume reflectance elevation in the green and red regions. The pivotal wavelength for which $\partial R / \partial C_{Chl}$ becomes zero has also moved to ~528 nm (for the selected Chl and SM cross section spectra), a consequence of an equilibrium having been established among the optical properties of three aquatic components. In this instance, the volume reflectance will be independent of chlorophyll concentration at that wavelength for which

$$\frac{(b_B)_{Chl}}{a_{Chl}} = \frac{(b_B)_w + C_{SM}(b_B)_{SM}}{a_w + C_{SM}a_{SM}} \tag{5.9}$$

The pivotal wavelength rapidly moves beyond 650 nm as the suspended

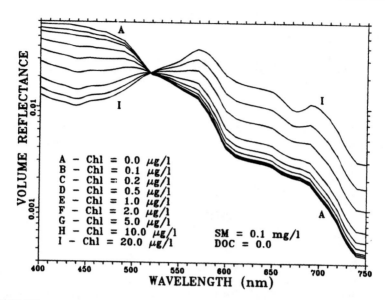

FIGURE 5.8

Volume reflectance spectra for various chlorophyll concentrations in a water column for which the DOC concentration is kept fixed at zero and the SM concentration is kept fixed at 0.10 mg/l.

mineral concentration increases. This dominance of significant concentrations of suspended minerals is evident from Figure 5.9 in which the family of volume reflectance spectra of Figure 5.8 have been recalculated for a fixed SM concentration of 10.0 mg/l. Notice that for $\lambda > 690$ nm the volume reflectance is independent of C_{Chl}. Thus, as evident from Figures 5.7, 5.8, and 5.9, non-Case 1 water bodies containing materials whose optical cross sections are comparable to those selected for this illustration, volume reflectance spectra with distinct maxima in the range 550–700 nm should be frequently encountered.

5.4 IMPACT OF SUSPENDED MINERALS ON VOLUME REFLECTANCE SPECTRA

Again using the same optical cross section spectra as in Section 5.3, Figure 5.10 illustrates the family of volume reflectance spectra just beneath the air-water interface $R(0^-,\lambda)$ for a hypothetical water column in which Chl$a$ and DOC concentrations are kept fixed at zero while the suspended mineral concentration C_{SM} is varied between 0 and 20 mg/l. The dramatic impact of suspended mineral concentrations on volume reflectance (and, hence, the dramatic impact of suspended mineral concentrations on aquatic color) is clearly evident. Even a small concentration of SM can substantially

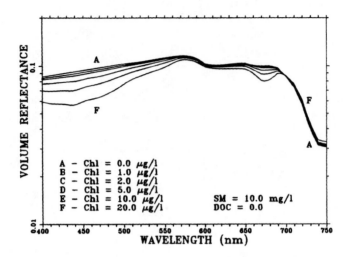

FIGURE 5.9

Volume reflectance spectra for various chlorophyll concentrations in a water column for which the DOC concentration is kept fixed at zero and the SM concentration is kept fixed at 10.0 mg/l.

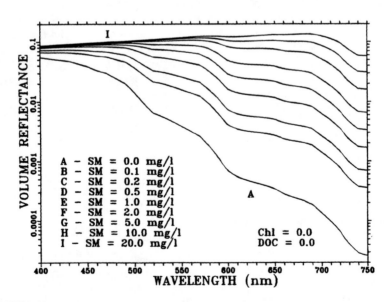

FIGURE 5.10

Volume reflectance spectra for various SM concentrations in a water column for which the chlorophyll and DOC concentrations are kept fixed at zero.

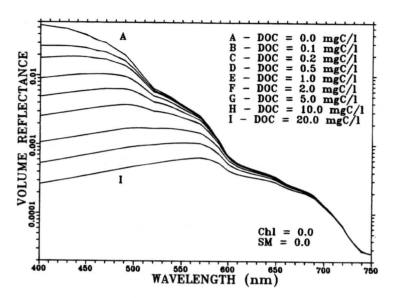

FIGURE 5.11
Volume reflectance spectra for various DOC concentrations in a water column for which the chlorophyll and SM concentrations are kept fixed at zero.

increase the volume reflectance in a manner that becomes more pronounced as the wavelength becomes longer. At very large concentrations of SM, the volume reflectance spectrum displays an asymptotic approach to a broad peak at ~700 nm. We shall return to this impact of suspended solids (predominantly SM and Chl$a$) on volume reflectance in subsequent discussions of chromaticity and aquatic color.

5.5 IMPACT OF DISSOLVED ORGANIC MATTER ON VOLUME REFLECTANCE SPECTRA

The addition of dissolved organic matter to the water column as a fourth component also affects the volume reflectance spectrum, but almost exclusively at the shorter wavelengths. This impact of DOM is displayed in Figure 5.11 wherein are plotted the family of volume reflectance spectra for a water column in which the suspended mineral and the chlorophyll a concentrations are kept fixed at zero, and the dissolved organic carbon concentration C_{DOC} varies in the range 0–20 mg C/l. For relatively clear waters (i.e., waters characterized by small concentrations of suspended minerals and small-to-moderate values of chlorophyll), the systematic addition of DOC results in a rapid depression of volume reflectance at wavelengths < ~600 nm (i.e., the blue and green regions of the visible spectrum) and a

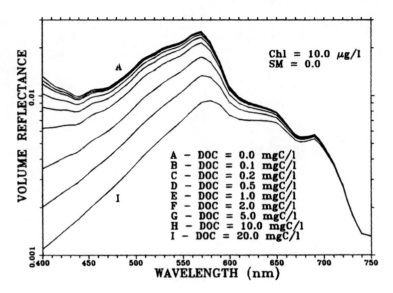

FIGURE 5.12
Volume reflectance spectra for various DOC concentrations in a water column for which the
SM concentration is kept fixed at zero and the chlorophyll concentration is kept fixed at 10.0
μg/l.

comparatively minimal depression of the volume reflectance at wavelengths
> ~600 nm (i.e., the red region of the visible spectrum). Beyond λ ~690
nm, DOC concentrations have little or no impact on the volume reflectance.
Waters containing high concentrations of DOC (and no SM or Chl*a*), as do
waters containing high concentrations of SM or Chl*a*, again display volume
reflectances that peak in the green and red wavelengths.

Figure 5.12 illustrates the impact of DOC on a water column containing
a large fixed concentration of Chl*a* (10.0 μg/l) and no SM. Figure 5.13 illustrates the impact of DOC on a water column containing a large fixed concentration of SM (10.0 mg/l) and no Chl*a*. From Figures 5.11, 5.12, and 5.13, it
is seen that:

1. Waters containing high concentrations of suspended minerals display volume reflectance spectra that are an order of magnitude higher than those
   displayed by waters containing high concentrations of chlorophyll.

2. The effect of DOC is more noticeable in waters containing high chlorophyll
   and low suspended mineral concentrations than in waters containing low
   chlorophyll and high suspended mineral concentrations.

3. The volume reflectance spectra of Figures 5.12 and 5.13 reinforce the ubiquitous occurrence of distinct maxima in the 550–600-nm wavelength interval
   suggested by Figures 5.8 and 5.9.

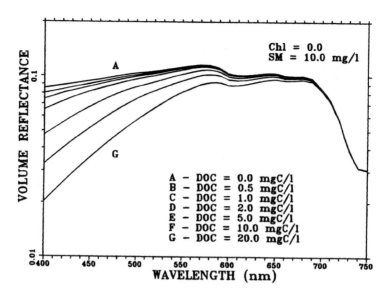

FIGURE 5.13

Volume reflectance spectra for various DOC concentrations in a water column for which the chlorophyll concentration is kept fixed at zero and the SM concentration is kept fixed at 10.0 mg/l.

The ubiquitous occurrence of sub-surface volume reflectance spectra with peak values in the green region of the visible spectrum is shown in Figure 5.14. Taken from Bukata et al.[46] (their Figure 4), representative examples of directly measured volume reflectance spectra are illustrated for both nearshore and mid-lake stations in Lake Ontario throughout the field season.

5.6 VOLUME REFLECTANCE IN NON-CASE I WATER BODIES CONTAINING MODERATE TO HIGH CONCENTRATIONS OF DISSOLVED ORGANIC MATTER

Although local values of DOM may vary both temporally and spatially, it has been observed from optical surveys of the lower Great Lakes[38] that a not unreasonable typical value for dissolved organic carbon concentrations in these waters is ~2.0 mg C/l. Optical surveys of major Russian lakes[239] have recorded values of ~9.0 mg C/l as representative of the DOC concentrations in the waters of Ladoga Lake. We will now consider the effect on observable volume reflectance of varying the suspended mineral and/or chlorophyll concentrations in waters containing such moderate and high DOC concentrations.

Figures 5.15 and 5.16 illustrate the anticipated changes in volume reflectance resulting from changes in chlorophyll a concentration for a water

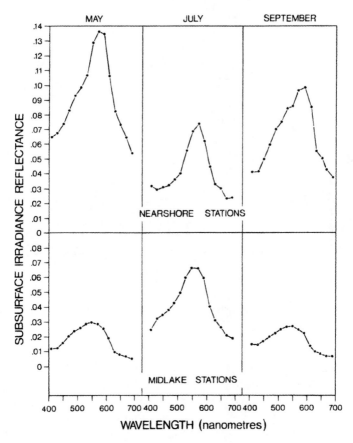

FIGURE 5.14
Representative examples of directly measured volume reflectance spectra obtained at nearshore and mid-lake stations in Lake Ontario throughout the field season. (From Bukata, R. P., Jerome, J. H., Kondratyev, K. Ya., and Pozdnyakov, D. V., *J. Great Lakes Res.*, 17, 461–469, 1991. With permission.)

column containing no suspended mineral concentration and fixed DOC concentrations of 2.0 and 9.0 mg C/l, respectively. Figures 5.17 and 5.18 illustrate the anticipated changes in volume reflectance resulting from changes in suspended mineral concentration for a water column containing no chlorophyll *a* concentration and fixed DOC concentrations of 2.0 and 9.0 mg C/l, respectively. The family of volume reflectance spectra in Figures 5.15 and 5.16 are constructed for chlorophyll concentrations in the range (0–20.0) µg/l. The family of volume reflectance spectra in Figures 5.17 and 5.18 are constructed for suspended mineral concentrations in the range (0–20.0) mg/l.

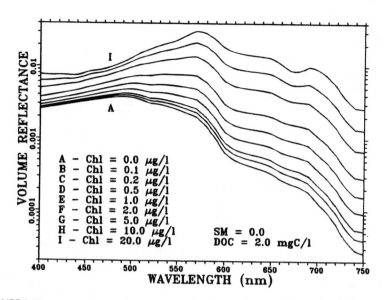

FIGURE 5.15
Volume reflectance spectra for various chlorophyll concentrations in a water column for which the SM concentration is kept fixed at zero and the DOC concentration is kept fixed at 2.0 mg C/l.

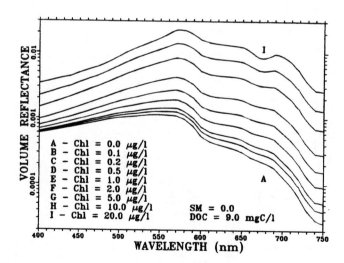

FIGURE 5.16
Volume reflectance spectra for various chlorophyll concentrations in a water column for which the SM concentration is kept fixed at zero and the DOC concentration is kept fixed at 9.0 mg C/l.

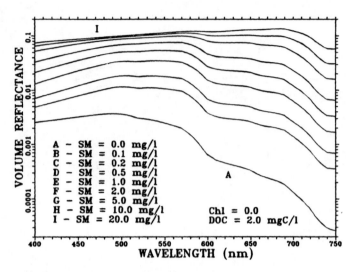

FIGURE 5.17
Volume reflectance spectra for various SM concentrations in a water column for which the chlorophyll concentration is kept fixed at zero and the DOC concentration is kept fixed at 2.0 mg C/l.

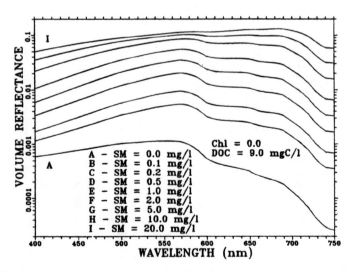

FIGURE 5.18
Volume reflectance spectra for various SM concentrations in a water column for which the chlorophyll concentration is kept fixed at zero and the DOC concentration is kept fixed at 9.0 mg C/l.

It is seen that increasing either the suspended mineral or chlorophyll *a* concentrations in a water body containing a DOC concentration of 2.0 mg C/l (Figures 5.15 and 5.17) results in a more rapid volume reflectance increase in the red end of the spectrum than in the blue. The basic shape of the spectrum resulting from increasing the concentration of chlorophyll in the absence of suspended minerals becomes very similar to the basic shape of the spectrum resulting from increasing the concentration of suspended mineral in the absence of chlorophyll (although the chlorophyll peaks in the green and red are more clearly resolved than are the SM peaks). Large concentrations of suspended minerals quite clearly result in volume reflectance values that are an order of magnitude greater than those resulting from large concentrations of chlorophyll *a*. However, it is also seen that for moderate values of DOC, comparable values of volume reflectance are observed for water columns containing either large concentrations of chlorophyll (and no suspended minerals) or relatively small concentrations of suspended minerals (and no chlorophyll).

Similar behavior is noted for water columns containing high concentrations of DOC (Figures 5.16 and 5.18). The effect of high concentrations of suspended minerals is much greater than the effect of large concentrations of chlorophyll in terms of volume reflectance magnitudes. In terms of spectral shape, however, we have seen that the presence of either chlorophyll or suspended minerals tends to produce volume reflectance spectra that peak at mid-wavelengths (increasing chlorophyll concentrations also generates a distinct peak near 700 nm). Substantial concentration of DOC, due to its large impact on shorter wavelengths, further accentuates this spectral shape. Thus, the presence of each of Chl, SM, and DOC contributes to the very often-encountered volume reflectance spectra displaying maxima in the 550–600-nm range.

The examples of subsurface volume reflectance spectra presented in this and preceding sections are certainly far from exhaustive, and by no means cover the myriad of spectra possible for inland and coastal waters. Nevertheless, they should serve to display the impact that variations in co-existing concentrations of chlorophyll, suspended minerals, and dissolved organic carbon will have on the volume reflectance spectra of natural waters. Thus, they emphasisize the care that must be taken in deconvolving these optically competitive aquatic components when measured or inferred values of subsurface volume reflectance spectra are used to monitor a desired parameter such as chlorophyll or carbon concentration in natural waters.

The volume reflectance spectra illustrated in this chapter all have been constructed using a selected set of optical cross section spectra. Indigenous optical cross section spectra are spatially and temporally variant. Natural water bodies would, therefore, require other specific optical cross section spectra, and these would result in volume reflectance spectra that would deviate from the illustrations we have presented. Appropriate absorption and scattering cross section spectra are mandatory inputs to any planned mission of remotely monitoring the compositions of natural waters. Thus, we **strongly** advocate the determination of local aquatic cross section spectra on a per water body basis in order to reliably **quantify** the concentrations of the co-existing concentrations of organic and inorganic matter comprising the water body in question. However, we are not uncomfortable with using a reasonably representative set of optical cross section spectra for illustrative purposes and feel that the impacts on volume reflectance spectra of varying the co-existing concentrations of Chl, SM, and DOC as depicted in Sections 5.3, 5.4, 5.5, and 5.6 are realistic.

5.7   VOLUME REFLECTANCE AT A SINGLE WAVELENGTH

To this point in our discussion we have considered volume reflectance as a spectrum and have tacitly assumed an operational capability for determining such an *in situ* spectrum. The volume reflectance spectra of Figure 5.11 demonstrate that the influence of dissolved organic carbon is most pronounced in the blue region of the spectrum and least pronounced in the red region. Therefore, a natural water body might be approximated, to some extent, by a three-component model (namely, pure water, chlorophyll, and suspended minerals) if only volume reflectance at red wavelengths were considered. It is therefore reasonable to assume that estimations of chlorophyll and/or suspended mineral concentrations from a single wavelength (rather than a complete spectrum) measurement of volume reflectance, if possible at all, might display the highest probability of success if the single wavelength were selected in the red region of the spectrum.

Figure 5.19 illustrates the family of curves that exist between the volume reflectance at 650 nm (red) and the chlorophyll concentration for the selected illustrative water body in which the DOC concentration is kept fixed at 2.0 mg C/l and the suspended mineral concentration assumes values between 0 and 20 mg/l. It is seen that although the red volume reflectance is quite sensitive to changes in chlorophyll concentrations in the absence of significant suspended mineral concentrations, this sensitivity rapidly vanishes once even a trace of suspended mineral (~0.10 mg/l) enters the aquatic system. Similar relationships, displaying even less sensitivity potential than shown in Figure 5.19, may be constructed for volume reflectance wavelengths in the blue and green spectral regions. Thus, reliable estimates of chlorophyll concentrations in non-Case I waters cannot result from a single wavelength measurement of sub-surface volume reflectance.

The sensitivity of the sub-surface volume reflectance at 650 nm (red) to changes in suspended mineral concentration for a water body containing a fixed DOC concentration of 2.0 mg C/l is shown in Figure 5.20 for chlorophyll *a* concentrations between 0 and 20.0 μg/l. The large impact of a variation

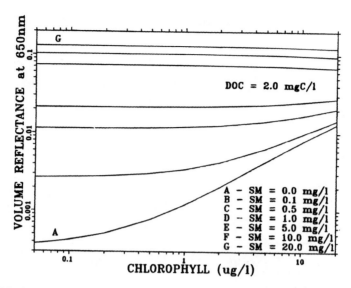

FIGURE 5.19
Relationship between volume reflectance at 650 nm (red) and chlorophyll concentration for a water column in which the DOC concentration is kept fixed at 2.0 mg C/l and the SM concentration assumes values between 0 and 20 mg/l.

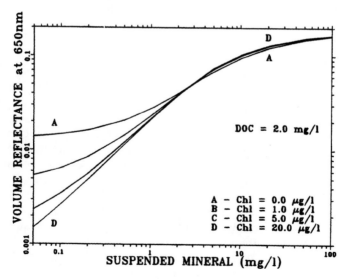

FIGURE 5.20
Relationship between volume reflectance at 650 nm (red) and SM concentration for a water column in which the DOC concentration is kept fixed at 2.0 mgC/l and the chlorophyll concentration assumes values between 0 and 20.0 μg/l.

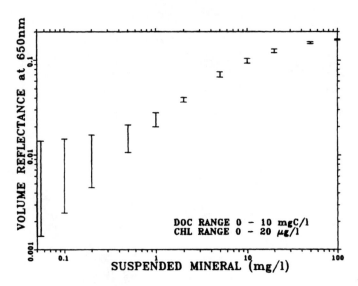

FIGURE 5.21

Relationship between volume reflectance at 650 nm (red) and SM concentration for a water body that may display a range of chlorophyll concentrations between 0 and 20.0 μg/l in addition to a range of DOC concentrations between 0 and 10.0 mg C/l.

in chlorophyll concentration on the volume reflectance at red wavelengths in waters containing small concentrations of suspended minerals is an obvious feature of Figure 5.20. However, for suspended mineral concentrations > ~0.10 mg/l, the impact of chlorophyll on the red volume reflectance is dramatically reduced.

When the minimal impact of DOC concentrations on the volume reflectance observable in the red region is also considered, the sensitivity curve of Figure 5.21 may be constructed. Herein are plotted the observable ranges of volume reflectance at 650 nm (red) as a function of suspended mineral concentration for a water body which may display a range of DOC concentrations between 0 and 10.0 mgC/l and a range of chlorophyll concentrations between 0 and 20.0 μg/l. It is seen that a reasonable sensitivity to suspended mineral concentration may be ascribed to sub-surface volume reflectance at 650 nm, particularly for suspended mineral concentrations > ~0.10 mg/l. Consequently, a single volume reflectance measurement at red wavelengths does appear to possess some capability for estimating suspended mineral concentrations in most inland and coastal waters.

The asymptotic nature of Figure 5.21 is also quite evident. This would render determinations of suspended mineral concentrations of more than ~100 mg/l very hazardous. However, in the general range 0.10 mg/l < C_{SM} < 100 mg/l, a single volume reflectance value at 650 nm may provide a reasonable estimate of suspended mineral concentration despite the competing optical activity resulting from the simultaneous presence of chlorophyll and dissolved organic matter within the water column.

The sensitivity of the subsurface volume reflectance at 650 nm to SM concentration is further improved for natural water bodies that do not display an excessive amount of biological activity (such as oceans and oligotrophic and some mesotrophic lakes and rivers). Reducing the chlorophyll and DOC concentration ranges will reduce the vertical bars of volume reflectance ranges of Figure 5.21. This will result in a very evident increased sensitivity (as compared to the sensitivity of Figure 5.21) for the use of $R(650$ nm$)$ as an estimate of suspended mineral concentration.

In capsule, the optical complexities of inland and coastal waters generally present overwhelming obstacles to the use of single-wavelength measurements of volume reflectance as a means of estimating aquatic component concentrations. It is unfeasible to attempt a reliable estimate of the chlorophyll concentration from such single-wavelength measurements, irrespective of the wavelength considered. However, despite the severe restrictions resulting from optical interference of chlorophyll concentrations (particularly at low concentrations of SM), single-wavelength volume reflectance values in the red region of the spectrum possess applicability to the estimation of small-to-moderate concentrations of suspended minerals. In the vast majority of instances, however, the entire volume reflectance spectrum $R(\lambda,z)$ will be required to deconvolve the concentrations of all aquatic components.

5.8 DECONVOLVING THE AQUATIC ORGANIC AND INORGANIC CONCENTRATIONS FROM THE VOLUME REFLECTANCE SPECTRA

Consider, as we have to this point, that the volume reflectance at any wavelength λ, $R(\lambda)$, can be mathematically expressed as a function of the optical cross section values of the co-existing concentrations of pure water, chlorophyll a (C_{Chl}), suspended minerals (C_{SM}), and dissolved organic carbon (C_{DOC}). That is:

$$R(\lambda) =$$

$$f[a_w, C_{Chl}a_{Chl}, C_{SM}a_{SM}, C_{DOC}a_{DOC}, (b_B)_w, C_{Chl}(b_B)_{Chl}, C_{SM}(b_B)_{SM}] \quad (5.10)$$

where the wavelength dependence of the optical cross sections has been dropped for simplicity of notation and the backscattering cross section of DOC has been taken to be zero.

The rate of change of volume reflectance with wavelength (i.e., the slope of the measured volume reflectance spectrum at wavelength λ), $dR(\lambda)/d\lambda$, is therefore defined as:

$$\frac{dR(\lambda)}{d\lambda} = \frac{\partial R}{\partial a_w} \cdot \frac{da_w}{d\lambda} + c_{Chl} \frac{\partial R}{\partial a_{Chl}} \cdot \frac{da_{Chl}}{d\lambda} + c_{SM} \frac{\partial R}{\partial a_{SM}} \cdot \frac{da_{SM}}{d\lambda}$$

$$+ c_{DOC} \frac{\partial R}{\partial a_{DOC}} \cdot \frac{da_{DOC}}{d\lambda} + \frac{\partial R}{\partial (b_B)_w} \cdot \frac{d(b_B)_w}{d\lambda} \qquad (5.11)$$

$$+ C_{Chl} \frac{\partial R}{\partial (b_B)_{Chl}} \cdot \frac{d(b_B)_{Chl}}{d\lambda} + C_{SM} \frac{\partial R}{\partial (b_B)_{SM}} \frac{d(b_B)_{SM}}{d\lambda}$$

where the partial derivatives of the optical cross sections are implicitly a function of λ.

Using the Monte Carlo results of Jerome et al.[204] expressed in equations (5.3) and (5.4), the volume reflectance $R(\lambda)$ may be written (neglecting the slopes and intercepts) in the form:

$$R(\lambda) \sim b_B(\lambda)/a(\lambda) \qquad (5.12)$$

where

$$b_B = (b_B)_w + C_{Chl}(b_B)_{Chl} + C_{SM}(b_B)_{SM}$$

and

$$a = a_w + C_{Chl}a_{Chl} + C_{SM}a_{SM} + C_{DOC}a_{DOC}$$

for each individual value of λ.

Making the debatable assumption that the spectral optical cross sections are mutually independent of one another, partially differentiating equation (5.12) with respect to each of the cross sections, yields:

$$\frac{\partial R}{\partial a_w} \sim -\frac{b_B(\lambda)}{[a(\lambda)]^2}$$

$$\frac{\partial R}{\partial a_{Chl}} \sim -\frac{b_B(\lambda)C_{Chl}}{[a(\lambda)]^2}$$

$$\frac{\partial R}{\partial a_{SM}} \sim -\frac{b_B(\lambda)C_{SM}}{[a(\lambda)]^2}$$

$$\frac{\partial R}{\partial a_{DOC}} \sim -\frac{b_B(\lambda)C_{DOC}}{[a(\lambda)]^2} \qquad (5.13)$$

$$\frac{\partial R}{\partial (b_B)_w} \sim \frac{1}{a(\lambda)}$$

$$\frac{\partial R}{\partial (b_B)_{Chl}} \sim \frac{C_{Chl}}{a(\lambda)}$$

$$\frac{\partial R}{\partial (b_B)_{SM}} \sim \frac{C_{SM}}{a(\lambda)}$$

for each individual value of λ.

Substituting equation set (5.13) into equation (5.11) yields:

$$\frac{dR(\lambda)}{d\lambda} \sim -\frac{b_B(\lambda)}{[a(\lambda)]^2}\left[\frac{da_w}{d\lambda} + (C_{Chl})^2\frac{da_{Chl}}{d\lambda} + (C_{SM})^2\frac{da_{SM}}{d\lambda} + (C_{DOC})^2\frac{da_{DOC}}{d\lambda}\right]$$

$$+ \frac{1}{a(\lambda)}\left[\frac{d(b_B)_w}{d\lambda} + (C_{Chl})^2\frac{d(b_B)_{Chl}}{d\lambda} + (C_{SM})^2\frac{d(b_B)_{SM}}{d\lambda}\right] \qquad (5.14)$$

Equation (5.14) then relates the slope of a directly measured or remotely estimated volume reflectance spectrum at any wavelength λ, $R(\lambda)$, to the co-existing concentrations of chlorophyll, suspended minerals, and dissolved organic carbon responsible for generating that observed spectrum. Thus, in principle, these co-existing concentrations of Chl, SM, and DOC could be estimated by simultaneously solving equation (5.14) at three discrete values of wavelength λ (higher order models would, of course, require additional discrete wavelengths). There are, however, distinct practical obstacles to deconvolving the volume reflectance spectra in such a manner. In addition to the requisite of a judicious selection of wavelengths, very accurate values of the volume reflectance slopes are mandatory at the selected λ values. As evident from some of the $R(\lambda)$ spectra discussed herein, such slopes can be numerically very small, as well as being either positive or negative. Similar statements apply to the slopes of the absorption and backscattering cross sections. Also, equation (5.14) includes the bulk optical properties $a(\lambda)$ and $b_B(\lambda)$, themselves functions of C_{Chl}, C_{SM}, and C_{DOC}, the concentrations being sought. Thus, reliable estimates of the total attenuation and total backscattering coefficients must also be obtained.

Therefore, equation (5.14) has, in principle, the ability to provide three equations which may be simultaneously solved to determine C_{Chl}, C_{SM}, and C_{DOC}. However, the severe onus of precise slope requirements in both measured and available optical cross section spectra adversely impacts the reliability of such mathematical calculations. The need to determine $\Sigma C_i a_i(\lambda)$ and $\Sigma C_i (b_B)_i(\lambda)$ for each *in situ* measurement of $R(\lambda)$ further renders the use of equation (5.14) unattractive. Ideally, a method of extracting the desired concentrations, C_{Chl}, C_{SM}, and C_{DOC} from a single measurement of the volume reflectance spectrum (once the pertinent optical cross section spectra are obtained) should be sought. Such a method is provided by multivariate optimization analyses (i.e., techniques that enable a continuous intercomparison of co-existing sets of independent variables until a satisfactory approximation to a measured dependent variable is obtained). This will be discussed in the next section.

5.9 MULTIVARIATE OPTIMIZATION AS A MEANS OF DECONVOLVING THE AQUATIC ORGANIC AND INORGANIC CONCENTRATIONS FROM THE VOLUME REFLECTANCE SPECTRUM

The fundamental problem to be solved is as follows: Given a directly measured subsurface volume reflectance spectrum $R(\lambda)$, determine those co-existing concentrations of Chl, SM, and DOC that will generate a volume reflectance spectrum most closely resembling the measured volume reflectance spectrum. It is also given that $R(\lambda)$ may be expressed in terms of the bulk aquatic optical properties [as illustrated by the $R(\lambda) = f\{a(\lambda), b_B(\lambda)\}$ expressions arising from Monte Carlo simulations as discussed in Section 3.6] and that the bulk optical properties themselves can be expressed as functions of the specific optical cross sections and the co-existing component concentrations. If the volume reflectance spectrum and the appropriate optical cross section spectra are known, therefore, it should be possible to simultaneously vary the aquatic concentrations C_i until a "best-fit" volume reflectance spectrum is generated in an iterative manner that will optimally replicate the observed volume reflectance spectrum. Such a multivariate optimization technique appropriate for estimating "best fit" concentrations of Chl, SM, and DOC should also be appropriate for solving the inverse problem, namely: Given directly measured volume reflectance spectra and the co-existing concentrations of Chl, SM, and DOC, determine the optical cross section spectra appropriate to the indigenous Chl, SM, and DOC.

Writing the volume reflectance [equation (5.12)] in vector form:

$$R[a(\lambda), b_B(\lambda), C] \sim b_B(\lambda) \cdot C / a(\lambda) \cdot C \qquad (5.15)$$

where $a(\lambda) = [a_w(\lambda), a_{Chl}(\lambda), a_{SM}(\lambda), a_{DOC}(\lambda)]$
        $b_B(\lambda) = [(b_B)_w(\lambda), (b_B)_{Chl}(\lambda), (b_B)_{SM}(\lambda), 0]$
        $C = (1, C_{Chl}, C_{SM}, C_{DOC})$

Given a measured spectrum $\{S_j\}$ consisting of a set of volume reflectance values at discrete values of λ_j, the weighted residuals between the measured $\{S_j\}$ and theoretical volume reflectances $R[a_j, (b_B)_j, C]$ can be written as:

$$g_j(C) = [S_j - R\{a_j, R\{a_j, (b_B)_j, C\}]/R\{a_j, (b_B)_j, C\} \qquad (5.16)$$

Here it is assumed that the absorption and backscatter vector functions **a** and $\mathbf{b}_B$ are known for the homogeneous water mass under consideration and that it is desired to find (at the wavelengths corresponding to the set $\{S_j\}$) the multidimensional least-squares solution by minimizing over **C** the function

$$f(\mathbf{C}) = \sum_j g_j^2(\mathbf{C}) \qquad (5.17)$$

The value of the concentration vector **C** for which $f(\mathbf{C})$ is the minimum will then define the parameter concentrations for the water column that produced the observed volume reflectance spectrum $\{S_j\}$.

Equations (5.16) and (5.17) are generally solved by means of finite-difference techniques (algorithms that minimize the differences between successive vectors in a mathematical phase space). One existing computer subroutine appropriate for such a finite-difference multivariate non-linear optimization analyses is a Levenberg-Marquardt algorithm [subroutine ZXSSQ] available in the International Mathematical and Statistical Library.[183] In-depth discussions of the finite-difference technique will not be presented here. Details of the Levenberg-Marquardt least-squares optimization method may be found in Levenberg[245] and Marquardt.[259] It should be noted, however, that the Levenberg-Marquardt (and other finite-difference algorithms) are not infallible. In general, such algorithms, given a suitable initial vector value $\mathbf{C}_o$, will systematically determine a local minimum of $f(\mathbf{C})$. This value of $f(\mathbf{C})$, however, may not be the smallest achievable over the valid range of **C**. Consequently, numerous starting points $\{\mathbf{C}_{oj}\}$ are chosen and the algorithm is permitted to determine the corresponding minima $\{f_j(\mathbf{C})\}$. The **C** associated with the minimum $f_j(\mathbf{C})$ of this set is then selected as the appropriate solution. Unfortunately, there is no guarantee that a particular starting point $\mathbf{C}_{oj}$ will result in the algorithm successfully finding any minimum for $f(\mathbf{C})$ since the algorithm may diverge, rather than converge, from that particular starting point. Furthermore, a **C** may be found which, although mathematically satisfactory, is physically meaningless (e.g., numerically negative concentrations). An additional complication could arise at certain wavelengths where the absorption and/or backscattering cross section of an aquatic component decreases to such an extent that the contribution of that component to the volume reflectance becomes comparable in magnitude to the uncertainties in the measured data. Unambiguous optimization is impossible under such conditions. There are, of course, techniques that have been developed to at least partially combat these and other difficulties encountered in multivariate optimization analyses, and the reader is directed to mathematical textbooks and computer libraries such as IMSL for details of such techniques.

Difficulties in the applications of multivariate finite difference optimization analyses notwithstanding, however, we strongly advocate their respect-

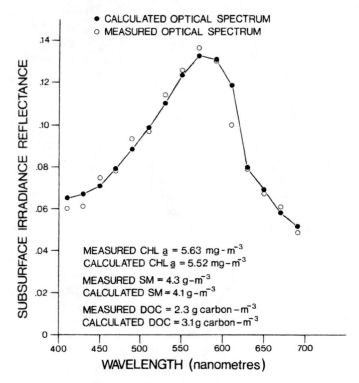

FIGURE 5.22
Comparison between calculated "best-fit" and directly measured sub-surface irradiance reflectance (volume reflectance) spectra for a Niagara River plume station in Lake Ontario. (From Bukata, R. P., Jerome, J. H., Kondratyev, K. Ya., and Pozdnyakov, D. V., *J. Great Lakes Res.*, 17, 461–469, 1991. With permission.)

fully cautious use in both (a) the determination of optical cross section spectra from measured volume reflectance spectra and known organic and inorganic aquatic component concentrations and (b) the determination of organic and inorganic aquatic component concentrations from measured volume reflectance spectra and known optical cross section spectra.

Figure 5.22 illustrates an example of the successes that can be realized in deconvolving the organic and inorganic aquatic component concentrations from a single measurement of volume reflectance spectra in a natural water body. Taken from Bukata *et al.*[46] (their Figure 5), it compares calculated "best-fit" with directly measured 15-point (λ increments of 20 nm) volume reflectance spectra for a Niagara River plume station in Lake Ontario. The co-existing concentrations of Chl$a$, SM, and DOC arising from an application of the Levenberg-Marquardt finite difference multivariate optimization algorithm (with the optical cross section spectra of Figures 5.5 and 5.6) are listed alongside directly measured values of these concentrations obtained from water samples collected in tandem with the spectroradiometric measurements. The agreement between the two sets of co-existing concentration determinations is extremely gratifying. Such remarkable agreement, however, cannot be expected in every occasion, a consequence of such factors as the limitations of the optimization process, an excessively optically complex and vertically layered natural water column, insufficient knowledge of the optical cross section spectra, statistically insufficient ranges of the co-existing concentrations to apply optimization analyses effectively (in the case of the inverse optimization process), among others. The limited number of times to date that this technique has been applied to inland waters has produced encouraging results, however. For waters as optically complex as those comprising the lower Great Lakes, it has been reported[41] that chlorophyll concentrations should be measurable to accuracies $\sim \pm 50\%$, suspended mineral concentrations to $\sim \pm 25\%$, and DOC concentrations to something in between. Such accuracies could improve as higher order bio-optical models are developed and a larger library and more understanding of optical cross section spectra become available.

Using multivariate optimization techniques to deconvolve co-existing aquatic concentrations from volume reflectance spectra, is necessary but not restrictive to non-Case I waters. Even though only a single-parameter concentration (e.g., chlorophyll a) is sought, the more optically complex the water column, the more essential becomes the need to determine the concentrations of the co-existing optically competitive (but not sought after) aquatic components. The less optically complex the water column, the less problematical becomes optical competitiveness. Consequently, applications of optimization analyses to Case I waters should anticipate greater accuracy than that anticipated for inland and coastal waters.

It must be stressed that the above discussions have considered directly measured values of the volume reflectance spectrum. The accuracies and intercomparisons with actual co-existing concentration values will, understandably, downgrade as remote sensing estimations of the volume reflectance are considered in place of the directly measured spectra.

We will end this chapter with a brief reiteration of the operations leading to estimating the co-existing aquatic component concentrations from either direct or remote estimates of volume reflectance spectra. Figure 5.23 illustrates these operations in flow diagram form. Multivariate optimization analyses has a dual role: (a) to utilize directly measured volume reflectance spectra, $R(\lambda,z)$, in concert with directly measured co-existing concentrations of Chl, SM, and DOC, to determine the optical cross section spectra $a_i(\lambda)$ and $(b_B)_i(\lambda)$, and (b) to utilize the optical cross section spectra $a_i(\lambda)$ and $(b_B)_i(\lambda)$ in concert with volume reflectance spectra inferred from upwelling radiance spectra recorded remotely to estimate the concentrations of Chl, SM, and DOC. The determination of optical cross sections necessitates a coordinated spectro-optical field program in which *in situ* sub-surface measurements

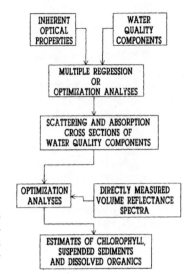

FIGURE 5.23

Flow diagram outlining operations in estimating the co-existing concentrations of aquatic chlorophyll, suspended sediment, and dissolved organic carbon from volume reflectance spectra, optical cross section spectra, and the multivariate optimization analyses.

of optical parameters (from transmissometers and spectroradiometers) are conducted in tandem with the collection of water samples for laboratory determinations of co-existing Chl*a*, SM, and DOC (and whatever additional components are deemed necessary to define a suitable water column model) concentrations. Atmospheric and air-water interface transference models are required to convert radiance spectra observable at the remote platform into volume reflectance spectra just below the air-water interface.

Chapter 6

CHROMATICITY AND THE
COLOR OF NATURAL
WATER

6.1 COLOR AND VOLUME REFLECTANCE

Water color has, throughout history, been perhaps as much a psychological as a physiological sensation. There is the inescapable tendency for an observer to associate a personal sense of aestheticism with water color and thereby ascribe perceived water quality criteria to a personal sensation of water color. Strong reactions to aquatic color have created both justifiable and unjustifiable emotions in people for centuries. Perceived aquatic color, in addition to being a function of the individual response characteristics of each observer's eyes, is also a function of the position of the sun, the location and orientation of the observer, the state of the water's surface, and the atmospheric conditions under which the viewing is performed. The perceived color of the same water body at the same time could quite easily become moot. It was, therefore, early recognized that means of recording water color were required that would retain the physiological nature while removing the subjective nature of direct visible observation. Chromaticity provides such a means.

The observed color of a natural water body is a direct consequence of the interaction of the incident downwelling solar and sky irradiances with whatever optically responsive organic and inorganic matter comprise the water column at the instant of observation. As we have seen, these materials absorb and scatter radiation in a spectrally selective manner. Thus, a measure of natural aquatic color is logically provided by the volume reflectance spectrum within the range of wavelengths to which the human eye is sensitive [i.e., $R(\lambda, z)$ for the broad spectral region 390–740 nm with a maximum sensitivity at ~555 nm for photopic vision]. It is this volume reflectance spectrum that we shall consider in our discussion of water color, and, in a manner similar to that of Chapter 5, we will investigate the impact on water color of varying the mix of co-existing concentrations of Chl$a$, SM, and DOC. However, effects of surface reflection of incident irradiance will be ignored. Although surface reflection influences the *perceived* color of a water body, it is not representative of the water body's *true* color.

As we have seen, the volume reflectance may be expressed as a function of bulk inherent optical properties such as the absorption coefficient $a(\lambda)$ and the backscattering coefficient (b_B) (λ), i.e.,

$$R(\lambda) = g\{a(\lambda), (b_B)(\lambda)\} \qquad (6.1)$$

Further, these bulk inherent optical properties may be expressed as the sum of the products of the specific inherent optical properties (optical cross sections) and the concentrations of the co-existing aquatic components, i.e.,

$$a(\lambda) = \sum a_i(\lambda)x_i \qquad (6.2)$$

$$(b_B)(\lambda) = \sum (b_B)_i(\lambda)x_i \qquad (6.3)$$

where x_i represents the concentration of the ith aquatic component.

Thus, the volume reflectance may be expressed as a function of the optical cross sections of the aquatic components and their concentrations, i.e.,

$$R(\lambda) = h\{a_i(\lambda), (b_B)_i(\lambda), x_i\} \qquad (6.4)$$

Water color may be expressed as a function of volume reflectance; thus, water color may also be considered as a function of the optical cross sections and concentrations of the co-existing aquatic components, i.e.,

$$\text{Water color} = f\{a_i(\lambda), (b_B)_i(\lambda), x_i\} \qquad (6.5)$$

The optical cross section spectra, of course, display temporal and spatial dependencies, representing the indigenous geological and biological character of the water body being studied. It is, therefore, possible that water columns possessing the same co-existing combinations of chlorophyll, suspended sediments, and dissolved organic matter, viewed under identical conditions of atmospheric composition, solar elevation and azimuth, and viewing orientation, may, in fact display different colors, even when dispassionately recorded by an optical sensor. This reinforces the requisite of obtaining the cross section spectra of indigenous organic and inorganic aquatic matter if remote measurements of water color are to be reliably related to aquatic composition. We will consider water color in the manner suggested by equation (6.5) in the upcoming sections of this chapter.

The relatively clear (Case I) waters defining mid-oceans have for centuries been observed to be blue. The considerably less clear (non-Case I) waters of lakes, rivers, and coastal areas have generally been observed to be a wide variety of colors, but most consistently to be green.

Bunsen[51] attributed the blue hue of Case I waters solely to spectrally selective absorption of incoming photons by the pure water itself. Other less appropriate theories at the time were founded solely upon spectrally selective scattering. Raman[340] illustrated that a combination of absorption and scattering was required to explain the color of ocean waters. A most comprehensive treatise of ocean color was advanced by Shuleikin,[368] who mathematically incorporated all color-producing agents (scattering by water molecules, air bubbles, suspended matter; absorption by water molecules and dissolved matter). Other notable ocean color work has been performed by Kalle,[208,209] Lenoble,[244] and many others.

Many theories have been proposed over the years regarding the changes in natural water color with changes in water turbidities. *Turbidity* is a term that has historically suffered from a lack of precise definition. Unfortunately, turbidity has been used in a somewhat cavalier manner to represent, at some time or other, virtually every combination of organic and/or inorganic material present in otherwise "pure" water. Understandably, therefore, much of the earlier works in relating water color to turbidity have not necessarily appeared to agree as well as they perhaps actually did. The generally accepted current definition of turbidity is that it is a consequence of the total suspended organic and inorganic matter residing within the water column and consequently includes sediments and chlorophyll-bearing biota. It does not include dissolved matter. While we prefer not to use the term in a quantitative manner, we do recognize its continued usage and, when we refer to turbidity, it shall be within the context of the above definition. Controversy over the term notwithstanding, however, it has always been acknowledged that absorption and scattering are the processes by which natural water color is determined.

6.2 PHOTOMETRIC UNITS

Photometry refers to the general measurement of the intensity, spectral distribution, color, and attenuation of visible light. To properly describe the psychophysical aspects of aquatic optics, a photometric nomenclature exists that is related to the physical optical nomenclature used to this point. Since such optical nomenclature is well documented in existing textbooks and since the main point of this chapter is to relate aquatic color to the responsible aquatic colorants, we will briefly discuss only two of the concepts that are of importance to the discussions to follow. These two concepts will illustrate

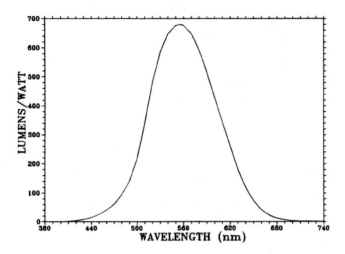

FIGURE 6.1
The luminosity function plotted throughout the visible spectral region.

that the system of photometric units parallels the system of radiometric units that are used in this book.

The first of these concepts is *luminance*. Identical to its radiometric counterpart radiance, $L(\lambda)$, luminance, $\mathscr{L}(\lambda)$, is defined as the luminous flux per unit of solid angle per unit of projected area. Luminance is generally measured in units of lumen m^{-2} s^{-1}, with a lumen being the photometric unit of luminous flux (corresponding to the radiometric unit of radiant flux, the watt). For completeness, the radiometric concepts of radiant energy (joule), radiant intensity (watt s^{-1}), and irradiance (watt m^{-2}), possess the photometric counterparts of luminous energy (talbot), luminous intensity (lumen s^{-1} or candela), and illuminance (lumen m^{-2} or lux), respectively.

The second of these concepts is the *standard luminosity curve*. The actual conversion between physical and psychophysical sensations depends upon such variables as the age, experience, and health of the observer, the orientation of the light beam with respect to the retina, the spectral shape of the incident beam, along with, of course, the ambient conditions under which the observations are made. Of these variables, perhaps the most important is the spectral shape of the incident beam, since sources of equal radiance will appear to an observer to possess different psychophysical luminances. This difference between physical and psychophysical sensations is strongly wavelength dependent and varies from observer to observer. Physiometric studies of numerous observers have generated a standard curve that plots an absolute *luminosity function* against wavelength throughout the visible spectral region. This curve (Figure 6.1) provides the wavelength dependent conversion factor relating photometric units to radiometric units. From Figure 6.1 it can be seen that, at $\lambda = 555$ nm, 1 watt of radiant flux is equivalent to about 680 lumens of photometric flux. At 410 nm and at 710 nm, however, 1 watt of radiant flux is equivalent to only 1 lumen of photometric flux.

6.3   COLORIMETRIC SYSTEM: CHROMATICITY

From the perspective of the human observer, color is the result of the interplay between the light spectrum reaching the eye and the spectral response of that eye. Color perception is most developed in the central region of the human retina. As is incorporated within the Young-Helmholtz theory, this region of the retina (for normal vision) is tri-chromatic (i.e., responsive to the three spectral regions red, green, and blue). Therefore, any perceived color can be created by appropriately proportioning red, green, and blue light. Such apportionment of light into its tristimulus red, green, and blue components forms the basis of *chromaticity* analyses.

A measured upwelling irradiance spectrum, $E(\lambda)$ (either above or below the air-water interface), may, therefore, be related to a perception of visual color through chromaticity analyses that integrate the sensitivity of the human eye with the irradiance spectrum impinging upon it.[65] Such an integration results in tristimulus values X', Y', and Z', from which the *chromaticity coordinates* X (red), Y (green), and Z (blue) may be determined.

Following the Commission Internationale de l'Éclairage (CIE) standard colorimetric system,[64] the tristimulus values of an upwelling irradiance spectrum $E(\lambda)$ are given by:

$$X' = \int E(\lambda)x(\lambda)d\lambda \qquad (6.6)$$

$$Y' = \int E(\lambda)y(\lambda)d\lambda \qquad (6.7)$$

$$Z' = \int E(\lambda)z(\lambda)d\lambda \qquad (6.8)$$

where $x(\lambda)$, $y(\lambda)$, and $z(\lambda)$ represent *CIE color mixture* data[64] for the red, green, and blue regions of the spectrum, respectively. These color mixture data are wavelength-dependent hypothetical standard values selected in such a manner that $y(\lambda)$ identically corresponds to the standard luminosity curve for photopic vision. The numerical values of $x(\lambda)$, $y(\lambda)$, and $z(\lambda)$ in equations (6.6) to (6.8) are those appropriate to an equal energy incident spectrum and are listed in Table 6.1.

The chromaticity coordinates X, Y, and Z (for red, green, and blue, respectively) are then obtained from:

TABLE 6.1

CIE Color Mixture Data for Equal Energy Spectrum

Wavelength (λ) (nm)	$x(\lambda)$	$y(\lambda)$	$z(\lambda)$
380	0.0023	0.0000	0.0106
390	0.0082	0.0002	0.0391
400	0.0283	0.0007	0.1343
410	0.0840	0.0023	0.4005
420	0.2740	0.0082	1.3164
430	0.5667	0.0232	2.7663
440	0.6965	0.0458	3.4939
450	0.6730	0.0761	3.5470
460	0.5824	0.1197	3.3426
470	0.3935	0.1824	2.5895
480	0.1897	0.2772	1.6193
490	0.0642	0.4162	0.9313
500	0.0097	0.6473	0.5455
510	0.0187	1.0077	0.3160
520	0.1264	1.4172	0.1569
530	0.3304	1.7243	0.0841
540	0.5810	1.9077	0.0408
550	0.8670	1.9906	0.0174
560	1.1887	1.9896	0.0077
570	1.5243	1.9041	0.0042
580	1.8320	1.7396	0.0032
590	2.0535	1.5144	0.0023
600	2.1255	1.2619	0.0016
610	2.0064	1.0066	0.0007
620	1.7065	0.7610	0.0003
630	1.2876	0.5311	0.0000
640	0.8945	0.3495	0.0000
650	0.5681	0.2143	0.0000
660	0.3292	0.1218	0.0000
670	0.1755	0.0643	0.0000
680	0.0927	0.0337	0.0000
690	0.0457	0.0165	0.0000
700	0.0225	0.0081	0.0000
710	0.0117	0.0042	0.0000
720	0.0057	0.0020	0.0000
730	0.0028	0.0010	0.0000
740	0.0014	0.0006	0.0000
750	0.0006	0.0002	0.0000
760	0.0003	0.0001	0.0000
770	0.0001	0.0000	0.0000
Total	21.3713	21.3714	21.3715

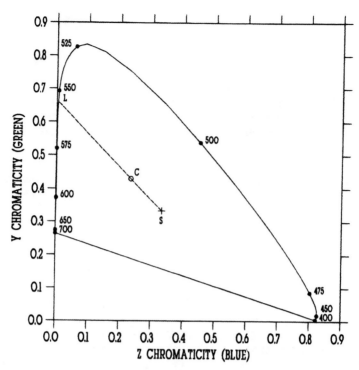

FIGURE 6.2
Chromaticity diagram using the chromaticity coordinates Y (green) and Z (blue).

$$X = X'/(X' + Y' + Z') \qquad (6.9)$$

$$Y = Y'/(X' + Y' + Z') \qquad (6.10)$$

$$Z = Z'/(X' + Y' + Z') \qquad (6.11)$$

where $X + Y + Z = 1$.

Thus, using the CIE color mixture values $x(\lambda)$, $y(\lambda)$, and $z(\lambda)$ and assuming monochromatic light of a given wavelength as the spectrum $E(\lambda)$, the CIE chromaticity coordinates may be obtained for that particular wavelength. By repeating this procedure, CIE chromaticity coordinates may be obtained for each λ throughout the entire visible spectrum. Since the sum of the three chromaticity coordinates is unity, only two chromaticity coordinates are required to construct a *chromaticity diagram* (i.e., a diagram in which corresponding pairs of the chromaticity coordinates X, Y, Z are plotted to illustrate the dependence of color on some independent variable or group of variables). Figure 6.2 shows such a chromaticity diagram [using chromaticity coordinates Y (green) and Z (blue)] constructed in the above manner for monochromatic light at each wavelength. The locus of the plotted (Z,Y) pairs defines an envelope which encompasses all possible chromaticity values. Comparable chromaticity diagrams could be constructed using X (red) and Y (green) or X (red) and Z (blue).

For a flat polychromatic or "white" irradiance spectrum (i.e., $E(\lambda)$ is invariant over the range of λ), $X = Y = Z = 0.333$. These values of the chromaticity coefficients define the *white point* or the *achromatic color S* in a chromaticity diagram, as is indicated in Figure 6.2. A numerical value oftimes ascribed to color is then obtained geometrically by drawing a line from this white point S through the plotted pair of chromaticity coefficients of a measured spectrum (indicated in the figure by the point C) and intersecting the chromaticity envelope. This intersection of the line SC with the encompassing chromaticity envelope (indicated by the point L) specifies the *dominant wavelength* of the observed irradiance spectrum, considered as a colorimetric definition of perceived color. Such a dominant wavelength definition of color applied to natural water bodies, however, encounters difficulties when an attempt is made to relate color to the concentrations of co-existing aquatic components. An infinite number of points C can accommodate the line SL. Thus, an infinite number of pairs of chromaticity coefficients (each pair a consequence of a particular combination of aquatic components) can result in the same dominant wavelength. The distinctiveness of this dominant wavelength within the irradiance spectrum $E(\lambda)$ for each of these combinations, however, is not identical. This distinctiveness of the dominant wavelength is termed *spectral purity* and is geometrically defined in Figure 6.2 as the ratio of the length of line segment CS to the length of the line LS. Thus, spectral purity is a measure of the magnitude of the contribution of the dominant wavelength to an observed optical spectrum. A spectral purity of 1.0 indicates a monochromatic spectrum at the dominant wavelength, while a spectral purity of 0 indicates a flat "white" spectrum. Together, therefore, the dominant wavelength and its associated spectral purity uniquely define an aquatic color.

An example of the use of chromaticity coordinates to indicate change of color with depth is illustrated in Figure 6.3 taken from Jerlov[198] (his Figure 94). Herein are plotted the loci of the corresponding Y (green) and X (red) chromaticity coordinates for directly measured downwelling (longer curve) and upwelling irradiances (shorter curve) in the Sargasso Sea. Also shown on the figure is the white point S and a segment of the Y-X chromaticity envelope for monochromatic light. It is seen that the downwelling light changes from a dominant wavelength of ~490 nm at the surface (bluish-green) to a dominant wavelength of ~465 nm (blue) at a depth of 100 m. The curvature of the chromaticity locus for the downwelling light illustrates a rapidly increasing spectral purity indicative of a tendency toward monochromatic blue light at greater depths. The upwelling light, however, is defined by a very short and nearly straight line representing a color change of only ~5 or 6 nm in the upper 100 m of the Sargasso Sea.

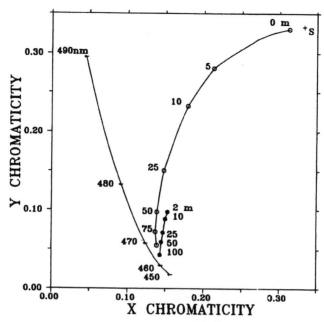

FIGURE 6.3
CIE chromaticity diagram illustrating change of aquatic color with depth (0 m to 100 m) in the Sargasso Sea. The longer curve illustrates the depth dependence of color of downwelling irradiance. The shorter curve illustrates the depth dependence of the color of upwelling irradiance. (From Jerlov, N. G., *Marine Optics*, Elsevier Oceanography Series 14, 1976. With permission.)

Recently, Jerome et al.[206,207] have utilized chromaticity coordinates in conjunction with the optical cross section model and direct measurements of subsurface volume reflectance spectra to effect a study of water color variability (in terms of the factors controlling the dominant wavelength) in both glacier-fed and non-glacier-fed river systems of the Canadian Cordillera in British Columbia. They observed that:

1. Waters with low concentrations of suspended minerals, chlorophyll, and dissolved organic carbon display dominant wavelengths in the color range blue to turquoise.

2. Waters with low concentrations of suspended mineral and chlorophyll but moderate concentrations of dissolved organic carbon display dominant wavelengths in the color range green to brown.

3. Waters with high concentrations of suspended mineral and/or chlorophyll and/or dissolved organic carbon display a restricted range of dominant wavelengths in the "end-point" color brown.

4. With increasing concentrations of suspended or dissolved matter, the spec-

tral purities as well as the dominant wavelengths of natural waters asymptotically approach "end-point" values. This "end-point" brown hue and "end-point" spectral purity are characteristic of all natural waters.

5. River water within simple basins (i.e., basins that can be represented by a single hydrograph) that are glacier-fed and meltwater-dominated generally appear blue to turquoise to green, a consequence of low to moderate turbidity predominantly comprised of suspended inorganic matter.

6. River water within simple basins that are non-glacier-fed and groundwater-dominated generally display dominant wavelengths in the color range green to brown, a consequence of low to high concentrations of suspended inorganic matter coupled with significant concentrations of dissolved organic matter.

7. River water within complex basins [i.e., basins that require the integration of hydrographs from a number of sources] generally display dominant wavelengths in the restricted range of "end-point" reddish-brown colors.

In the upcoming discussions, however, we will restrict our attention to chromaticity coordinates and not convert them into dominant wavelengths and spectral purities.

In the manner discussed in Chapter 5, and briefly recapped at the start of this chapter, we will now construct a number of volume reflectance spectra for hypothetical water columns containing various combinations of co-existing concentrations of chlorophyll a, suspended minerals, and dissolved organic carbon. Standard chromaticity analyses will then be performed on these spectra to determine the coordinates X, Y, and Z.

From equation (6.5) water color (chromaticity) is a function of the optical cross section spectra of the indigenous aquatic components. To illustrate the impact of aquatic composition on the color of natural water, we will use the same optical cross section spectra that we used in Chapter 5. Thus, the hypothetical water body that we will consider is one whose optical properties are determined by co-existing concentrations of pure water as given in Smith and Baker,[375] chlorophyll-bearing biota of the type described by Hoepffner and Sathyandrenath,[169] suspended red clay of the type measured by Witte et al.,[428] and dissolved organic carbon of the type we have observed in Lake Ontario. We fully realize the inappropriateness of extrapolating these spectrally selective specific absorption and scattering coefficients to natural water bodies. However, the lack of universality in the optical cross section spectra should not detract significantly from the general nature of the chromaticity analyses to follow, since similar conclusions could be drawn from other regionally appropriate spectra. In fact, the illustrative material to follow is not at variance with our earlier work utilizing the optical cross section spectra pertinent to Lake Ontario waters.[40]

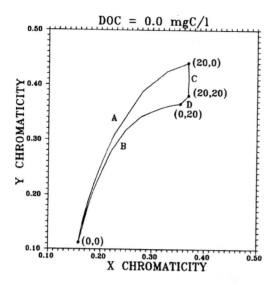

FIGURE 6.4
The loci of the relationships between X (red) and Y (green) for water columns in which the chlorophyll and SM concentrations are considered as independent variables while the DOC concentration remains fixed at zero.

6.4 CHROMATICITY AS A MEANS OF ESTIMATING AQUATIC COMPONENT CONCENTRATIONS

Consider a three-component water column in which the water can be represented by a homogeneous assemblage of pure water molecules, suspended minerals, and suspended chlorophyll a. No dissolved matter is present. Figure 6.4 illustrates the loci of the interrelationships among the chromaticity coordinates X (red) and Y (green) for water columns in which the chlorophyll and suspended mineral concentrations are considered as independent variables while the dissolved organic component remains fixed at a concentration of zero. Figure 6.4 is plotted in duoisoplethic form, with each point on the diagram representing a co-existing concentration of chlorophyll a and suspended minerals recorded in the numerical order (Chl$a$,SM). Curve A illustrates the locus of X and Y values for water columns containing no suspended mineral concentration and a range of chlorophyll concentrations between 0 (at the lowest values of X and Y) to 20 μg/l (the highest values of X and Y). Curve B illustrates the locus of X and Y values for water columns containing no chlorophyll concentration and a range of suspended mineral concentrations between 0 and 20 mg/l. Curve C illustrates the locus of X and Y values for a fixed Chl$a$ concentration of 20 μg/l and various SM concentrations, while Curve D illustrates the locus for a fixed SM concentration of 20 mg/l and various Chl$a$ concentrations.

The entire ranges of chlorophyll and suspended mineral concentrations plotted onto Figure 6.4 are observable in inland and coastal waters. Consequently, the chromaticities that are measured above non-Case I waters would generally be bounded by the Curves A, B, C, and D. Non-Case I waters with components characterized by optical cross section spectra substantially different from those used in this illustration would result in chromaticity loci that would be numerically dissimilar to the curves of Figure 6.4, but not significantly different in shape and position. Relatively clear waters (low concentrations of chlorophyll and suspended minerals) would fall within the lower regions bounded by Curves A and B. All but the most transparent of lake water, however, would fall within the upper region of X and Y values bounded by Curves A, B, C, and D, since this region contains the entire concentration range from low Chl$a$ coupled with high SM to low SM coupled with high Chl$a$. The restricted ranges of X and Y defining this bounded region places an unreasonable onus on the accuracy to which the chromaticity coordinates must be measured to reliably separate the contributions of Chl$a$ and SM to water color.

The maximum benefit of chromaticity to relating remote estimates of water color to the responsible colorants might lie in the monitoring of clear natural waters (concentrations in the ranges $0 \leq$ Chl$a \leq 2$ μg/l and $0 \leq$ SM ≤ 1 mg/l) since Curves A and B are the most easily defined of the loci in Figure 6.4. However, while curves A and B are readily definable, they are not readily distinguishable from one another. Thus, even if the natural water column were devoid of dissolved organic matter, the successful application of chromaticity analysis to estimate chlorophyll concentrations would require either the further absence of SM or that the SM concentration be both known and invariant. Similarly, to estimate SM concentrations would require either the absence of Chl$a$ or its presence in a known and invariant concentration. In two component water masses (i.e., natural waters for which there is only one irrefutably known principal colorant), chromaticity could be a valuable tool for assessing water color. Two-component water masses provide a most convenient basis for distinguishing between Curves A and B and by this distinction expanding the concentration ranges of reliable remote estimates. Therefore, chromaticity can be more successfully applied to the remote sensing of oceans than to the remote sensing of most inland and coastal waters.

The interference of chlorophyll and suspended sediments becomes even more pronounced when dissolved organics are added to the aquatic mix. Figure 6.5 illustrates the impact on Figure 6.4 of the addition of a moderate fixed concentration of 2.0 mg C/l of dissolved organic carbon to the water column. The range of observable chromaticities is dramatically compressed, thereby restricting the reliability of chromaticity-based predictions even further. Chromaticity loci such as depicted in Figure 6.5 are representative of some of the natural waters comprising the Laurentian Great Lakes. Lakes of high DOC concentrations (such as Ladoga Lake in Russia) experience an even more compressed range of available chromaticity coordinates with which to record aquatic color. Figure 6.6 illustrates the loci of X and Y values (in the manner of Figures 6.4 and 6.5) for a water column containing a fixed

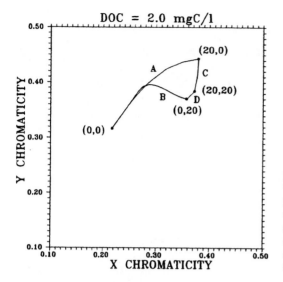

FIGURE 6.5
The loci of the relationships
between X (red) and Y (green) for
water columns in which the chlorophyll and SM concentrations are
considered as independent variables while the DOC concentration remains fixed at 2.0 mg C/l.

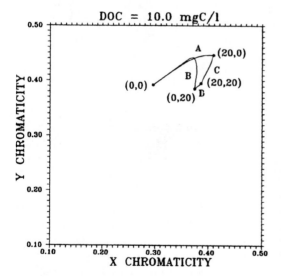

FIGURE 6.6
The loci of the relationships
between X (red) and Y (green) for
water columns in which the chlorophyll and SM concentrations are
considered as independent variables while the DOC concentration remains fixed at 10 mg C/l.

substantial DOC concentration of 10 mg C/l. The onus on the required
accuracy in recording the upwelling spectrum is readily apparent.

Similar conclusions regarding the ambiguities inherent to chromaticity
as a means of extracting aquatic colorant concentrations apply to the use of
the Y (green) and Z (blue) chromaticity coordinates as apply to the use of
the X (red) and Y (green) coordinates.

6.5 CHROMATICITY AND REMOTE SENSING

The earliest chromaticity studies of natural water bodies from satellite altitudes were performed using the multispectral scanning spectrometer (MSS) mounted aboard the earth-orbiting Landsat vehicles (Munday,[282] Munday and Alföldi,[283] and others). These MSS bands (initially numbered Bands 4, 5, 6, and 7 but currently renumbered as Bands 1, 2, 3, and 4) possess the broadband wavelength values:

Band 1: 500–600 nm
Band 2: 600–700 nm
Band 3: 700–800 nm
Band 4: 800–1100 nm

It is seen that values of $\lambda < 500$ nm are omitted from the observed wavelengths and that Band 3 is a hybrid wavelength band straddling the upper red and lower near-infrared regions of the electromagnetic spectrum. The restrictions in both spectral range and spectral sensitivity prevent a direct application of the type of CIE chromaticity analyses described above. As a result a Landsat chromaticity system (LCS) was developed (Munday[282]). The chromaticty coefficients of LCS are defined as:

$$X_1 = L_1 / \sum L_i \qquad (6.12)$$

$$X_2 = L_2 / \sum L_i \qquad (6.13)$$

where L_1, L_2, and L_i are the radiances recorded in Bands 1, 2, and i of the Landsat MSS, respectively. Since the infrared radiance of Band 4 is usually zero over all but the most turbid of natural waters, the summation is invariably over Bands 1, 2, and 3. Thus:

$$\sum L_i = L_1 + L_2 + L_3 \qquad (6.14)$$

Equation (6.14) serves as a definition of *aquatic brightness,* namely the total spectral radiance recorded over the water body by the Landsat MSS. The LCS chromaticity coordinates are, therefore, normalized to the aquatic brightness. Although not directly comparable to the X, Y, and Z CIE chromaticity coordinates, the LCS chromaticity coordinates X_1 and X_2 can be considered as distantly related to the CIE chromaticity coordinates Y (green) and X (red), respectively.

Alföldi and Munday[6] and Munday and Alföldi[284] used chromaticity analyses of upwelling radiance recorded in the visible channels of Landsat to measure suspended sediment concentrations in coastal waters with a reported accuracy of $\pm 44\%$ over the range (1–1000) mg/l (Munday et al.[285]).

Chromaticity loci curves based on three MSS channels were generated and regressed with suspended solids concentrations taken from natural waters found in the Bay of Fundy and the James River and off the eastern shore of Virginia. The suggestion in this and other contemporary work was that comparable results could be achieved for estimating chlorophyll concentrations. Theoretical work by Bukata et al.,[40] however, based on optical cross section spectra, showed that chromaticity analyses result in total ambiguity if the natural water contains optically significant concentrations of both suspended inorganic matter and chlorophyll, with additional complications arising from the presence of dissolved organic matter within the water column. This difficulty in using chromaticity techniques to remotely monitor the component concentrations of non-Case I waters was subsequently observed by Lindell et al.[247] for several Swedish lakes and by Gallie and Murtha[121] for Chilko Lake, British Columbia.

Gallie and Murtha,[121] performing LCS chromaticity analyses on Bands 1 (green), 2 (red), and 3 (red/near-infrared) of the Landsat MSS, observed the optical interference among organic and inorganic substances to be even more profound than does the CIE chromaticity analyses described in Section 6.3. Part of this added confusion is undoubtedly a consequence of the broadband spectral characteristics of the Landsat multispectral scanning device. In an attempt to reduce this optical interference, they treated brightness (the numerical sum of the three band radiances) as a chromaticity coordinate and found that it did improve the separability of SM and Chl$a$, but that Chl$a$ and dissolved organic matter generated near-identical (X_1,X_2) loci. The contribution of brightness to the distinction between SM and Chl$a$ arises from the fact that aquatic brightness and aquatic color are not directly correlated in optically complex water columns. Aquatic brightness increases with increases in SM concentration (due to the dominance of scattering associated with the presence of SM), while aquatic brightness decreases with increases in Chl$a$ and dissolved organic matter concentration (due to the dominance of absorption by these aquatic components).

Gallie and Murtha also observed a distinctive V-shaped curve in the (X_1,X_2) chromaticity locus appropriate to Chilko Lake, an oligotrophic (low concentrations of chlorophyll and dissolved organic matter) elongated north-south water body in British Columbia characterized by a pronounced sediment gradient varying from ~20 mg/l at the northern end to ~0.5 mg/l at the southern end. They interpreted this curve-back feature of the chromaticity loci (which occurs at low concentrations of suspended minerals) as a natural consequence of an aquatic SM gradient disappearing when atmospheric radiance is included within the recorded satellite signal. Since atmospheric radiance was not included in the CIE chromaticity analyses described above, this inversion in the chromaticity loci does not appear. Such an LCS loci inversion at low values of SM concentrations should, however, be a feature of the chromaticity loci resulting from remotely acquired data over all low-turbidity natural water bodies. The distinctness of this inversion and the SM concentrations at which this inversion will occur, is, of course, dependent upon the optical cross section spectra of the indigenous aquatic matter comprising the water body.

Recapping, chromaticity as a means of estimating aquatic component concentrations based on the CIE chromaticity analyses of Bukata *et al.*[40] and the LCS chromaticity analyses of Gallie and Murtha[121] show that:

1. Chromaticity analyses yield totally ambiguous results for water columns containing large concentrations of suspended sediments, large concentrations of chlorophyll, or large concentrations of both.

2. Chromaticity analyses can distinguish relatively clear water (small concentrations of SM and Chl*a*) from more turbid water if the dissolved organic component were either negligible or of a known concentration. The component creating the turbidity (Chl*a* or SM), however, would be in doubt. The presence of large and/or variable concentrations of dissolved organics could render even the distinction between turbid and clear water hazardous.

3. Part of the doubt of point 2 above could be alleviated and some separability of SM from Chl*a* could be invoked by a judicious use of aquatic brightness as a pseudo-chromaticity coordinate.

4. Chromaticity analyses provide a reasonable approach to extracting *one of* SM, Chl*a*, or DOC concentrations from a water column whose volume reflectance can be inarguably taken to be the consequence of a known single component colorant. For such a water column, chromaticity loci can be confidently ascribed to this known principal colorant.

The optically competitive organic and inorganic material comprising inland and coastal waters, therefore, prohibits the use of chromaticity as a single-component monitoring aid. Even if the spectral limitations of existing satellite imaging systems were overcome, chromaticity analyses of the upwelling radiance spectrum generally would be of little value to any but the least optically complex of natural waters.

Chapter 7

OBSERVATIONS OF OPTICAL PROPERTIES OF NATURAL WATERS (THE LAURENTIAN GREAT LAKES)

7.1 INTRODUCTORY REMARKS

At this point in the discussion of optical properties of non-Case I waters, it might be of interest to take a break from the mathematical formulism and theoretical discussions we have been pursuing and show some examples of directly measured optical parameters of inland waters. Perhaps one of the most familiar limnological systems is the Laurentian Great Lakes of central North America, and we will use these lakes liberally in presenting illustrative examples of *in situ* aquatic optical data. To provide a geographic orientation, Figure 7.1 is a schematic representation of the five lakes (Superior, Huron with Georgian Bay, Michigan, Erie, and Ontario) composing the Great Lakes Basin. The Canada-U.S. border is indicated, as are the adjacent American states and the Canadian province of Ontario. While the Laurentian Great Lakes will be prominently featured in this chapter, the optical observations need not necessarily be considered to be atypical of other inland water bodies.

7.2 THE SECCHI DISK AND ATTENUATION OF SUBSURFACE IRRADIANCE

The attenuation of light in natural water bodies has fascinated seafaring and non-seafaring observers alike for many centuries. One of the earliest and simplest attempts to quantify such attenuation was to lower an object (something white or otherwise distinctive enough to facilitate its recognition by the human eye) into deep waters and to track that object visually until

FIGURE 7.1
Schematic map of the Laurentian Great Lakes Basin.

it disappeared from the view of an observer stationed above the air-water interface. One such object is the Secchi disk, named after its generally acclaimed founder.[361] Estimates of water clarity have evolved from the use of Secchi disks to the use of submersible optical sensors such as transmissometers and spectroradiometers. The depth at which the lowered disk vanishes from the view of the above-water observer is termed the *Secchi depth* of the water being monitored. Despite the very apparent conflict between the highly subjective nature of the Secchi disk method and the considerably more objective nature of the method pertinent to the interpretation of data gathered by *in situ* optical devices, the convenience, deceptive simplicity, and longevity of the database associated with Secchi disk usage have resulted in the Secchi disk having become an integral component of ocean and large lake surveillance strategies. The Secchi disk is widely used today as an attractive means of obtaining an instant evaluation of water clarity and, in many instances, observations of Secchi depth form the only major recorded "optical history" for water bodies on a global scale.

It is extremely easy to criticize the use of Secchi depths as an indicator of water clarity and even easier to criticize their use as an indicator of water quality. Over the years the disks themselves have not necessarily been standardized, varying not only in size and shape (although most have been round), but also in component material (generally plastic or metal) and color (from white to off-white to patterns of alternating white and black segments). Further, the disks have been viewed by numerous observers possessing non-identical eyesights, and readings have been obtained under non-identical environmental viewing conditions and from non-identical viewing positions. These criticisms notwithstanding, however, the reality of the Secchi disk's popularity must be recognized. Also, it must be recognized that this simple and subjective data-gathering procedure can, through appropriate physical and physiological considerations, be converted into quantitative expressions for the depth rate of decay of natural light in natural waters (Preisendorfer[327,328,333] and many others). In fact, the excellent five-volume series of lecture notes by Preisendorfer[328-332] is mandatory reading for any serious student of aquatic optics.

Recall that the attenuation of downwelling irradiance in an aquatic medium is a consequence of absorption and scattering processes proceeding in tandem as the photons encounter organic and inorganic matter in their subsurface propagation. Recall further that this irradiance attenuation is described in terms of the *irradiance attenuation coefficient* K_d, which is an *apparent* optical property (i.e., a property dependent upon the spatial distribution of the downwelling radiation). When photons comprise a beam of light, the attenuation is described in terms of the *beam attenuation coefficient c*, which is an *inherent* optical property (i.e., a property independent of the spatial distribution of the radiation).

The Secchi depth S, therefore, is indeed related to such optical parameters as the attenuation length (τ), the optical depth (ζ), the beam attenuation coefficient (c), the irradiance attenuation coefficient (K_d), the scattering albedo (ω_0), and the forwardscattering and backscattering coefficients (b_F and b_B). The efforts of such workers as Preisendorfer, Beeton,[22] Tyler,[401] Højerslev,[171] Gordon and Wouters,[137] and others have shown that the Secchi depth, S, generally expressed in meters, may be used to provide a quantitative estimate of the optical property ($c + K_d$). Thus, the Secchi depth results in the estimation of a *hybrid* optical property comprised, in part, by an *inherent* optical property of the water column (c) and in part by an *apparent* optical property of the water column (K_d). On the basis of empirical relationships obtained from marine water measurements, Poole and Atkins[323] had earlier suggested that the Secchi depth was inversely proportional to the downwelling irradiance attenuation coefficient, K_d. Therefore, by measuring S the apparent optical property K_d could be inferred. Using the contrast transmittance theory developed by Duntley and Preisendorfer,[93] however, Tyler[401] concluded, on theoretical grounds, that Secchi depth was more appropriately represented as a function of the hybrid optical property ($c + K_d$) than solely as a function of the apparent optical property K_d, and derived the relationship

$$S = 8.69/(c + K_d). \tag{7.1}$$

Holmes,[174] from measurements performed in coastal waters, obtained the empirical relationship

$$S = 9.42/(c + K_d). \tag{7.2}$$

In most natural waters the numerical value of c is substantially greater than the numerical value of K_d. In general, therefore, the Secchi depth is more a consequence of c than of K_d. Beam transmissometry is the measurement of the propagation of a beam of light through a given medium over a known path length. A transmissometer, therefore, may be used to provide a reliable estimate of c in a direct and dispassionate manner without the added complexities of the influence of K_d. By profiling the transmissometer, the total attenuation coefficient may be obtained as a function of subsurface depth z. *In situ* spectral irradiance profiles are readily obtained by directly submerging irradiance meters such as scanning spectroradiometers. These profiles yield reliable estimates of not only the downwelling irradiance coefficient, $K_d(z)$, but also of the subsurface irradiance reflectance ratio (*volume reflectance*), $R(z)$.

It is, therefore, easy to advocate the extensive use of submerged optical devices in research-oriented and surveillance-oriented aquatic activities as a means of obtaining independent estimates of c and K_d. Indeed, we unhesitatingly *do* advocate the continued use of such submersible optical devices. It is also easy to condemn the use of Secchi disk depths to visually estimate the hybrid $(c + K_d)$. The historical universal popularity of the simplistic Secchi disk determinations is, however, due in part to lack of access, by many observers, to sophisticated submersible optical instrumentation. Consequently, we must stop considerably short of condemning the use of Secchi disks and recognize the convenience and simplicity that are so appropriately ascribed to them.

The National Water Research Institute (NWRI) in Ontario, Canada, has, as part of its lake optics program, collected Secchi disk, transmission, and spectral irradiance data since 1973 in four of the Laurentian Great Lakes (no direct measurements were performed in Lake Michigan). These data were used to relate, through statistical regressions, Secchi depth values S to each of the optical parameters c and K_{PAR}, as well as to use these relationships to intercompare the Great Lakes themselves.

Tables 7.1 and 7.2 are taken from Bukata et al.[43] Table 7.1 illustrates the per lake or lake region relationships between the total attenuation coefficient c and the inverse Secchi depth S^{-1} and Table 7.2 the relationships between the irradiance attenuation coefficient K_{PAR} and the inverse Secchi depth S^{-1} for the Great Lakes waters. These relationships are graphically displayed in Figures 7.2 and 7.3, respectively.

TABLE 7.1

Relationships between Total Attenuation Coefficient c and Secchi Depth S for Great Lakes Waters

Great Lake	Number of (c, S) Data Pairs	Mathematical Relationship	Correlation Coefficient
Superior	291	$c = 2.85\ (S^{-1})^{0.80}$	0.98
Huron	184	$c = 4.55\ (S^{-1})^{0.95}$	0.99
Georgian Bay	171	$c = 3.90\ (S^{-1})^{0.99}$	0.99
Erie	347	$c = 5.85\ (S^{-1})^{1.00}$	0.99
Ontario	1442	$c = 4.35\ (S^{-1})^{0.90}$	0.99

TABLE 7.2

Linear Relationships between the Irradiance Attenuation Coefficient K_{PAR} and Secchi Depth S for Great Lakes Waters

Great Lake	Mathematical Relationship	
Superior	$K_{PAR} = 0.67S^{-1} + 0.10$	for 2m $\leq$ S $\leq$ 20m
Huron	$K_{PAR} = 0.74S^{-1} + 0.07$	for 1m $\leq$ S $\leq$ 20m
Georgian Bay	$K_{PAR} = 0.81S^{-1} + 0.07$	for 2m $\leq$ S $\leq$ 20m
Erie	$K_{PAR} = 1.28S^{-1}$	for 0.5m $\leq$ S $\leq$ 10m
Ontario	$K_{PAR} = 0.76S^{-1} + 0.06$	for 1m $\leq$ S $\leq$ 3m
Ontario	$K_{PAR} = 0.86S^{-1} + 0.03$	for 3m $\leq$ S $\leq$ 10m

The number of individual data pairs entering into the linear relationship between c and S of Figure 7.2 for each water body is listed in Table 7.1. Since the number of data pairs for each lake or lake region varied from 171 to 1,442, only the calculated regression curves are shown in Figure 7.1. The scatter of the individual members of the data pair complement are not shown. Significant scatter does, indeed, occur, but as seen from the correlation coefficients also listed in Table 7.1, the existing individual statistical scatter is overcome by the sheer volume of collected data pairs.

It is seen from Table 7.1 and Figure 7.2 that the relationship between the total attenuation coefficient c and the inverse Secchi depth S^{-1} varies with location, and that in progressing from Lake Superior to Georgian Bay to Lake Ontario to Lake Huron to Lake Erie, a higher ratio of c to S^{-1} is generally encountered.

Kirk[222] has shown that the depth profile of PAR does not, throughout the water column, follow the simple exponential form that generally is tacitly assumed in discussions of this sort. Consequently, the irradiance attenuation coefficients K_{PAR} reported in Bukata et al.[43] were determined by least-square fits to each irradiance profile from the air-water interface to the depth of the 1% irradiance level (i.e., to the bottom of the *euphotic zone*).

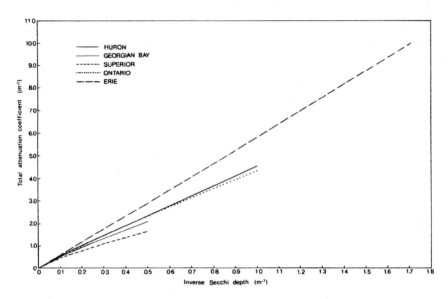

FIGURE 7.2
Relationships between the total attenuation coefficient c and the inverse Secchi depth S^{-1} for various Great Lakes waters. (Adapted from Bukata, R. P., Jerome, J. H., and Bruton, J. E., *J. Great Lakes Res.*, 14, 347–355, 1988.)

It is seen from Table 7.2 and Figure 7.3 that Georgian Bay and Lakes Huron, Superior, and Ontario display distinctly similar (K_{PAR}, S^{-1}) regressions, while Lake Erie displays a markedly different regression with a slope that is 50% greater than the slopes observed for the other water bodies. Figure 7.4, also taken from Bukata *et al.*,[43] illustrates the effect of averaging the linear (K_{PAR}, S^{-1}) regressions of Georgian Bay and Lakes Huron, Superior, and Ontario into a single relationship over the common range of Secchi depths 2 m $\leq$ S $\leq$ 10 m. Consequently, with the exclusion of Lake Erie, the Great Lakes waters monitored by the NWRI surveillance program may be approximately defined by the single relationship

$$K_{PAR} = 0.757S^{-1} + 0.07 \qquad (7.3)$$

over the range 2 m $\leq$ S $\leq$ 10 m. The average percent difference between the use of this single equation and the actual regressions (listed in Table 7.2) is 5% for Georgian Bay, 2% for Lake Huron, 5% for Lake Superior, and 7% for Lake Ontario.

For the Great Lakes system, it therefore appears that while the relationship between the Secchi depth and the inherent optical property total attenuation coefficient c varies from lake to lake, the relationships between the

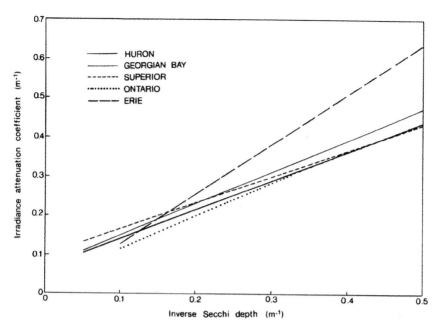

FIGURE 7.3

Relationships between the irradiance attenuation coefficient K_{PAR} and the inverse Secchi depth S^{-1} for various Great Lakes waters. (Adapted from Bukata, R. P., Jerome, J. H., and Bruton, J. E., *J. Great Lakes Res.*, 14, 347–355, 1988.)

Secchi depth and the apparent optical property downwelling irradiance attenuation coefficient K_{PAR} display a remarkable similarity for much of the lake waters. This would further reinforce the suggestion that spatial departures from constancy in the local relationships between Secchi depth and the hybrid optical property $(c + K_d)$ may be predominantly due to departures from constancy of the *inherent* optical properties of the aquatic systems, an important consideration when contemplating the remote measurements of radiance upwelling from aquatic targets.

The greatest hazard in relating, through statistical regression techniques, the Secchi depth observations to direct optical measurements, lies in the tendency, once such relationships are established, to utilize Secchi depth readings to infer optical parameters which are much more appropriately obtained from other, more sophisticated techniques. Such relationships tend to smooth out the effects of seasonal and spatial variations in the physical, chemical, and biological activity occurring within the aquatic system. These effects, along with the physiological subjectiveness of the Secchi disk user and the variety of environmental conditions under which the observations are made, generate a significant degree of statistical scatter among all pairs

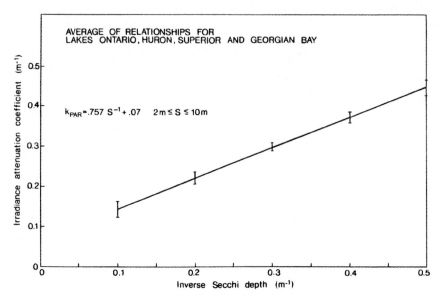

FIGURE 7.4

The averaged single linear relationship relating the irradiance attenuation coefficient K_{PAR} to the inverse Secchi depth S^{-1} over the common range of Secchi depths (2 m $\leq$ S $\leq$ 10 m) in Lakes Huron, Superior, and Ontario as well as Georgian Bay. (Adapted from Bukata, R. P., Jerome, J. H., and Bruton, J. E., *J. Great Lakes Res.*, 14, 347–355, 1988.)

of optical data sets (particularly those data sets involving S). On an individual surveillance basis, this limited number of data sets further aggravates the statistical scatter. However, in the absence of more rigorous data, and in recognition of the fact that Secchi depths represent the only existing historical record for many global waters, the cautious and careful utilization of Secchi depth relationships with inherent and apparent optical properties could perhaps be of beneficial consequence.

Ergo, the Secchi disk remains a bittersweet reality of contemporary optical surveillance of water bodies on a global scale.

7.3   BEAM TRANSMISSION, $T(\lambda, z)$

As we have seen, the transmission of light through a natural water column is determined by the specific complement of absorption and scattering centers comprising the water column at the time of observation. The *beam transmittance*, $T(\lambda, z)$, also commonly referred to as the *beam transmission* of a water mass at depth z, is defined as the ratio of the radiant flux emerging in a beam from an infinitesimal aquatic layer, ϕ_{trans}, to the incident radiant flux impinging in a beam on that infinitesimal layer, ϕ_{inc}, i.e., $T = \phi_{trans}/$

ϕ_{inc}. A comparable term, *beam attenuance*, $C(\lambda, z)$, or simply *beam attenuation*, is used to define the ratio of radiant flux lost from the beam within the infinitesimal layer to the beam of incident radiant flux, i.e., $C = (\phi_{inc} - \phi_{trans})/\phi_{inc}$. Clearly, $T + C = 1$. The beam transmission, T, of a given water mass at depth z over a path length r is given from Beer's Law as:

$$T(\lambda, z) = \exp[-c(\lambda, z)r] \qquad (7.4)$$

where c = beam attenuation coefficient $[c = a + b]$
$\quad\quad a$ = absorption coefficient
$\quad\quad b$ = scattering coefficient

Equation (7.4), therefore, represents the fractional transmission of a light beam through a distance r in the water column, while the fractional attenuation of light by the water mass is given as:

$$C(\lambda, z) = 1 - \exp[-c(\lambda, z)r] \qquad (7.5)$$

A beam transmissometer (sometimes referred to as a beam attenuation meter) is a device that enables *in situ* determinations of the beam attenuation $C(\lambda, z)$ within a prescribed instrument pathlength r. The beam attenuation coefficient $c(\lambda)$, as seen from equation (7.5), is then obtained from:

$$c(\lambda, z) = -\ln[1 - C(\lambda, z)]/r \qquad (7.6)$$

Three instrument pathlengths r commercially available are 1.0 meter, 0.25, and 0.10 meter, although specialized variable pathlength transmissometers are often utilized as research instruments. The transmissometer generally consists of [see the discussion of design principles along with their limitations presented by Austin and Petzold[12]] a light source with associated lens system to produce a collimated light beam, separated by a pathlength distance r from a receiver lens/detector system. This light source/receiver configuration is then used to observe an intervening water column. To obtain an accurate estimation of $c(\lambda, z)$ (as discussed in Chapter 1) requires an ability to estimate a rectilinear loss in photon energy as the photons propagate through the attenuating medium. Such a rectilinear loss restriction, while not unduly problematical to absorptive interactions, can be distinctly problematical to scattering interactions. This problem arises from the fact that most of the scattering in natural waters occurs at small angles to the direction of pre-scattered photon propagation. This requires that the acceptance angle of the transmissometer detector be of the order of 1° or less.[308] A large acceptance angle allows scattered photons to essentially remain in the beam, and, consequently, failure to provide a sufficiently small acceptance cone can result in a significant underestimation of the beam attenuation.

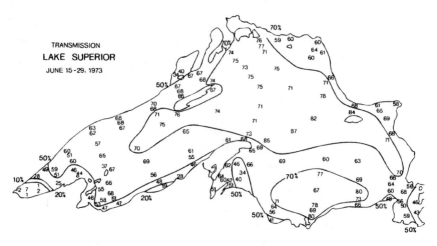

FIGURE 7.5
Near-surface transmission contours of Lake Superior, June 1973.

If a collimated beam of light is passed through natural water and the light scattered from an intervening fixed water mass of known composition is measured at various angles between 0° and 180°, then the volume-scattering function $\beta(\theta)$ can be determined. In reality, however, such measurements are most difficult to execute (again largely due to the need to measure small angle scattering and the difficulties this imposes upon the alignment of the source and detector). Consequently, the $\beta(\theta)$ values for very few natural waters have been determined in this manner.

Transmission profiles on four of the Laurentian Great Lakes have been routinely obtained by the National Water Research Institute since 1973 utilizing Martek XMS transmissometers of 1 meter or 0.25-meter pathlengths fitted with a Wratten 45 optical filter. These Great Lake data will be used to illustrate briefly the nature of the optical measurements and the contributions the routine use of submersible optical devices continue to provide to large lake surveillance protocols.

Figure 7.5 illustrates the near-surface (depth of 1 meter) unpublished transmission contours of Lake Superior taken during an NWRI surveillance cruise in late June 1973. In general Lake Superior displays a high degree of clarity (transmission values $> \sim 60\%$ and peaks well over 85% for a 1-m pathlength). From Figure 7.5 and comparable contours obtained throughout the field season, it has been seen that the lowest lake-wide transmission values are observed in the autumn, while the highest lake-wide transmission values are observed in mid-summer. Areas of generally low transmission throughout the year include the tip of the western basin (near Duluth, Minnesota), the north shore area (Thunder, Black, and Nipigon Bays, Ontario), and the south shore (west of the Keweenaw Peninsula).

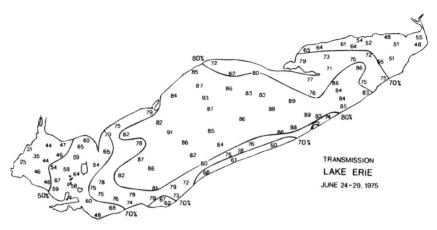

FIGURE 7.6
Near-surface transmission contours of Lake Erie, June 1975.

While not as optically transparent as Lake Superior, Lake Huron and Georgian Bay generally display comparable transmission values > 55% and peaks of > 70% for a 1-meter pathlength.

In general Lake Erie displays a low degree of optical clarity combined with a high degree of spatial variability. Figure 7.6 illustrates the near-surface (depth of 1 meter) unpublished transmission contours of Lake Erie taken during an NWRI surveillance cruise in late June 1975. Throughout the year, measured values of transmission vary from as low as 1% for a 0.25-meter pathlength (corresponding to a transmission of $\sim 1 \times 10^{-6}\%$ for a 1-meter pathlength) to as high as 90% for a 0.25-meter pathlength (corresponding to a transmission of $\sim 66\%$ for a 1-meter pathlength). While a seasonal variation is in evidence (lowest transmission in spring and autumn; highest transmission in mid-summer), the most notable feature of the Lake Erie transmission contours is the transmission gradient along the lake from east to west. Lake Erie is generally considered to comprise three distinct sub-basins. The shallowest western basin of Lake Erie is characterized by the least clarity, while the intermediate depth central basin is most often characterized by a clarity intermediate to those characterizing the shallow western and the deep eastern basins.

Lake Ontario generally displays a beam transmission intermediate to the beam transmissions displayed by Lake Erie and the upper Great Lakes. Lake Ontario transmission values usually lie in the range 50% to 85% for a 0.25-meter pathlength ($\sim 6\%$ to 52% for a 1-meter pathlength). The lowest lake-wide transmission values occur during the summer months, the highest during spring and winter. The areas of generally lowest transmission values throughout the year include Hamilton Harbour, the Niagara River plume region, the Toronto area, Black Bay, and in general the nearshore zone around

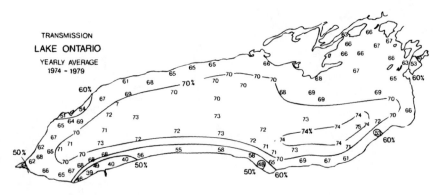

FIGURE 7.7
Near-surface transmission contours of Lake Ontario, yearly average 1974–1979.

the lake. These regions are particularly evident in Figure 7.7 wherein are plotted the unpublished NWRI yearly averages of the near-surface (1-meter depth) transmission contours of Lake Ontario using data collected continually between 1974 and 1979.

Aquatic areas that have for short or prolonged periods of time been subjected to natural and/or anthropogenic stress can, depending upon the nature of the environmental stress, display transmissometer readings indicative of the entire range of beam transmission from high optical transparency to high optical opacity. Since the inherent optical property $c(\lambda)$ responds to the totality of absorption and scattering centres residing within the water column, transmission measurements cannot unambiguously identify these absorption and scattering centres, let alone provide an estimation of their possible deleterious natures. Certainly problem areas are known to exist, and effluents from both well- and ill-defined point sources and extended sources are known to disrupt the natural evolution of unperturbed ecosystems.

Hamilton Harbour, an enclosed body of water located at the western end of Lake Ontario and designated as an Area of Concern by the Canada-United States International Joint Commission, is one of the most polluted sites in the Great Lakes system (Ontario Ministry of the Environment, MOE[300,301]). Being a receiving body for a number of municipal and industrial effluents, Hamilton Harbour represents a distinct water management problem and has been designated a Remedial Action Plan (RAP) site. As such, it has been the focal point for concentrated multidisciplinary studies for at least the past two decades. [These studies have produced invaluable information on the descriptive limnology of the harbour, the mechanisms through which the turbulence spectrum of the harbour is maintained by means of water exchange between the harbour and its adjacent waters, the severe near-bottom anoxia, the light regime aspects of the harbour, the impacts of nutrient loadings on eutrophication, as well as the impacts of toxicants on the biological character of the harbour. A woefully incomplete but representative bibliography of principal research results would include the works by Harris et al.,[159,160] Klapwijk and Snodgrass,[230] Kohli,[234] Barica et al.,[19] Harris and Piccinin,[161] Polak and Haffner,[321] Haffner et al.,[152] Poulton,[324] and Janus[187]].

Consistently low transmission values are recorded in Hamilton Harbour and its adjacent waters, and it would indeed be tempting to ascribe this opacity to the known pollution inputs that have plagued Hamilton Harbour for many years. While suspended particulate contaminants could produce a significant impact on light attenuation, virtually every chemical contaminant in aquatic solution, irrespective of concentrations high enough to warrant their serious consideration, cannot impact optical attenuation in a manner that would compete with the scattering and absorption produced by chlorophyll, suspended sediments, and dissolved organic matter. Thus, with or without sources of municipal or industrial effluents, near-shore regions such as harbours, bays, deltas, and marshes would generally display transmissions distinguishably subordinate to that of open waters. This additional opacity of nearshore zones, of course, can modulate aquatic vulnerability to contaminant impact since the suspended organic and inorganic material can act as vehicles for contaminant transport.

Just as a low value of optical transparency does not necessarily indicate an aquatic resource under siege, a high value of optical transparency does not necessarily indicate a condition of environmental harmony. It is well known that the chlorophyll content of a water mass is a major indication of its biological productivity, and since acid stress is generally known to have an adverse impact on aquatic life, it would be reasonable to expect that low pH levels (e.g., pH < 5.0) should be accompanied by low chlorophyll concentrations. Acid stressed waters sufficiently distant from nearshore suspended inorganic matter would therefore display a very high degree of optical transparency. However, it is a very serious error in fundamental logic to assume that lakes having low chlorophyll concentrations are by necessity acid stressed. Thus, it is evident that naturally oligotrophic (see Section 7.10) inland waters and acidified lakes could conceivably display identical transmissometer readings.

Transmissometer depth profiles can be a valuable tool in assessing the limnological processes governing the behavior of inland water bodies. Figures 7.8 and 7.9 illustrate unpublished NWRI transmissometer depth profiles obtained from west-east transects of Lake Ontario taken in May 1982 and July 1982, respectively. Such depth profiles provide a snapshot of the vertical gradient in aquatic clarity and therefore represent a "layering" of aquatic regimes according to their respective attenuation characteristics.

Immediately after ice break-up in the spring, lakes in intermediate to higher northern latitudes are characterized by isothermal conditions

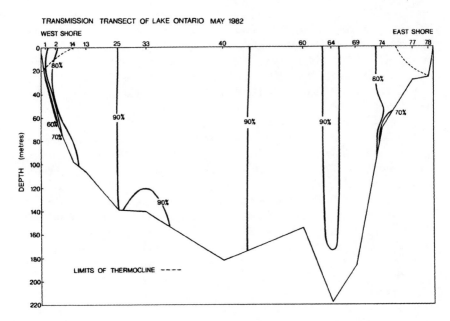

FIGURE 7.8
Transmission depth profiles of Lake Ontario, May 1982.

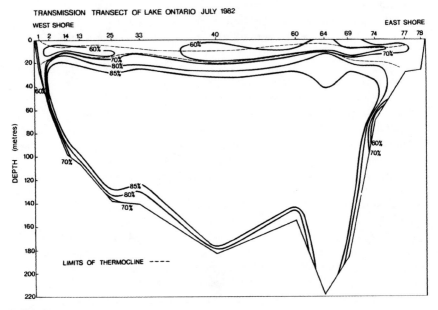

FIGURE 7.9
Transmission depth profiles of Lake Ontario, July 1982.

(between 0° and ~4° C). At this time the transmission transects across these lakes generally indicate a similar tendency toward an isotropic condition. No vertical gradient in transmission is in evidence. A slight horizontal gradient, however, can be seen in Figure 7.8, which was taken shortly after the onset of spring warming. As the season progresses and the lakes approach thermal stratification (the development of two distinctive thermal regimes, a warmer upper layer known as the *epilimnion* and a cooler lower layer known as the *hypolimnion*, with the region of demarcation between these thermal regimes being termed the *thermocline*), distinct layering develops among the transmission profiles. In general, a significant amount of this layering is associated with the actual location of the thermocline. The thermal evolution of inland lakes proceeds generally as an "onion-skin" effect. In its simplest terms it is a consequence of deeper (cooler) water upwelling and undergoing surface solar warming. A *thermal bar* is thus established around the lake, which progressively advances offshore until the central regions of the lake are incorporated into the elevated temperature. It is the initial development and advance of the thermal bar that is reflected in the horizontal transmission gradient seen in Figure 7.8. By mid-summer the lake is in full stratification, and the duality of the Lake Ontario thermal regime is quite evident from the transmission profile of Figure 7.9.

In Lakes Superior and Huron (and Georgian Bay), the thermocline is often characterized as the layer of minimum mid-lake transmission. Above and below the thermocline the waters display a higher degree of clarity.

Figure 7.10 illustrates representative depth profiles of transmission and temperature for a mid-lake station in Lake Huron in the summer (unpublished NWRI data collected in August 1974). The optical layering in the vicinity of the thermocline is the most obvious feature of the transmission profile.

The triple-basin nature of Lake Erie produces a considerably more complex vertical transmission gradient than that observed in the upper Great Lakes. Figure 7.11 illustrates unpublished NWRI transmission profiles collected during a June 1975 west-east transect of Lake Erie. The shallow western basin of Lake Erie is characterized by both a horizontal and a vertical transmission gradient with extremely low values of transmission being observed at lake-bed. The intermediate depth central basin is generally characterized by higher transmission values above the thermocline and lower transmission values below the thermocline. The deeper eastern basin tends to display the transmission/depth profile of the upper Great Lakes (i.e., clearer waters both above and below the thermocline).

In early summer Lake Ontario displays a vertical transmission gradient similar to that of the upper Great Lakes (thermocline as the region of minimal transmission). However, as the summer progresses, the epilimnion becomes the most turbid (lowest value of transmission) and the hypolimnion becomes the clearest (highest values of transmission). This summer situation is illus-

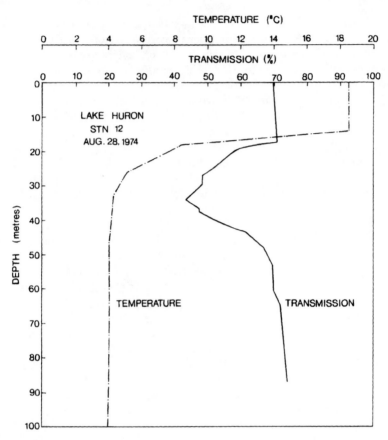

FIGURE 7.10
Vertical transmission and temperature depth profiles for a mid-lake station in Lake Huron, August 1974.

trated in Figure 7.12 wherein are plotted the unpublished NWRI transmission and temperature profiles of Lake Ontario obtained in August 1982.

During the summer, therefore, the deep eastern basin of Lake Erie displays a transmission transect structure similar to those of the upper Great Lakes, while Lake Ontario displays an inverse transmission transect structure to that of the central basin of Lake Erie. In autumn, as the lakes return to an isothermal condition, the transmission structure also collapses and returns to the spring conditions.

The ubiquitous occurrence of minimal transmission associated with the thermocline of Great Lakes waters, and the possible organic and/or inorganic origins of this layered turbidity have been discussed by Thomson and Jerome[398] and Jerome et al.[201]

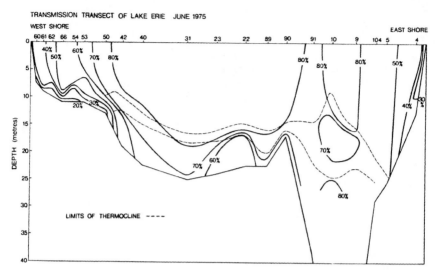

FIGURE 7.11
Transmission depth profiles of Lake Erie, June 1975.

An interesting variation in transmissometer design and operation has been presented by Fournier et al.,[109,110] who have used the device to gather intrinsic optical information on Canadian waters. Their transmissometer incorporates a wavelength scanning capability which displays the same unfortunate dependence upon the perverse variability of the volume scattering function, $\beta(\theta)$, that plagues conventional transmissometry. However, they found that a simple analytic form for the phase function, $P(\theta)$, namely,

$$P(\theta) = C_\mu \exp[-\alpha(\sin\theta/2)^{1/2}]/(\sin\theta/2)^\mu \qquad (7.7)$$

where C_μ is a normalization constant and α and μ are adjustable fitting parameters ($\alpha = 2\pi/\mu$ for $0.6 < \mu < 1.9$), gave a good fit to scattering over the first 90°. Use of the empirical equation (7.7) enabled realistic separations of the absorption and scattering phenomena in the waters of Cabot Strait and Labrador Sea.

7.4 THE NEPHELOID LAYER

As seen from the Great Lakes transmission/depth profiles such as Figures 7.9 and 7.11, an obvious feature of Lakes Erie and Ontario is the region of low transmission near the lake bottom. This region, referred to as the *nepheloid layer*, is generally attributed to resuspension of very fine bottom

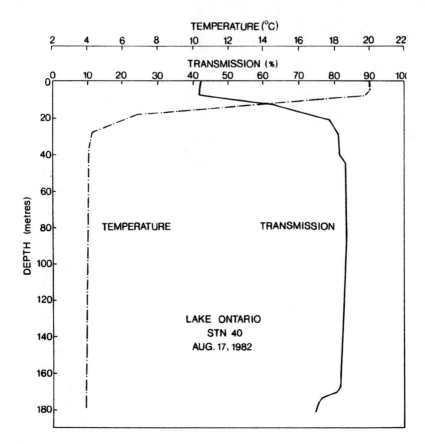

FIGURE 7.12
Transmission and temperature depth profiles of Lake Ontario, August 1982.

sediments due to currents that exist deep in ocean or lake basins [Hunkins et al.,[178] Eittreim et al.,[100] Ewing and Connary[101]]. There are several possible sources for nepheloid material: matter transported from continental slopes; ocean or lake bed matter eroded by bottom currents; suspended matter provided by river inputs; organic and inorganic matter settling or vertically transported from epilimnetic waters.[303] Light transmission measurements by Sandilands and Mudroch[348] indicated that a turbid layer at the bottom of Lake Ontario was a consistent feature of the entire lake at water depths greater than 60 meters, and that the thickness of this nepheloid layer averaged about 22 meters throughout the field season, but doubled this thickness during the late summer and early autumn months. While transmissometry is capable of locating and determining the extent in space and time of the nepheloid layer, the identification of the organic and/or inorganic matter comprising this layer requires the invoking of other techniques, such as the direct sampling of the water column for laboratory analyses. Both particle size and type are best determined from actual samples, and direct (although limited in number) observations suggest that the nepheloid layers in the Great Lakes contain small particles of amorphous or biogenic silica, presumably fragments of diatom fustules, in addition to calcite and clay minerals.[348,276] Such a *benthic* (pertaining to the plant and animal life that live in or is associated with the bottom of a water body) nepheloid layer represents a large surface area that is available for the absorption of chemical compounds, and thus can play a significant role in the transport of toxic contaminants either horizontally along the lake proper or vertically upwards into the euphotic zone [as discussed by Glover,[126] Sandilands and Mudroch,[348] Schelske,[352] and others].

The nepheloid layers in the western and central basins of Lake Erie are most prominent during the summer. In fact, the entire hypolimnion of central Lake Erie could be considered as comprising the nepheloid layer. The nepheloid layer of Georgian Bay, while most prominent during the summer, is neither as clearly defined nor as physically extensive as its lower Great Lakes counterparts. Chambers and Eadie[58] have shown a persistent benthic nepheloid layer with high total suspended matter to also be a widespread feature of the bottom waters of Lake Michigan. Our inability to monitor transmission at depths $> \sim 100$ meters from conventional ship transects have, to date, prevented direct observation of nepheloid layering in Lakes Superior and Huron.

7.5   IRRADIANCE ATTENUATION COEFFICIENT FOR PAR

Submersible irradiance spectrometers such as the Techtum QSM scanning quantaspectrometer used for the past two decades by the National Water Research Institute, are capable of directly providing a depth profile of the subsurface irradiance levels.

Recall that the downwelling irradiance at depth z, $E_d(\lambda,z)$, is given by:

$$E_d(\lambda, z) = E_d(\lambda, 0)\exp[-K_d(\lambda)z] \qquad (7.8)$$

where $K_d(\lambda)$ is the average value of the downwelling irradiance attenuation coefficient over the depth interval 0 to z.

The 1% irradiance level then satisfies the equation:

$$0.01 = \exp(-K_d(\lambda)z_{.01})$$

where $z_{.01}$ is the depth of the 1% irradiance level and from which

$$K_d(\lambda) = 4.605/z_{.01} \qquad (7.9)$$

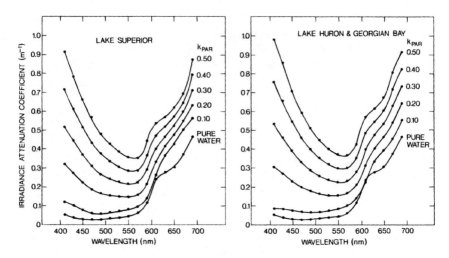

FIGURE 7.13
Spectral irradiance attenuation coefficients $K(\lambda)$ for five values of K_{PAR} for (a) Lake Superior and (b) Lake Huron and Georgian Bay. (From Jerome, J. H., Bukata, R. P., and Bruton, J. E., *J. Great Lakes Res.*, 9, 60–68, 1983. With permission.)

Scanning irradiance quantaspectrometers provide a record of the irradiance E versus wavelength λ from 400 nm to 700 nm in units of quanta m^{-2} sec^{-1}. For any spectral band the irradiance can then be converted into units of einsteins m^{-2} sec^{-1} where 1 einstein is equivalent to 6.023×10^{23} quanta. The total radiation in the 400 nm to 700 nm band is the photosynthetically available radiation (PAR) expressed in units of quanta irradiance (μ einsteins m^{-2} sec^{-1}).

For radiation within the PAR interval, therefore, we may write the downwelling irradiance attenuation as K_{PAR} and determine its value from equation (7.9). However, as discussed in Section 7.2, Kirk[222] has shown that the depth profile of PAR does not adhere to the simple exponential form expressed in equations (7.8) and (7.9). Consequently, the spectral irradiance attenuation coefficients $K(\lambda)$ and the irradiance attenuation coefficient for PAR, K_{PAR}, shown in the upcoming figures were determined by a least-squares fit to each directly measured quanta irradiance profile.

Figure 7.13, taken from Jerome *et al.*[203] (their Figure 3), illustrates the spectral irradiance attenuation coefficients $K(\lambda)$ for five values of K_{PAR} (covering the entire range of mid-lake values) for (a) Lake Superior and (b) Lake Huron/Georgian Bay. The irradiance attenuation curve for pure water[177] is also shown. The similarities between the two upper Great Lakes is immediately apparent.

Figure 7.14, also taken from Jerome *et al.*[203] (their Figure 4), illustrates the spectral irradiance attenuation coefficients $K(\lambda)$ for six values of K_{PAR} (covering the entire range of mid-lake values) for (a) Lake Erie and (b) Lake

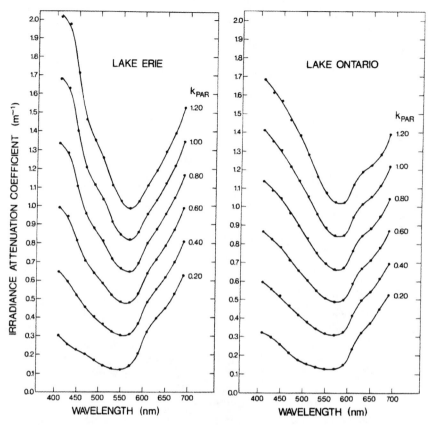

FIGURE 7.14
Spectral irradiance attenuation coefficients $K(\lambda)$ for six values of K_{PAR}: (a) Lake Erie and (b) Lake Ontario. (From Jerome, J. H., Bukata, R. P., and Bruton, J. E., *J. Great Lakes Res.*, 9, 60–68, 1983. With permission.)

Ontario. The two lower Great Lakes display similar properties except for the much greater attenuation at shorter wavelengths in Lake Erie for high values of K_{PAR}.

As an indication of the spectral distribution of PAR as a function of depth, Figures 7.15 and 7.16, again taken from Jerome *et al.*[203] (their Figures 5 and 6), illustrate the PAR spectrum at the surface (incident PAR) and at the 10% and 1% irradiance levels (expressed as a percentage of the incident PAR for Lakes Erie and Huron, respectively). The two extreme mid-lake values of observed K_{PAR} for each lake are considered. Lake Superior displays spectral distribution curves similar to Lake Huron (Figure 7.16) and Lake Ontario displays spectral distribution curves similar to Lake Erie (Figure 7.15). It is seen from Figures 7.15 and 7.16 that, in general, as the attenuation

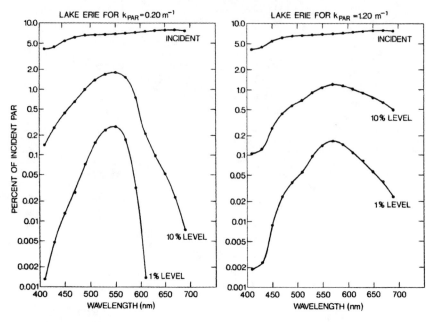

FIGURE 7.15
Spectral distribution of PAR at the surface and at the 10% and 1% irradiance levels of Lake
Erie. Values of K_{PAR} of 0.20 m^{-1} and 1.20 m^{-1} are considered. (From Jerome, J. H., Bukata, R.
P., and Bruton, J. E., *J. Great Lakes Res.*, 9, 60–68, 1983. With permission.)

of PAR, K_{PAR}, increases, the radiation in the blue (shorter wavelengths) region
of the PAR spectrum becomes less significant (the percentage comprising
the PAR spectrum becomes smaller) while the radiation in the red (longer
wavelengths) becomes more significant (the percentage comprising the PAR
spectrum becomes larger). This will impact the light availability for primary
production since shorter wavelength radiation plays a greater role in the
primary production process than does longer wavelength radiation.

This spectral distribution of PAR as a function of depth will be referred
to again in Section 7.8 which deals with the irradiance attenuation coefficients
$K(\lambda)$ of the spectral components of PAR in inland waters.

7.6 PHOTOSYNTHETIC USABLE RADIATION, PUR

The photosynthetic available radiation, PAR, represents the totality of
radiation in the 400-nm to 700-nm wavelength interval that is available for
photosynthetic activity. However, since the absorption spectrum of algae is
not invariant over the PAR wavelength interval, PAR does not transform

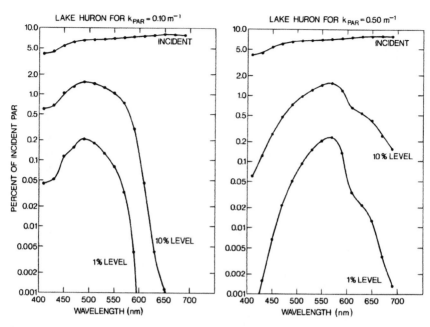

FIGURE 7.16
Spectral distribution of PAR at the surface and at the 10% and 1% irradiance levels of Lake Huron. Values of K_{PAR} of 0.10 m^{-1} and 0.50 m^{-1} are considered. (From Jerome, J. H., Bukata, R. P., and Bruton, J. E., *J. Great Lakes Res.*, 9, 60–68, 1983. With permission.)

directly into the amount of radiation that can be used for photosynthesis. The amount of PAR which is pertinent to the photosynthetic process is termed the *Photosynthetic Usable Radiation* (PUR) and is defined by:

$$PUR(z) = \int_{400}^{700} a(\lambda)E_Q(z, \lambda)d\lambda \qquad (7.10)$$

where $a(\lambda)$ = specific absorption for chlorophyll at wavelength λ in m^2
                 per mg Chl and
    $E_Q(z,\lambda)$ = spectral quanta irradiance at depth z and wavelength λ
                 in quanta per m^2-sec.

    PUR, therefore, is a value of PAR that is weighted according to the absorption capabilities of chlorophyll. If PAR rather than PUR were used in estimating the primary production of an aquatic resource, it is therefore probable that the primary productivity could be significantly over- or underestimated.

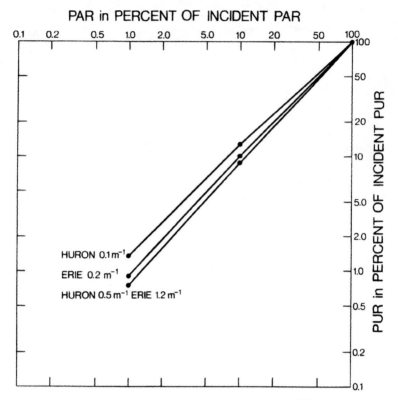

FIGURE 7.17
Relationship between PAR and PUR for Great Lakes waters. (From Jerome, J. H., Bukata, R. P., and Bruton, J. E., *J. Great Lakes Res.*, 9, 60–68, 1983. With permission.)

Using a specific absorption spectrum taken from Prieur and Sathyendranath,[337] along with directly measured spectral distributions of incident PAR at the air-water interface, Jerome *et al.*[203] determined the relationships between PUR and PAR for Great Lakes waters that are shown in Figure 7.17.

For the clearest waters ($K_{PAR} = 0.10$ m^{-1}) of the upper Great Lakes, the attenuation of PAR is less than the attenuation of PUR. Thus, for a given percentage of PAR in the downwelling irradiance, the percentage of PUR would be higher and the amount of primary production, if calculated on the basis of PAR, would be underestimated (by a factor of 0.73 at the 1% level of PAR).

For the clearest waters ($K_{PAR} = 0.20$ m^{-1}) of the lower Great Lakes, the percentage of PAR and PUR are almost equivalent at the 10% and 1% levels of PAR. Estimates of primary production based on PAR for these waters would, therefore, be quite accurate.

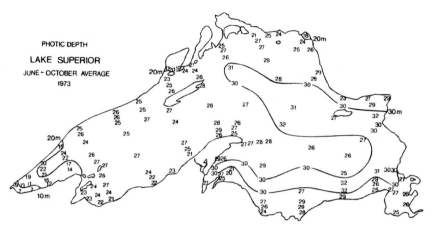

FIGURE 7.18
Photic depth contours of Lake Superior, averaged June–October 1973.

For the most turbid waters ($K_{PAR} = 0.50$ m^{-1} for the upper Great Lakes and $K_{PAR} = 1.20$ m^{-1} for the lower Great Lakes), the PUR is attenuated more rapidly than the PAR. Use of PAR rather than PUR would, therefore, result in an overestimation of primary production (by a factor of 1.3 at the 1% level of PAR).

Chlorophyll-bearing biota, however, have been observed to naturally adapt to their surroundings, and the above conclusions, based upon a specific chlorophyll absorption spectrum, does not take into consideration possible chromatic adaptation.

7.7   PHOTIC DEPTH

The photic zone is generally taken to represent the aquatic region bounded by the 100% and the 1% irradiance levels of the subsurface photosynthetic available radiation, PAR. The depths of subsurface irradiance levels are either directly measurable by means of submersible spectroradiometers or calculable from equation (7.8) if the downwelling irradiance just below the air-water interface and the appropriate K_{PAR} are both known. The *photic depth* is considered to be the depth from the air-water interface to the bottom of the photic zone (i.e., $z_{.01}$), the depth of the 1% irradiance level. There is, understandably, a similarity between the photic depth contours of a water body and its beam transmission contours.

Figure 7.18 illustrates unpublished NWRI summer averaged (June–October) photic depth contours of Lake Superior collected in 1973. Figure 7.19 illustrates unpublished May–October 1975 averaged NWRI photic depth contours for Lake Erie. Figure 7.20 illustrates unpublished NWRI photic

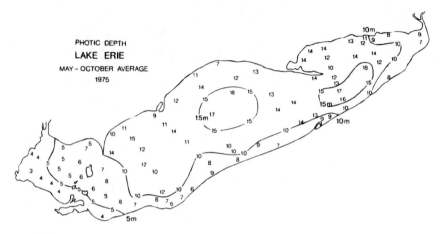

FIGURE 7.19
Photic depth contours of Lake Erie, averaged May–October 1975.

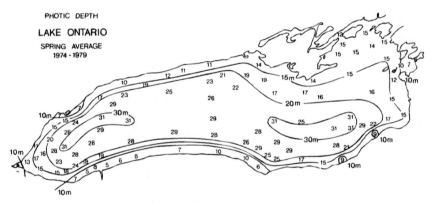

FIGURE 7.20
Photic depth contours of Lake Ontario, five-year averaged spring months 1974–1979.

depth contours of Lake Ontario averaging five years of spring data collected during 1974 to 1979. Figure 7.21 illustrates the five-year summer average of Lake Ontario photic depth contours during this same period.

From Great Lakes photic depth contours such as those illustrated in Figures 7.18 to 7.21, it can be seen that:

1. During the summer months the upper Great Lakes (including Georgian Bay) are characterized by consistently large values of photic depths (>20 meters with values >30 meters not uncommon).

2. During the summer months the lower Great Lakes are characterized by consistently smaller values of photic depths (<20 meters with values <15 meters not uncommon).

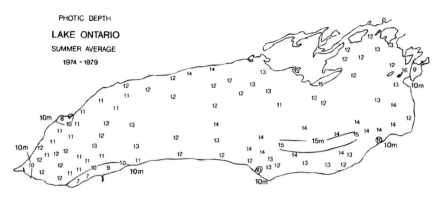

FIGURE 7.21
Photic depth contours of Lake Ontario, five-year averaged summer months 1974–1979.

3. In general, as expected, the areas of low surface transmission (as illustrated in Section 7.3) are reflected as areas of small photic depths, while the areas of high beam transmission in each lake are reflected as areas of large photic depths.

4. The "onion skin" pattern in photic depth contours of Lake Ontario (Figure 7.20) is distinctly prominent in the spring and winter months, very indistinct in the autumn months, and totally absent in the summer months, during which time the entire lake is characterized by photic depths of ~11 to ~14 meters (Figure 7.21).

The photic depths (like most optical characteristics of natural waters) are a direct consequence of the concentrations and the spectral absorption and scattering cross sections of the organic and inorganic matter in solution and in suspension within the water column. Therefore, photic depths generally observable in Case I waters are considerably larger than photic depths observable in inland and coastal waters. The depth of the 1% irradiance level is dependent upon the wavelength of the downwelling irradiance, with the 1% level for downwelling "blue" light in Case I waters being at a greater depth than the 1% level for downwelling "green" light (as dictated by the optical cross section spectra for near-pure water columns). The photic depth of downwelling PAR, therefore, accounts for the exponential approach to the 1% level of all the component wavelengths incorporated into the broadband PAR. Direct measurement of the photic depth of broadband PAR, however, results in an effective depth which, while possibly inappropriate to any specific monochromatic light, represents a compromise amongst the spectrum of visible irradiance wavelengths characterized by a variety of attenuation coefficients. On this basis, a theoretical ocean comprised solely of pure water would contain the 1% PAR irradiance level at a depth of ~140 meters. This hypothetical photic depth is just slightly larger than the 120-meter photic zone characterizing the Sargasso Sea, generally regarded as one of the most transparent regions of the Earth's oceans. Typical values of photic depths observable in Case I waters range from 40 meters to 90 meters, with observable photic depths diminishing in magnitude as land is approached.

Since plankton are the initiators of primary production, the presence of algae in natural waters will, by virtue of their absorption and scattering properties reduce the photic depth, and by virtue of their photosynthetic activity increase the primary productivity within the water column. It would, therefore, be logical to anticipate an inverse relationship as existing between primary production and photic depth. Such is indeed observed to be the case for mid-oceanic waters. In Case I waters plankton (and their generally accepted surrogate, chlorophyllous pigments) are the prominent aquatic absorption and scattering centres. Thus, in mid oceanic waters, plankton-free can be considered not only as production-free but also as essentially turbidity-free. In inland and coastal regimes, however, the presence of suspended inorganic and dissolved organic matter can independently or collectively become the prominent aquatic absorption and scattering centres, oftimes in the relative or total absence of plankton. Thus, in non-Case I waters an inverse relationship between primary production and photic depth may not exist, and is, therefore, of very little practical relevance insofar as providing an estimate of the primary productivity of a natural water body. Low to moderately high values of limnological and/or coastal photic depths could be indicative of either significant or very insignificant aquatic primary production. High values of limnological and/or coastal photic depths, however, can still be confidently assumed to represent waters of relatively low primary production.

7.8   SPECTRAL BAND VALUES OF IRRADIANCE ATTENUATION COEFFICIENTS

When the downwelling irradiance in the PAR broadband range 400 nm to 700 nm is considered, the irradiance attenuation coefficient, K_{PAR}, so determined will understandably be at variance with the irradiance attenuation coefficient $K(\lambda)$ determined for downwelling irradiance at any specific wavelength λ. Such variances observed in Great Lakes waters (Bukata et al.[41]) are illustrated in Table 7.3. Herein are listed the results of regressing (between 40 and 110 data sets for each of four of the Laurentian Great Lakes) the irradiance attenuation coefficients appropriate for each of the spectral bands 400–500 nm (blue), 500–600 nm (green), and 600–700 nm (red) against the irradiance attenuation coefficient for the spectral band 400–700 nm (PAR). In Table 7.3 these attenuation coefficients are designated as K_{blue}, K_{green}, K_{red}, and K_{PAR}, respectively. Along with each regressed mathematical relationship is a value of the correlation coefficient r indicative of the statistical scatter

TABLE 7.3

Relationships among Spectral Irradiance Band Attenuation Coefficients and Irradiance Attenuation Coefficients of PAR for Great Lake Waters

Lake	Per-Band Mathematical Relationship			r
Superior	K_{blue}	=	$1.24\ K_{PAR} - 0.04$	0.98
Superior	K_{green}	=	$0.76\ K_{PAR} + 0.01$	0.98
Superior	K_{red}	=	$0.74\ K_{PAR} + 0.27$	0.68
Huron and Georgian Bay	K_{blue}	=	$1.31\ K_{PAR} - 0.05$	0.96
Huron and Georgian Bay	K_{green}	=	$0.76\ K_{PAR} + 0.01$	0.98
Huron and Georgian Bay	K_{red}	=	$0.84\ K_{PAR} + 0.24$	0.74
Erie	K_{blue}	=	$1.31\ K_{PAR} - 0.05$	0.99
Erie	K_{green}	=	$0.92\ K_{PAR} - 0.04$	0.99
Erie	K_{red}	=	$0.86\ K_{PAR} + 0.24$	0.96
Ontario	K_{blue}	=	$1.23\ K_{PAR} + 0.04$	0.91
Ontario	K_{green}	=	$0.82\ K_{PAR}$	0.99
Ontario	K_{red}	=	$0.77\ K_{PAR} + 0.28$	0.92

encountered in establishing each of the listed relationships. From the entries in Table 7.3 it is seen that:

1. For the upper Great Lakes the statistical scatter between K_{blue} and K_{PAR} and between K_{green} and K_{PAR} is quite small. The statistical scatter between K_{red} and K_{PAR}, however, is quite large.

2. For the lower Great Lakes the statistical scatter between K_{PAR} and any of the three spectral irradiance band attenuation coefficient values is quite small.

3. The large intercept value of the regressions between K_{red} and K_{PAR} apparent in all the lakes is due to the high absorption of pure water in this wavelength interval.

4. For each lake the slopes between K_{green} and K_{PAR} and between K_{red} and K_{PAR} are generally comparable (and less than unity) but significantly lower than the slope between K_{blue} and K_{PAR} (which is more than unity).

The spectral dependencies of point 4 above, which characterize the attenuation of downwelling PAR in inland waters, are consistent with the spectral distribution of PAR discussed in Section 7.5 and illustrated in Figures 7.15 and 7.16. Therein it was seen that as the value of K_{PAR} increased (or the level of the subsurface irradiance decreased), there was a greater reduction in the blue component of the PAR spectrum than in the red component. This preferential attenuation at blue wavelengths is reflected in the equation entries of Table 7.3 and is a direct consequence of the absorption and scattering cross sections of the suspended inorganic and dissolved organic matter indigenous to non-Case I waters (see Chapter 5). Thus, the additional optical complexity inherent to inland and coastal water regimes tends to modify the spectral penetration from that observed in Case I waters. In oceanic waters maximum penetration of downwelling PAR irradiance occurs at blue wavelengths. In limnological and coastal waters, however, maximum penetration of downwelling PAR irradiance generally occurs at longer wavelengths. This is, of course, consistent with the predominantly blue hue of mid-oceanic waters and the wider range of green to red colors characterizing inland and coastal waters.

7.9   SUBSURFACE SIGHTING RANGE

The *subsurface sighting range* is generally taken to represent the maximum distance at which an object of a given size may be detected underwater by the human eye. As such, the term is highly subjective in nature, and, apart from the eyesight of the submerged viewer, the major factors governing the numerical value of the subsurface sighting range are the optical properties of the water column. Large values of subsurface sighting range are associated with clear waters, while small values are associated with turbid waters. The subsurface sighting range is analogous to the Secchi disk depth, the main distinction being the position of the human observer relative to the air-water interface. Understandably, therefore, the approaches taken to relate sighting ranges to the optical properties of natural waters parallel the approaches taken to relate Secchi disk depths to these optical properties.

Recall from Section 7.2 that the Secchi depth can be related to the hybrid optical property given by the sum of the total attenuation coefficient c and the irradiance attenuation coefficient K. This hybrid optical property ($c + K$) is also most readily related to subsurface sighting range.[93,328,401] Consequently, if the sighting range is reckoned as being measured vertically downward from the air-water interface, the sighting range so determined should be a zeroth-order approximation to the Secchi depth at that location. Such an intercomparison, however, while reasonable, considers non-identical physiological scenarios. The sighting range is usually designed for swimmers and/or divers engaged in visual searches for objects whose presence is anticipated but whose precise whereabouts are unknown (wreckage, shoals, lost articles, etc.). The Secchi disk user, however, is fully aware of the presence of the disk, and faithfully tracks its descent into oblivion. Therefore, the sighting range is physiologically conceptualized for the sudden appearance of a foreign (to the ambient water) object into a field of view, while the Secchi disk is physiologically conceptualized for the gradual disappearance of a foreign object from a field of view.

In addition to the optical properties of the water influencing the sighting range associated with a submerged target, the properties of the submerged object itself influence its ability to be detected.[91,93,94,328,333] Of major importance is the *contrast* of the submerged object to its surroundings. This contrast is a function of both the *reflectivity* of the target, R_T, and the *reflectivity* of the

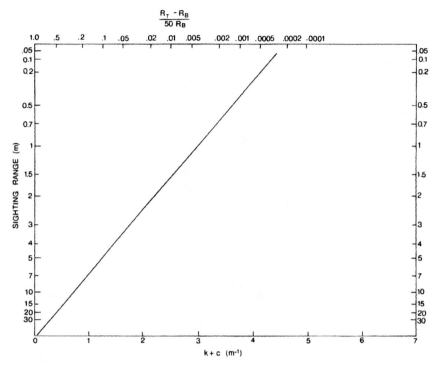

FIGURE 7.22
Nomograph for determination of near-surface, vertically downward sighting ranges for physical objects ≥ 100 cm^2 and reflectivity R_T. (From Bukata, R. P., Jerome, J. H., and Bruton, J. E., *J. Great Lakes Res.*, 14, 347–355, 1988. With permission.)

background, R_B (this background reflectivity is satisfactorily approximated by either the bottom reflectance if the object is at or near the ocean, lake, or river bottom or the volume reflectance if the target is far removed from the bottom). Contrast C_T is then defined simply as:

$$C_T = (R_T - R_B)/R_B \qquad (7.11)$$

Other factors that influence the sighting range associated with a submerged object are the physical size and shape of the object, the direction from which the object is viewed, and the availability of subsurface light (this latter factor being a direct function of the incident radiation).

Duntley[91] and Preisendorfer[328] illustrate how to construct nomographs from which the sighting range appropriate to a specific submerged object may be determined from the optical parameters c, K, R_T, and R_B. Such a nomograph is illustrated in Figure 7.22 for physical objects of projected area ≥ 100 cm^2 viewed vertically downward from the air-water interface. Figure

7.22 is constructed for all lighting conditions between 1 hour subsequent to sunrise and 1 hour prior to sunset. A determination of the surface sighting range, therefore, requires direct measurements of the total attenuation coefficient c and the irradiance attenuation coefficient K. Appropriate values of R_B may be obtained from direct spectroradiometric measurements of the subsurface volume reflectance. The reflectivity of an object, R_T, may be estimated as a number between 0 and 1, depending upon its color and finish. Secchi disks are generally taken to be defined by $R_T \sim 0.70$. Thus, the parameters $(c + K)$ and $(R_T - R_B)/50R_B$ may be readily determined, and a straight line drawn between these values of Figure 7.22 will yield the sighting range (as measured vertically downward from the air-water interface, i.e., a zeroth-order approximation to a Secchi depth if an R_T of ~ 0.70 is considered) as its interception point with the nomograph curve. For objects smaller than 100 cm^2, for non-vertical viewing directions, and for differing conditions of incident radiation, other nomographs suitable to those conditions would be required.

Using the nomograph of Figure 7.22, Bukata et al.[43] estimated the zeroth-order Secchi depths for Lakes Ontario, Erie, Superior, Huron, and for Georgian Bay. Figure 7.23 (taken from their Figure 7) displays these calculated vertical sighting ranges as a function of beam transmission values T (percent transmission for a 1-m pathlength) appropriate to the offshore, near-surface waters of each of the five Great Lakes regions during the summer months. The sighting ranges are seen to vary from as high as 20 m in Lake Superior to as low as 1 or 2 m in Lakes Erie and Ontario.

7.10   THE TROPHIC STATUS OF NATURAL WATER

The suffix -trophy refers to nutrition as well as growth related to nourishment. When applied to natural water bodies the suffix implies a condition resulting from the interplay of aquatic nutrients with aquatic life at a particular place at a particular time. Such an interplay involves a series of biophysical phenomena that are not easy to mathematically formulate let alone enable the establishment of an universal set of criteria from which to classify entire or even segments of natural water bodies. There have been extensive treatises written on classifying the trophic status of natural water [see, for example, the discussions in Hutchinson[181,182] and Vollenweider[416,417]]. These have resulted in a somewhat qualitative lake trophic typology comprised of the terms eutrophy, mesotrophy, and oligotrophy, along with the extended boundary terms hypertrophy and ultra-oligotrophy. Various attempts to ascribe quantitative criteria to this trophic typology, while generating considerable agreement among bio-limnologists, have, nonetheless, not removed the distinctly qualitative aspect of the typology.

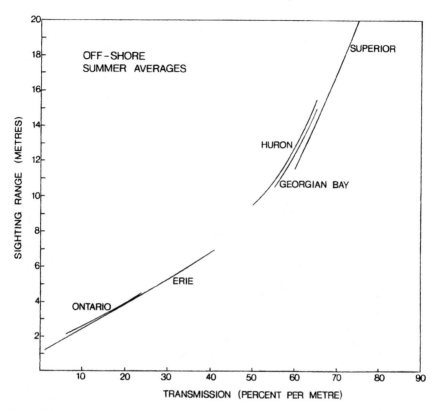

FIGURE 7.23
Near-surface vertical sighting ranges as a function of transmission for summer conditions in offshore waters of the Laurentian Great Lakes. (From Bukata, R. P., Jerome, J. H., and Bruton, J. E., *J. Great Lakes Res.*, 14, 347–355, 1988. With permission.)

Suffice here to say that -trophy is generally (although not universally) considered as a complex duality of epilimnetic water quality and hypolimnetic dissolved oxygen.[180] Near-surface water quality and subsurface dissolved oxygen content are themselves functions of a variety of independent variables such as phosphorus, nitrogen, and chlorophyllous pigment concentrations, phytoplankton, zooplankton, and fish and other aquatic denizen populations, thermal depth profiles, primary production, and the underwater light regime. *Eutrophication* is then defined as the enrichment of waters with plant nutrients (primarily phosphorus and nitrogen) leading to enhanced plant growth (algae and macrophytes) that generally results in some combination of visible surficial algal blooms, floating and/or submerged patches of algae or macrophytes, and increased benthic algae. Decay of this enhanced plant growth can be accompanied not only by the depletion of dissolved aquatic oxygen, but also by the release of a variety of undesirable

TABLE 7.4

Boundary Value Quantification of Trophic Status of Lakes and Reservoirs

Trophic Category	Total P (mg per m³)	Total Chl (mg per m³)	Secchi Depth (m)	Oxygen (% saturation)
Ultra-oligotrophic	0–3.5	0–0.8	>10	90
Oligotrophic	3.5–10	0.8–2.5	6–10	80
Mesotrophic	10–30	2.5–6.0	3–6	40–80
Eutrophic	30–90	6.0–18	1–3	0–40
Hypertrophic	>90	>18	<1	0–10

substances such as CO_2, H_2S, CH_4, corrosive gases, and toxins. The threats of eutrophication to the plant, animal, and human life dependent upon an aquatic basin for sustenance, therefore, can be quite dramatic. A water body displaying such an enhanced nutrient and concomitant enhanced plant growth is termed *eutrophic*, a water body displaying low nutrient concentrations and/or plant growth is termed *oligotrophic*, and the term *mesotrophic* describes an intermediate scenario. Although the impacts and reaction rates of nutrient dynamics are dependent upon climatological, limnological, and oceanological parameters and processes, the symptoms and manifestations of eutrophy as sketched above apply to all waters marine or fresh.

Nutrient/plant growth criteria demarcating these water bodies rely upon subjective assessment as to what parameters should be considered as eutrophy-limiting. Thus, such criteria can be somewhat arbitrary. Nevertheless, criteria suggested by various workers display a remarkable degree of agreement for encapsulated or mid-oceanic waters (i.e., waters with relatively long residence times. No criteria have, as yet, been established for running waters). Criteria based upon concentrations of total phosphorus and total chlorophyll, Secchi depth, and hypolimnetic oxygen have been suggested and/or reviewed by Dobson,[86] Chapra and Dobson,[60] and Vollenweider,[416,418] among others. Such criteria have been, of necessity, quantified on the basis of attempts to mathematically define the complex inter-relationships amongst the possible eutrophy indicators. In many cases, these relationships have been postulated on the basis of anticipated behaviour patterns, in other cases on the basis of mathematical regressions among parameters whose interrelationships might be more complex than such regressions would suggest. These trophic classification criteria are summarized in Table 7.4.

Despite the above caveats, the tabulated criteria have been independently assessed in terms of data obtained from hundreds of lakes and reservoirs in North American, European, Asian, Australian, South and Central American countries gathered from the late 1960s to the early 1990s, and serve as a reasonable guide to the trophic stature of inland waters, at least insofar as the phosphorus and chlorophyll limits are concerned. As we have been discussing in this chapter, the limitations of Secchi depths as an indicator of water quality must be fully recognized when suggesting such readings be a global criterion of eutrophy (this point will be briefly touched upon in Section 7.11). Two very important eutrophy parameters are excluded from Table 7.4, namely primary production and epilimnetic particulate nitrogen concentrations. Primary production, as will be discussed in Chapter 9, is related to the epilimnetic chlorophyll profile and the use of primary production as a trophic status criterion has been discussed by Chapra and Dobson,[60] Dillon and Rigler,[83] Vallentyne,[407] Vollenweider et al.,[419] and others, the general consensus being that primary productions $< \sim 145$ g Carbon per m^2 per yr characterize oligotrophic lakes, while primary productions $> \sim 240$ g Carbon per m^2 per yr characterize eutrophic lakes. Over the past several decades, scientific and public focus has been directed toward the nutrient phosphorus, a consequence of well-documented excesses of phosphates having been injected into inland and coastal water systems and the fact that phosphorus is generally the limiting factor in offshore algal growth. Schindler[353] has speculated that as a lake is increasingly enriched with phosphorus it will ultimately reach a state at which it becomes nitrogen limited. In such a state the biota indigenous to the lake may change (i.e., nitrogen-fixing algae may gain ascendancy over phosphorus-fixing algae in the algal hierarchy) to accommodate such a nutrient excess/shortage. Thus, the nitrogen-to-phosphorus ratio assumes prominence as a lake eutrophy modulator. If phosphorus abatement programs are successfully implemented, however, the threshold of this ratio is of minor importance to most eutrophication concerns. Thus, despite the fact that each of the Laurentian Great Lakes at present contains an excess of nitrates, this excess does not necessarily translate into a stimulus for excessive aquatic plant growth. Understandably, therefore, considerably less attention has been directed toward the removal of nitrogen from natural waters. Dobson,[87] however, suggests that the approximate range of particulate nitrogen for mesotrophic waters is 50 to 150 mg per m^3.

The above brief discussion of trophic status of inland water bodies has been consistent with the simplistic view that eutrophy is bad and oligotrophy is good. Indeed, such a view does possess credence. With increased productivity comes increased oxygen depletion and increased probability of the emergence of a variety of pernicious by-products. Irrefutable evidence exists on a global scale of the impairments such eutrophy has imposed upon drinking water, recreational areas, human and animal health, and ambient environmental ecosystems. However, as in all environmental situations, there are many shades of grey separating black from white. Oligotrophy (and even more dramatically ultra-oligotrophy) very often represents a near-barren aquatic resource, and it is quite easy to defend increases in lake productivity, in terms of increased fish populations, under such conditions. Lake Superior and Lake Tahoe are examples of such barren lakes that could perhaps benefit from controlled increased productivity. The artificial enrichment of ponds and small lakes to increase fish production has been an accepted practice in Asia and Europe, and to a somewhat lesser extent in North and Central America for decades [see Hasler[163]], such fertilization being implemented either through direct injection of appropriate plant nutrients or through basin drainage avenues. Of course, such an intentional enhancement of eutrophy above the ambient edaphic eutrophy can produce the same deleterious environmental impacts as an unintentional enhancement. Natural eutrophication is a slow process and is largely a function of morphometric change. Cultural eutrophication proceeds more rapidly and is largely dictated by the increases in nutrient load. Thus, predicting the long-term effects of a controlled increase in lake productivity can become a complex problem in applied bio-limnology.

Since the subsurface light regime plays a principal role in the monitoring of near-surface water quality and a governing role in aquatic primary production, one of the directives of both *in situ* and remote optical measurements has been to monitor and assess spatial and temporal variations in aquatic productivity, and therefore, corresponding changes in the trophic status of natural water bodies. To date there exist no algorithms or methodologies that will retrieve phosphorus, nitrogen, or oxygen concentrations from a recorded remote optical spectrum. Work is in progress at a number of oceanographic institutions to remotely distinguish algal types indigenous to Case I waters in terms of their chlorophyllous pigments. Such algal distinction for remote sensing of non-Case I waters, however, while perhaps theoretically possible, is as yet an unconquered challenge. Thus, to assist in the evaluation of changes in the trophic status of inland and coastal waters, remote sensing must content itself with monitoring aspects of near-surface water quality (e.g., simultaneous estimates of the organic and inorganic component concentrations). Subsurface oxygen conditions would have to be obtained either from direct sampling techniques, from historical trend data, or somehow inferred, if possible, through knowledge of the organic/inorganic component concentrations of the near surface layer.

7.11   OPTICS AND THE STATUS OF THE GREAT LAKES

The data presented in this chapter have illustrated that a time sequence of optical measurements can provide cryptic clues as to the evolving status of a natural water body. It is in this regard that, despite its obvious shortcomings in scientific rigor, the Secchi disk must with some trepidation and some reluctance be given its qualified due. The continuous Secchi depth history coupled with the less continuous transmissometry history and the somewhat sporadic spectroradiometric history of the Great Lakes has documented an optical evolution which, when used in conjunction with direct sampling and physical, chemical, and biological models, generates a relatively unambiguous documentary on the natural and anthropogenic uses and abuses of the Great Lakes waters, as well as the consequences of attempts to restore these waters to a pre-stressed state. Optical measurements have contributed to defining the trophic status of inland waters, the primary productivity and irradiation characteristics of natural waters, as well as temporal changes in aquatic composition. Unanticipated changes in the clarity of Great Lakes waters have resulted in scientific focus being quickly directed toward aquatic regions at risk. Such risk agents vary from point source contaminant injection and long-range transport of atmospheric pollutants circa 1950s and 1960s, to the direct invasion of exotic nuisance species such as the zebra mussel (*Dreissena polymorpha*) circa 1986 (Hebert et al.,[166] Mackie et al.,[253]).

The optical history of Lake Erie over the past three decades or so (Charlton et al.[62]) illustrates not only the evolution of Lake Erie, but also the cryptic nature of measurements of aquatic clarity. Water transparency in the western basin of Lake Erie as suggested from Secchi depth measurements performed routinely by NWRI corresponded to Secchi depths < 1 meter circa 1970. These low transparencies (along with data obtained from laboratory analyses of collected water samples) were readily interpreted as a consequence of both the suspended sediments and the highly *eutrophic* conditions anthropogenically induced within the Lake Erie ecosystem.

Vigorous bi-national programs of phosphorus loading reduction have enabled Lake Erie to recover to a mesotrophic ecosystem (as indicated by western basin Secchi depths of 2–3 meters circa 1985). With this increase in aquatic clarity have come the beneficial environmental impacts of decreased nuisance algae and increased healthy fish communities. A general direct relationship, therefore, appears to exist between an increasing Secchi depth and a decreasing trophic status.

About the time of the invasion of the exotic zebra mussel (native to the Caspian Sea), the western basin of Lake Erie had regained mesotrophic status. The zebra mussels confined their presence to the upper 20 meters of the water column, and by 1990 had colonized hard substrate substances such as rocks, piers, boat hulls, and municipal and industrial intake pipes. In 1991 it was discovered that a second similar looking but genetically distinct quagga mussel (*Dreissena bugensis*) had also invaded the Lake Erie system. Unlike the zebra mussel, however, the quagga mussel colonized sand, silt, and mud. As a consequence of the combined presence of the mussels at essentially all depths, vast quantities of suspended sediments and phytoplankton have been and are being removed from the water column and some benthic species have disappeared. Changes in the food chain have manifested as reductions in zooplankton populations and changes in indigenous fish stocks. Secchi depths in some areas of the western basin of Lake Erie have increased from pre-invasion values of (2–3) meters to post 1991 values of (3–5) meters. With this enhanced clarity and impending oligotrophy in Lake Erie comes the fears of the negative impacts associated with increased abundances of nuisance algae and diminished health of fish and crustacean communities.

Similar changes are being observed in Lake St. Clair, which, due to its heavily sediment-laden character consistently displays Secchi depths in the eutrophic to mesotrophic range. As the waters become clearer, aquatic macrophytes return, and the possibility of adverse effects on the fish stocks (e.g., turbidity-loving walleye) exists.

The above problem of exotic species disruption of the Great Lakes ecosystem is currently a priority concern of the International Joint Commission Council of Great Lakes Research, which is encouraging cooperative studies to define the specific nature of the aquatic conditions and thereby inaugurate appropriate remedial action.

Chapter 8

REMOTE SENSING OVER NATURAL WATER

8.1 UPWELLING RADIATION THROUGH THE AIR-WATER INTERFACE

As we have suggested in the *Gedanken* experiment of Section 1.7, the remote sensing of natural water bodies can be likened unto the mental transference of the remote sensing device in a systematic manner to each of the three locations: (1) an orbit just beneath the air-water interface; (2) an orbit just above the air-water interface; (3) an orbit at the satellite or aircraft altitude. Immediately beneath the air-water interface the key parameter is the subsurface volume reflectance spectrum $R(0^-,\lambda)$. Immediately above the air-water interface the key parameter is the upwelling radiance spectrum $L_u(0^+, \lambda, \theta_V, \phi_V)$. θ_V and ϕ_V are the are the zenith and azimuth viewing angles of the remote sensing device. At the remote sensor (altitude h above the air-water interface), the key parameter is the recorded upwelling radiance spectrum $L_u(h, \lambda, \theta_V, \phi_V)$. Interpretation of this radiance spectrum depends upon converting $L_u(h, \lambda, \theta_V, \phi_V)$ into $R(0^-,\lambda)$ and $R(0^-,\lambda)$ into co-existing aquatic concentrations.

The objective of a remote sensing mission (utilizing the optical range of wavelengths) is generally considered to encompass one or more of the following goals:

1. Delineation, at a fixed moment in time, of the spatial aquatic patterns that manifest from the recorded upwelling radiance data.
2. Collection and storage of such spatial data and relation of near-surface aquatic patterns to known or assumed patterns of ancillary sea-state and/or meteorological data.
3. Extraction from the recorded radiance spectra one or more of the co-existing organic and inorganic aquatic components residing within the penetration depth of the aquatic regime.
4. Use of these near-surface component concentrations to infer water column processes such as primary production and carbon budget information in Case I and non-Case I waters (as will be discussed in Chapter 9).

A portion of goals 1 and 2 may be approached from a purely descriptive and/or statistical point-of-view and, thus, can be accommodated by remotely sensed imagery that might be normalized by photometric techniques. Goals 3 and 4, however, require precise absolute (as opposed to relative) numerical values of the radiance and volume reflectance spectra, plus precise spectral values of the atmospheric and aquatic apparent and inherent optical properties. As we have discussed throughout the previous chapters, the transference of volume reflectance and radiance spectra to various altitudes and/or depths of attenuating media involve the concentration distributions and specific absorption and scattering properties of the constituents of the attenuating media.

Consider the transference of an upwelling subsurface irradiance spectrum, $E_u(0^-,\lambda)$, through the air-water interface, and its contribution to the upwelling radiance spectrum, $L_u(0^+, \lambda, \theta_V, \phi_V)$.

$$L_u(0^+, \lambda, \theta_V, \phi_V) = f_1 E_{sky}(\lambda) + f_2 E_{sun}(\lambda, \theta) + E_u(0^-, \lambda)T_{surf}/Q \quad (8.1)$$

where

$E_{sun}(\lambda,\theta)$ = the downwelling direct irradiance from the sun located at a solar zenith angle θ,

$E_{sky}(\lambda)$ = the downwelling diffuse irradiance from the sky,

Q = ratio of the upwelling irradiance below the water surface to the upwelling nadir radiance below the water surface, i.e., $Q = E_u(0^-, \lambda)/L_u(0^-, \lambda, 0°, \phi_V)$,

T_{surf} = transmission of nadir radiance through the air-water interface including correction for the n^2 radiance law (n is the relative index of refraction),

f_1 = ratio of the upwelling radiance entering the field-of-view of the remote sensing device (originating from diffuse skylight reflected from the surface) to the downwelling sky irradiance, and

f_2 = ratio of the upwelling radiance entering the field-of-view of the remote sensing device (originating from direct sunlight reflected from the surface) to the downwelling solar irradiance.

For simplicity of notation we will on occasion dispense with the λ, θ, and ϕ dependencies of the atmospheric and aquatic terms. Such dependencies, however, are implicit. Rearranging equation (8.1) to solve for the upwelling subsurface irradiance, $E_u(0^-)$, yields:

$$E_u(0^-) = Q[L_u(0^+) - f_1 E_{sky} - f_2 E_{sun}]/T_{surf} \quad (8.2)$$

The downwelling irradiance just below the air-water interface, $E_d(0^-)$, is given by:

$$E_d(0^-) = f_3 E_{sky} + f_4 E_{sun} \tag{8.3}$$

where f_3 = fraction of the downwelling diffuse sky irradiance that is transmitted into the water, and

f_4 = fraction of the downwelling direct solar irradiance that is transmitted into the water.

The volume reflectance spectrum just below the air-water interface is then given as the ratio of equations (8.2) and (8.3):

$$R(0^-) = E_u(0^-)/E_d(0^-)$$

$$= \frac{Q[L_u - f_1 E_{sky} - f_2 E_{sun}]}{T_{surf}[f_3 E_{sky} + f_4 E_{sun}]} \tag{8.4}$$

The fraction f_1 may be expressed as:

$$f_1 = \alpha \rho_0 + \beta_1 \tag{8.5}$$

where α = the spectrally dependent ratio of downwelling zenith sky radiance to downwelling sky irradiance, i.e., $L_{sky}(0°)/E_{sky}$,

ρ_0 = Fresnel reflectivity for vertical incidence, and

β_1 = ratio of the upwelling radiance entering the field-of-view of the remote sensing device (as a result of sky irradiance being reflected by surficial waves) to the downwelling diffuse sky irradiance.

For wind speeds in the range $(0 - 10)$ m s^{-1}, the value of β_1 can be considered as zero.[142] ρ_0 for a relative index of refraction of 1.341 has a numerical value of 0.0212. Thus, equation (8.5) can for most situations be expressed as:

$$f_1 = 0.0212\alpha \tag{8.6}$$

From Maul,[262] the radiance reflected from a water surface due to wave action $(f_2 E_{sun})$ may be written:

$$f_2 E_{sun} = \frac{\rho(\Omega)E_{sun}\exp[-\tan^2\delta/S^2]}{4\pi S^2 \cos\psi\cos^4\delta}$$

from which the fraction f_2 may be expressed as:

$$f_2 = \frac{\rho(\Omega)\exp[-\tan^2\delta/S^2]}{4\pi S^2 \cos\psi\cos^4\delta} \tag{8.7}$$

where Ω = the angle of incidence resulting in E_{sun} being reflected into
               the field-of-view of the remote sensing device,

   $\rho(\Omega)$ = Fresnel reflectivity for incident angle Ω,

   ψ = zenith angle of the remote sensing device ($0°$ for a sensor
               that is looking vertically downward),

   δ = wave slope resulting in observable sunglint by the remote
               sensing device, and

   S^2 = mean square wave slope

       = $0.003 + (0.512 \times 10^{-2} \times$ wind speed).

For ambient logistical monitoring conditions such as a range of solar zenith angles $40°–65°$ (corresponding, for nadir viewing, to Ω values in the range $20°–33°$ for wind speeds up to 5 m s^{-1}), values of f_2 are in the range 8×10^{-8} to 7×10^{-4} (i.e., f_2 is negligible and sunglint is not a problem). As discussed at length in Chapter 3, the judicious choice of sensor geometry, time of overflight, sea-state conditions, and direction of viewing can combat the purely geometric problem of mirror reflection (sunglint) from the air-water interface.

For waters displaying small to moderate concentrations of suspended sediments, it is reasonable to assume that the volume reflectance in the near-infrared wavelength band is near-zero. For this condition the above-surface upwelling infrared radiance $L_u(ir)$ is written:

$$L_u(ir) = f_1 E_{sky}(ir) + f_2 E_{sun}(ir) \tag{8.8}$$

from which the wave fraction f_2 may be expressed as:

$$f_2 = [L_u(ir) - E_{sky}(ir)\rho_0\alpha(ir)]/E_{sun}(ir) \tag{8.9}$$

where $L_u(ir)$, $E_{sky}(ir)$, $E_{sun}(ir)$, and $\alpha(ir)$ are the appropriate near-infrared band values for these parameters. However, equation (8.9) is not valid for heavily sediment laden water bodies such as most rivers, estuaries, littoral zones, and lakes such as Lake St. Clair and western Lake Erie.

The fraction f_3 may be expressed as:

$$f_3 = 1 - (\rho_{sky} + \beta_3) \tag{8.10}$$

where ρ_{sky} = Fresnel reflectivity of a uniformly diffuse sky irradiance from
               a flat air-water interface and

   β_3 = fraction of sky irradiance that is reflected by surficial waves.

Values of the sky irradiance fraction β_3 are related to the wind speed, e.g., for a wind speed of 4 m s^{-1}, $\beta_3 = -0.010$ (Payne[305]) and for a wind speed of 7 m s^{-1}, $\beta_3 = -0.014$ (Cox and Munk[68]). From Jerlov[198], $\rho_{sky} = 0.066$. For many remote sensing situations $0.93 \leq f_3 \leq 0.95$.

The fraction f_4 may be expressed as:

$$f_4 = 1 - [\rho(\theta) + \beta_4] \tag{8.11}$$

where $\rho(\theta)$ = Fresnel reflectivity of direct solar irradiance (solar zenith angle θ) from a flat air-water interface and

β_4 = fraction of direct solar irradiance that is reflected due to surficial waves.

Cox and Munk[68] have expanded the term $[\rho(\theta) + \beta_4]$ in a mathematical series, $\rho_w(\theta)$, of the form:

$$\rho_w(\theta) = \rho(\theta) + \beta_4$$

$$= \rho(\theta)\{{}^1/_2[1 + I(k)] + {}^1/_2\pi^{-1/2}aS \cdot \exp(-k^2)$$

$$+ {}^1/_4 bS^2[1 + I(k)] - 2\pi^{-1/2}k \cdot \exp(-k^2) + {}^1/_4 cS^2[1 + I(k) \tag{8.12}$$

$$+ \cdots\cdots$$

where

$$S^2 = \text{mean square wave slope,}$$

$$k = (2S)^{-1}\cot\theta,$$

$$I(k) = 2\pi^{-1/2} \int_0^k \exp(-t^2)dt,$$

$$F = \rho(\theta)\cos\theta,$$

$$a = \frac{1}{F}\frac{dF}{d\theta},$$

$$b = \frac{1}{2} + \frac{1}{2F}\frac{d^2F}{d\theta^2},$$

and

$$c = \frac{1}{2} + \frac{\cot\theta}{2F}\frac{dF}{d\theta}.$$

Thus,

$$f_4 = 1 - \rho_w(\theta) \tag{8.13}$$

Substituting the expressions for the fractions $f_1, f_2, f_3,$ and f_4 into the subsurface volume reflectance expression of equation (8.4) yields:

$$R(0^-) = \frac{Q\left[L_u - (\alpha\rho_0 + \beta_1)E_{sky} - \dfrac{\rho(\Omega)\exp(-\tan^2\delta/S^2)E_{sun}}{4\pi S^2\cos\psi\cos^4\delta}\right]}{T_{surf}\{[1 - \rho_{sky} - \beta_3]E_{sky} + [1 - \rho_w(\theta)]E_{sun}\}} \tag{8.14}$$

Equation (8.14) is the general expression for the volume reflectance spectrum R immediately beneath the air-water interface (depth $z = 0^-$) in terms of the Fresnel coefficients, sea-state, global radiation, time-of-day, and the upwelling radiance spectrum just above the air-water interface (depth $z = 0^+$). Equation (8.14) thus defines the transference of an upwelling radiance spectrum just above the air-water interface into a subsurface volume reflectance spectrum or *vice versa*. To use equation (8.14), however, requires either reliable previously determined or extant values of the essential parameters Q, α, and $\rho_w(\theta)$. Recapping, α is the spectrally dependent ratio of the downwelling zenith sky radiance to the downwelling sky irradiance. In a remote sensing mission, therefore, both an upward-viewing radiance sensor as well as an all-sky (exclusive of the direction of the sun, θ) irradiance sensor should constitute part of the suite of optical sensors. The parameter $\rho_w(\theta)$ is the solar irradiance reflection term given by the series expansion of equation (8.12). The parameter Q is the ratio of the upwelling irradiance to the upwelling nadir radiance immediately below the air-water interface. Bukata et al.[44] utilized Monte Carlo simulations of radiative transfer to estimate Q as a function of solar zenith angle θ over a range of absorption and scattering coefficients applicable to non-Case I waters. Figure 8.1 is taken from Bukata et al.[44] (their Figure 2) and illustrates their estimated Q values (upwelling irradiance $E_u(0^-)$ divided by the upwelling radiance $L_u(0^-)$ for an 11° field-of-view) as a function of θ. Curve A represents a water body with a relatively high backscattering probability of 0.044. Curve B represents a water body with a relatively low backscattering probability of 0.013. Curve C is the arithmetic mean of Curves A and B. The ratio of backscattering to absorption was observed to not have a significant impact on Figure 8.1. The hazard of selecting a fixed value of Q for a remote sensing mission that covers a substantial period of the day is evident.

A value of T_{surf} of 0.544 for relative refractive index of 1.341 is given in Duntley et al.[95] For many remote sensing situations and locations, the general equation (8.14) can reduce, with minimal impact, to the simplified equation

$$R(0^-) = \frac{Q[L_u(0^+) - 0.0212\alpha E_{sky} - f_2 E_{sun}]}{0.544(0.944E_{sky} + f_4 E_{sun})} \tag{8.15}$$

The composition of global radiation varies throughout the day (i,.e., the time dependency of F, the fraction of diffuse sky E_{sky} to total global irradiance). Thus, both remotely collected continuous and contiguous data streams contain data sets that are not directly comparable. To achieve internal consistency

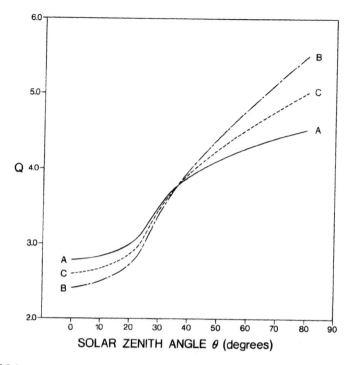

FIGURE 8.1

Monte Carlo estimations of Q (ratio of upwelling irradiance immediately below the air-water interface to upwelling nadir radiance immediately below the air-water interface) as a function of solar zenith angle λ for large backscattering probability (Curve A) and for low backscattering probability (Curve B), with Curve C being the arithmetic mean. (From Bukata, R. P., Jerome, J. H., and Bruton, J. E., *Remote Sens. Environ.*, 25, 201–229, 1988. With permission.)

within the data stream, it is necessary to normalize the subsurface volume reflectance data inferred from a remotely measured upwelling radiance spectrum to a standard subsurface volume reflectance spectrum observed under a standard global radiation spectrum. If the diffuse fraction of the global radiation is F, the direct fraction of the global radiation is $(1 - F)$, and the subsurface volume reflectance just beneath the air-water interface, $R(0^-, \theta, F)$, may be written:

$$R(0^-, \theta, F) = (1 - F)R_{sun}(0^-, \theta) + FR_{sky}(0^-) \qquad (8.16)$$

where $R_{sun}(0^-, \theta, F)$ = the volume reflectance just beneath the air-water interface resulting from the direct fraction of the global radiation and

$R_{sky}(0^-)$ = the volume reflectance just beneath the air-water interface resulting from the diffuse fraction of the global radiation.

Bukata et $al.,$[44] using Monte Carlo simulations, illustrate that the volume reflectance $R_{sun}(0^-, \theta)$ for direct solar radiation downwelling at solar zenith angle θ is related to the volume reflectance $R_{sun}(0^-, 0^\circ)$ for direct solar radiation from a zenith sun ($\theta = 0^\circ$) via the cosine of the in-water refracted angle θ_w, viz.,

$$R_{sun}(0^-, \theta) = R_{sun}(0^-, 0^\circ)/\cos\theta_w \qquad (8.17)$$

where the in-water refraction angle θ_w is the angle whose sine is given from Snell's Law as $(\sin\theta)/n$.

Further, they determine $R_{sky}(0^-)$ and $R_{sun}(0^-, 0^\circ)$ to be related via the linear equation

$$R_{sky}(0^-) = 1.165R_{sun}(0^-, 0^\circ). \qquad (8.18)$$

Substituting equations (8.17) and (8.18) into equation (8.16) and rearranging terms yields:

$$R_{sun}(0^-, 0^\circ) = R(0^-, \theta, F)/[1.165F + (1 - F)(\cos\theta_w)^{-1}]. \qquad (8.19)$$

Equation (8.19) then determines the subsurface volume reflectance due to a zenith sun ($\theta = 0^\circ$) as a function of the volume reflectance for the extant global radiation.

As discussed in Bukata et $al.$[39] for the usual sun and viewing directions encountered in Landsat overpasses of the Canadian terrain [solar zenith angle $\theta < 40^\circ$, nadir angle of viewing $\theta_v < 40^\circ$ resulting in Fresnel reflectances $\rho \sim 0.02$] and assuming a flat air-water interface, equation (8.15) can be further approximated. For this case the relationship between the upwelling radiance L_u immediately above the air-water interface and the volume reflectance R immediately below the air-water interface can be expressed at any wavelength λ as:

$$R(0^-) = \frac{L_u - 0.02L_{sky}}{0.48L_u - 0.0096L_{sky} + (0.9604E_{sky} + 0.9153E_{sun})/\pi n^2} \qquad (8.20)$$

where once again it is seen that (due to the parameter Q) to transform the upwelling radiance spectrum immediately above the air-water interface into the subsurface volume reflectance spectrum that would be recorded immediately beneath the air-water interface, values of solar and sky irradiance spectra [$E_{sun}(\lambda)$ and $E_{sky}(\lambda)$] are required as well as the downwelling zenith sky radiance spectrum $L_{sky}(\lambda)$ at the time of the remote sensing overpass.

To transform the radiance spectrum observed at the satellite or airborne sensor at altitude h [i.e., $L_u(h, \lambda, \theta_v, \phi_v)$] into the radiance spectrum observed just above the air-water interface [i.e., $L_u(0^+, \lambda, \theta_V, \phi_V)$] requires removal of the attenuation and amplification produced by the intervening atmosphere.

8.2   ACCOUNTING FOR ATMOSPHERIC INTERVENTION

In Section 8.1 we have essentially illustrated the manner in which the first two components of our *Gedanken* approach to remote sensing over natural water could be accommodated. The subsurface volume reflectance spectrum R (to which a remote sensing device orbiting just beneath the air-water interface would respond) can be converted into the upwelling radiance spectrum $L_{up}(0^+)$ (to which a remote sensing device orbiting just above the air-water interface would respond). The third component, however, namely the relating of $L_{up}(0^+)$ to the upwelling radiance spectrum recorded at the height h of the remote sensing device $L_{up}(h)$ is extremely problematical.

Many conscientious efforts to remotely sense aquatic targets have been stymied by atmospheric intervention. Scientists have been aware since prior to the pre-planning stages of ERTS-1 that the removal of a varying atmospheric signal from a varying (and generally low intensity) optical return from natural water bodies is not a trivial task. In fact, we must admit, somewhat shamefully, that in much of our own research work we have deliberately avoided the need to atmospherically correct the collected data by performing remote measurements close to the air-water interface. One example of such low level remote sensing is described in Bukata et al.[44] wherein a shipboom-mounted dual radiometer reflectance system (recording spectral upwelling aquatic radiation and spectral downwelling global radiation) was used to conduct a continuous remote lakewater quality survey from a moving research vessel.

As discussed in Chapter 2, upon encountering the Earth's atmosphere the extraterrestrial solar radiation abruptly departs from its constant irradiance value E_{solar}. Absorption and scattering by the atmospheric gases, water vapor, aerosols, and clouds result in a global radiation being observed at a point on the Earth's surface that is comprised of a direct component E_{sun} and a diffuse component E_{sky}. Both E_{sun} and E_{sky} display temporal and spatial variations that are governed by temporal and spatial variabilities within the local atmosphere. Further, the upwelling radiation (as a consequence of both surficial and volumetric returns of E_{sun} and E_{sky}) encounters the atmosphere once again—an atmosphere that may well be governed by a set of component distributions different from that which attenuated the downwelling global radiation.

Not only does the atmosphere reduce the radiance return from the aquatic target, it also contributes backscattered radiation of its own to the upwelling radiance spectrum recorded at the remote sensing platform. In cases of dense aerosol concentrations, the atmospheric radiation can comprise > 95% of the recorded radiance over water at satellite altitudes.

To properly account for the impact of the atmosphere on both the downwelling global radiation and the upwelling reflected radiation requires knowledge of the atmospheric conditions existing at the time and location of the remote sensing mission. As has been mentioned earlier, direct measurements of global radiation, where possible, should be included within the remote sensing protocol.

One early (and very popular) attempt at compensating for atmospheric intervention was to ascribe all the radiation return recorded in a specific satellite wavelength band to atmospheric intervention. The Landsat band thus selected was the (700–800) nm band (also duplicated on the Nimbus-7 Coastal Zone Color Scanner CZCS). The CZCS also possessed a 20-nm wide spectral band centered on the red wavelength 670 nm. This band was also widely considered as the basis for atmospheric correction algorithms, based upon the premise that $R(0^-,670nm)$ = zero, and therefore, all the radiance contained within that band was simply due to surface reflection and molecular and aerosol scattering occurring within the atmosphere. Successes enjoyed by such an atmospheric algorithm approach over Case I waters, however, could not be duplicated over non-Case I waters, since, as shown by Bukata et al.,[37] non-zero subsurface volume reflectances in the range 0.01–0.10 abound within the Great Lakes system. Figure 8.2 illustrates their observations of Lake Ontario volume reflectance $R(0^-,670nm)$ plotted against total suspended minerals during the 1979 field season. Wrigley and Klooster[432] showed that an assumption of zero light exiting from the water at 670 nm was also inappropriate for coastal waters. They observed, however, that the wavelength band centered on 1012 nm was not responsive to a heavily laden sediment plume at the mouth of the Mississippi River. In upcoming satellite programs such as SeaWiFS (Sea-viewing Wide Field-of-View Sensor), bands in the near-infrared wavelength region are being considered for atmospheric correction. This consideration, while certainly alleviating the problem of attributing to the atmosphere an optical return that is actually attributable to the water mass, does not completely remove the ambiguity. Very high concentrations of near-surface suspended solids and/or the presence of extensive near-surface chlorophyll blooms can result in a substantial aquatic return being recorded at near-infrared wavelengths.

Numerous approaches have been considered to provide atmospheric corrections to remotely acquired spectroradiometric data. Some are based upon physical principles, while others are based upon statistical evidence. Since we must derive physical units such as radiance and irradiance from the remotely sensed optical data, we shall here consider the physical approach. To quantify the estimates of atmospheric scattering and absorption using either type of approach, however, requires knowledge of the local atmospheric conditions at or very close to the time of the remote sensing

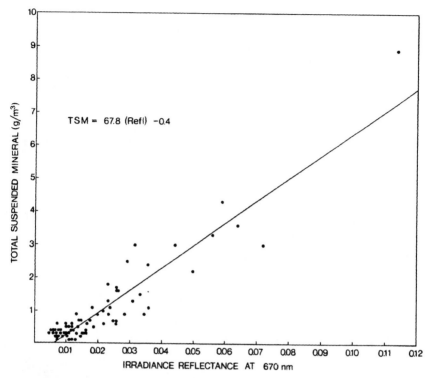

$$TSM = 67.8 \ (Refl) \ -0.4$$

FIGURE 8.2

Correlation between observed irradiance reflectance (volume reflectance) at 670 nm and measured concentrations of total suspended mineral in Lake Ontario. (From Bukata, R. P., Jerome, J. H., and Bruton, J. E., *Remote Sens. Environ.*, 25, 201–229, 1988. With permission.)

activity. Impacts of the Rayleigh scattering from atmospheric gases can be readily predicted since (a) the principles of Rayleigh scattering are well established and (b) nitrogen and oxygen comprise 99% of the atmosphere, the gases are homogeneously mixed, and their concentration profiles with altitude are known. Impacts of Mie (particulate) scattering, however, are highly variable due to the wide variety of aerosols present in the atmosphere and the intractable patchiness of their spatial and temporal distributions. Because of this intractable patchiness, it has become incumbent to measure both aerosol distributions and their optical properties at the time of the remote sensing overpass or flight line. Size, shape, refractive index, and concentration of the aerosol particles determine such radiative transfer parameters as the *aerosol optical depth* (an estimate of the total aerosol concentration, defined by the integration of the vertical atmospheric beam attenuation coefficient over altitude h); the *single-scattering albedo*, defined, as for scattering albedo ω_0 in aquatic optics, by the ratio of the atmospheric scattering coefficient to the atmospheric beam attenuation coefficient); and the *aerosol phase function* (a measure of light scattered by an aerosol particle as a function of angle reckoned from the direction of the incident beam, defined, as in aquatic optics, by the ratio of the volume scattering function to the atmospheric scattering coefficient). Aerosol optical depth and single-scattering albedo are dimensionless quantities, while the phase function has units of sterad^{-1}.

Spanner *et al.*[382] describe a technique for estimating aerosol optical depths using an airborne tracking sunphotometer to calculate total atmospheric optical depth from which is subtracted best-estimates of the optical depths due to Rayleigh scattering and absorption due to ozone and nitrogen dioxide. Mie theory is then used to estimate columnar aerosol densities, single-scattering albedo, and phase functions. The atmospheric optical properties so determined can then be utilized with radiative transfer models [Wrigley *et al.*[431]] to atmospherically correct Landsat Thematic Mapper data and high altitude C-130 aircraft data. Before proceeding with a discussion of some of the existing atmospheric correction models and/or techniques, we will briefly discuss radiative transfer models.

8.3   RADIATIVE TRANSFER MODELS

As mentioned in Section 3.5, detailed discussions of the *radiative transfer equation* which mathematically describes the transfer of energy between an electromagnetic field and its confining medium, are beyond the scope of this book [see Chandrasekhar,[59] however, for an excellent treatise]. Each atmospheric correction technique, by necessity, employs some approximation to or some simulation of the intricacies of radiative transfer.

From Fraser *et al.*,[112] and consistent with the discussions in Section 1.7, the radiance L, at each wavelength λ, that would be recorded by a remote sensor over a surface of uniform reflectance may be written:

$$L = L_p + (T_p E_0 R_s)/(1 - R_s S)\pi \qquad (8.21)$$

where L_p = the atmospheric path radiance, i.e., radiance that does not interact with the air-water interface [the term $L_a(\theta, \phi, \lambda)$ of equation (1.71)]. In discussions to follow, however, L_a will refer to radiance from atmospheric aerosols.

$\quad\;\; T_p$ = the direct transmittance of the atmosphere between the sensor and the air-water interface,

$\quad\;\; E_0$ = the downwelling irradiance above a surface of zero reflectance,

$\quad\;\; R_s$ = the reflectance of the water body, which is assumed uniform, and

S = the fraction of the reflected irradiance that is reflected by the atmosphere back to the surface.

Equation (8.21), while exact, contains terms that are difficult to measure and/or estimate exactly due in large measure to multiple-scattering events. These difficulties make application of equation (8.21) to remote sensing data comprised of millions of pixels precarious. Particularly problematical are the terms S and E_0, as they require measurements from a surface of zero reflectance. Further, equation (8.21) technically is only valid for a Lambertian reflecting surface. These limitations notwithstanding, however, Fraser et al.[112] presented a lookup table for specific sites (pixels), specific viewing geometries (solar, look, and field-of-view angles), and specific atmospheric conditions (optical depths) that could be interpolated for intervening pixels.

Gordon[132] suggested that (a) a single-scattering approximation be used to separate the Rayleigh from the Mie scattering effects, (b) a spectral band containing little or no return from or from below the air-water interface be used to provide a measure of aerosol radiance in that band, and (c) the spectral value of aerosol radiance in that band be used to estimate the aerosol radiance at other wavelengths. A major advantage to this approach is that direct measurements of the aerosol optical depths at the time of remote observations are not required, and such an approach is very often successful for mid-oceanic (Case I) waters. A major disadvantage, as illustrated in Figure 8.2, results from the ubiquitous presence, in inland and coastal (non-Case I) waters, of light of subsurface origin in the wavelength bands generally designated for the solely aerosol contribution.

From Gordon et al.,[141] a single-scattering approach to the Nimbus-7 Coastal Zone Color Scanner is given as:

$$L = L_R + L_a + T_{p'}L_s \qquad (8.22)$$

where L_R = the Rayleigh scattering component of the recorded radiance L,
L_a = the aerosol Mie scattering component of L,
$T_{p'}$ = modified atmospheric transmittance referred to as diffuse transmittance (see below), and
L_s = the reflected surface radiance component of L.

The modified (diffuse) atmospheric transmittance term of equation (8.22) empirically attempts to consider forwardscattering from neighboring pixels into the radiance registered in a sensor above a particular pixel. Direct transmittance, T_p, is given by the atmospheric transparency coefficient expression of equation (2.6):

$$T_p = \exp(-\tau/\mu) \qquad (8.23)$$

where τ = the total optical depth, i.e., the sum of the optical depths associated with Rayleigh and aerosol scattering, and absorption due to water vapor, ozone, and nitrogen dioxide, and

μ = cosine of the zenith angle from the air-water interface to the remote sensor.

The diffuse transmittance $T_{p'}$ of equation (8.22) is similar to T_p except it considers a 50% reduction in the optical depths asociated with the scattering optical depths, i.e.,

$$T_{p'} = \exp\{-[(\tau_R + \tau_a)/2 + \tau_{wv} + \tau_{oz} + \tau_{no}]/\mu\} \qquad (8.24)$$

where τ_R, τ_a, τ_{wv}, τ_{oz}, and τ_{no} are the separated Rayleigh and aerosol scattering optical depths and the separated water vapor, ozone, and nitrogen dioxide absorption optical depths, respectively. The wavelength bands of the Landsat Thematic Mapper were selected to avoid water vapor absorption in its visible bands. However, water vapor absorption impacts near infra-red and thermal bands. The optical depth of nitrogen dioxide is only of consequence in the blue spectral band.

The Rayleigh path radiance L_R can be written[141] as:

$$L_R = (\tau_R E_{solar\lambda} T_{oz} P_R)/(4\pi\mu) \qquad (8.25)$$

where τ_R = the optical depth associated with Rayleigh scattering,

$E_{solar\lambda}$ = the extraterrestrial irradiance at the spectral wavelength or band λ considered,

T_{oz} = the ozone transmittance at the wavelength λ, included to account for the reduction of both the downwelling irradiance and upwelling radiance due to ozone, and

P_R = the phase function appropriate to atmospheric Rayleigh scattering.

The Rayleigh phase function (normalized to 4π) is defined by:

$$P_R = 3(1 + \cos^2\gamma)/4 \qquad (8.26)$$

where γ is the scattering angle between the direction of incoming direct solar radiation and the direction of light scattered into the remote sensing device. Similarly, the Mie scattering aerosol path radiance L_a can be written:

$$L_a = (\omega_a \tau_a E_{solar\lambda} T_{oz} P_a)/4\pi\mu \qquad (8.27)$$

where the subscript a refers to the aerosol equivalent of the Rayleigh parameters of equation (8.25). ω_a is the aerosol single scattering albedo.

Wrigley et al.[431] presented a simplified radiative transfer model using a modified single-scattering approximation wherein the Rayleigh scattering due to atmospheric molecules are distinguishable from the Mie scattering of aerosols. However, the contribution of scattered skylight to the global radiation was ignored. In their work, Wrigley et al. utilized an airborne tracking sunphotometer to derive the aerosol optical depth τ_a. The aerosol scattering albedo ω_o and the aerosol phase function P_a were calculated using the technique of King et al.,[218] whereby aerosol particle size distribution estimates are inferred from multispectral aerosol optical depth measurements. Once an appropriate particle size distribution is obtained and reasonable assumptions (generally based upon either extant observations or a priori knowledge) as to the aerosol particle shape (spherical or elliptical) and refractive index, Mie theory can yield reasonable estimates of ω_o and P_a. The authors showed that their model yielded atmospherically-corrected surface radiances for the green, red, and near-infrared Landsat Thematic Mapper bands that were within a few percent of direct readings from a surface radiometer. The highly susceptible (to atmospheric vagaries) atmospherically corrected blue Thematic Mapper radiances agreed to within 15%. Similar agreements were observed when the simplified radiative transfer model was applied to spectral data collected over water by a Thematic Mapper simulator flown aboard a high altitude C-130 aircraft.

8.4 EXISTING ATMOSPHERIC CORRECTION MODELS

As discussed above, the impacts of the intervening atmosphere upon the radiance spectrum remotely recorded at altitude can be considered through the proper combination of mathematical representations of Rayleigh molecular scattering and Mie aerosol scattering. Single-scattering algorithms, due to their relative simplicity, are in general use. Multiple-scattering algorithms, by virtue of the statistical (albeit within the physical confines of conservation principles) nature of multiple encounters, can involve curve-fitting to Monte Carlo-type simulations. Guzzi et al.[150] incorporated such multiple-scattering considerations into their single-scattering model. The Gordon et al.[141] single-scattering atmospheric correction approach to CZCS data was combined with the Guzzi et al.[150] multiple-scattering approach to atmospherically correct high altitude Airborne Ocean Color Imager (AOCI) data utilizing airborne-tracking sunphotometer measurements [Wrigley et al.[432]].

Perhaps the best known atmospheric correction model is one that has been and is being developed as an on-going updated series termed LOW-TRAN (Kneizys et al.[231]). Utilizing detailed radiative transfer algorithms, the LOWTRAN computer coded program calculates atmospheric transmittance, background atmospheric radiance, single-scattered solar (and lunar) radiance, multiple-scattered solar radiance, direct solar irradiance, and thermal radiance. Molecular spectral line absorption, molecular continuum absorption, molecular scattering, water vapour absorption and scattering, aerosol absorption and scattering, atmospheric refraction, extraterrestrial radiation, Earth curvature, cloud cover and cloud type are among the parameters considered in the estimation of atmospheric impacts on upwelling and downwelling radiation. One of the later in the series, LOWTRAN-7 (Kneizys *et al.*[232]), in widespread use requires input values of air temperature, solar azimuth, land elevation, aerosol density, day of year, cloud profiles, rainfall, atmospheric composition, wind speed, and various other obligatory and optional atmospheric parameters including one called *visibility* and another called *meteorological range*.

The terms *visibility* and *meteorological range*, like the terms *albedo*, *turbidity*, and *clarity*, among others that we have encountered in our discussions, are terms that have, over the years, suffered a variety of obscure and often self-contradictory usages. In the LOWTRAN atmospheric correction models, Kneizys *et al.* use the definitions given in Gordon,[143] namely that *visibility* V is qualitatively defined as the greatest distance at which it is just possible to see and identify an object of a given size with the unaided eye (a definition that harks back to our discussions of Secchi depth), and *meteorological range* V_M is quantitatively defined as a mathematical formulism of visibility given by the relationship:

$$V_M = [\ln(1/\epsilon)]/\beta = 3.912/\beta \qquad (8.28)$$

where β = the total atmospheric extinction coefficient given as the sum of molecular and aerosol extinction and

ϵ = a quantification of threshhold contrast of an object with its background, fixed in the LOWTRAN code at the value 0.02.

The LOWTRAN model utilizes equation (8.28) in its operation. However, most often only the observer's visibility V is available and the LOWTRAN-7 estimates the meteorological range V_M as $(1.3 \pm 0.3)V$. As a further complication, visibility as utilized in the LOWTRAN code is defined horizontally along the ground. Thus, daytime visibility would refer to viewing a dark object against a horizon sky while nighttime visibility would refer to viewing a moderately intense source of known light against a dark horizon sky. This also indicates that horizontal visibility or meteorological range is used to specify aerosol density. Fraser *et al.*[111] argue that concentrations of columnar aerosol and surficial aerosol are not uniquely correlatable for visibilities exceeding 5 km. It is, of course, columnar aerosol density that is required for atmospheric corrections.

Although many of the atmospheric parameters required as inputs to the LOWTRAN model may be either directly measured (as specific targets in controlled research activities) or routinely monitored (components of regional monitoring network activities), many others are not always readily available. This is particularly problematic when historical satellite or airborne data are used in an attempt to obtain temporal information on a particular target site. Consequently, LOWTRAN-7 incorporates a user's choice of six generic model atmospheres based upon specified selections of atmospheric parameters.[7,232] These generic atmospheres include a tropical atmosphere (15°N), a midlatitude summer atmosphere (45°N, July), a midlatitude winter atmosphere (45°N, January), a subarctic summer atmosphere (60°N, July), and a subarctic winter atmosphere (60°N, January), in addition to a 1976 U.S. Standard atmosphere generated from data obtained from the NASA US Standard Atmosphere Supplements of that year. Provision also exists within LOWTRAN-7 for generating custom atmospheric profiles or for modifying the basic generic atmospheres to accommodate specific conditions.

The Full-Width Half-Maximum (FWHM) spectral resolution of the LOW-TRAN model is 20 cm^{-1} in steps of 5 cm^{-1}, from 0 to 5 $\times$ 10^4 cm^{-1} (from 200 nm to infinity). Thus, LOWTRAN-7, when utilized for the case of a remote sensing device looking vertically downward can yield values (within the limitations imposed by horizontal rather than vertical visibility) of the atmospheric transmission T_p between the sensor and the air-water interface; the upwelling radiance due to atmospheric aerosols, L_a; the upwelling radiance due to Rayleigh scattering, L_R; the direct solar irradiance, E_{sun}; and the background sky radiance L_{sky} for any value of λ. This enables conversion of the upwelling radiance recorded at the remote sensing device [L_{up} (h)] into the upwelling radiance that would be recorded just above the air-water interface [L_u (0$^+$)]. However, as seen from equation (8.20), in order to convert L_u (0$^+$) into the volume reflectance just beneath the air-water interface, $R(0^-)$, values of E_{sun}, L_{sky}, and E_{sky} are required. The sky irradiance, E_{sky}, does not emerge from the LOWTRAN model and, hence, remains to be determined by some other means. E_{sky} does not seem an unreasonable parameter to be determined from LOWTRAN-7, and it is certainly possible that LOWTRAN-7 may, at some future time, be modified to yield this parameter. At this time, however, LOWTRAN-7 possesses no such capability (except by integration of sky radiance).

Models such as those presented by Dave et al.[77] and Hatfield et al.[164] readily compute direct and diffuse spectral fluxes at the Earth's surface for custom and/or generic atmospheres. Therefore, values of $E_{sky}(0^+)$ and $E_{sun}(0^+)$ may be determined for any location and any set of atmospheric conditions. Calculations described in Dave[75] form the basis of the user-oriented lookup tables determined from the algorithm of Fraser et al.[112]. As with the LOW-TRAN model, however, the algorithms of the above models calculate the atmospheric impact along a prescribed single path through the atmosphere, and do not consider the literally millions of paths that are associated with a remotely sensed aquatic or terrestrial image.

Due to the complexities that are continually encountered in trying to remove the impacts of a dynamic and perversely anisotropic atmosphere from the comparatively low intensity return from an aquatic surface, water researchers have attempted to simplify the atmospheric correction algorithms. A popular approach to atmospheric correction has been to use deep, transparent lakes to locate a series of pixels (ground areas corresponding to the spatial resolution of the remote sensing device—the remote sensing image is the totality of such pixels) from which the upwelling spectral radiance could be assumed negligible.[2,3] This philosophy is the extended limit of the zero $R(670$ nm) philosophy discussed earlier since it assumes $R(\lambda)$ is zero at all the visible wavelengths. Such an approach then uses the path radiances in these wavelength bands to invert a radiative transfer model to obtain atmospheric effects. The difficulties with such an approach are obvious. First, such completely barren, infertile lakes must exist in the vicinity of the water body being remotely sensed. Second, these lakes must be found and directly verified to be deep (absence of bottom reflectance) and transparent (absence of volume reflectance). Third, unjustifiably ignoring the energy return from the water column can lead to significant errors in both the determination of atmospheric correction over the designated barren lake and the determination of subsurface volume reflectance for the targetted (non-barren) water bodies within the remotely sensed scene.

For several decades considerable effort has been directed toward evaluating the role of remote sensing in estimating both natural and cultivated vegetative crop vigor. It is well known that chlorophyll strongly absorbs visible light and strongly reflects near-infrared light (as illustrated in the optical cross section spectra of chlorophyll). Thus, if a scatter diagram of simultaneously recorded red and near-infrared radiances from a vegetative canopy were plotted, highly vigorous vegetation would cluster into a region of low red radiance return coupled with high near-infrared radiance return. Vegetation of lesser vigor would cluster into a region of higher red radiance return coupled with lower near-infrared return. Such visible/near-infrared clustering techniques have been in widespread use for monitoring agricultural biomass since the "lushness parabola" determinations of terrestrial ground-water flow pathways suggested by Bukata et al.[35,48] and Bobba et al.[25] and the "tasseled-cap" transformations of the four Landsat bands first proposed by Kauth and Thomas.[214] These orthogonal linear transformations yielded sets of brightness, yellowness, greenness, and nonesuch classification indices which were extended by Richardson and Weigand,[342] Dave,[76] Jackson,[186] Gallo and Daughtry,[122] and others to include sets of vegetation indices which could be utilized for computer-based analyses such as soil distinction, crop identification, and terrestrial biomass assays. Such "lushness parabola" relationships between visible and near-infrared radiance returns are applicable to aquatic as well as terrestrial vegetation, and have been instrumental in remotely identifying and monitoring oceanic red tides, seasonal surficial chlorophyll blooms, and emergent vegetation in shallow coastal waters in a quasi-operational manner.

Modifying the "tasseled-cap" brightness indices, Hall et al.[153] selected dark and bright areas of minimum greenness within a satellite image to serve as a basis for defining regions of no or unchanging vegetation. The differences in recorded radiance over such areas on different dates were then assumed to be due to differences in the atmospheric path radiances on those dates. This provided a means of radiometrically normalizing a sequence of satellite visits to a specific target by applying a pixel-to-pixel linear normalization to each scene. Although such normalization would provide a relative as opposed to an absolute intercomparison of satellite observations, the Hall et al. technique could provide a means of intercomparing archival data collected in the absence of ancillary atmospheric information. Natural waters (generally the darkest regions in a satellite scene) that display scene-to-scene changes in subsurface volume reflectance (due to variable sediment loading, near-surface rock outcrops, or reflections from a shallow bottom) can very easily compromise such a normalization technique. However, a major advantage of such a technique [as with the Wrigley et al.[431] approach referred to earlier] is that it provides a treatment for an entire image, and thus incorporates the myriad of downwelling radiation paths converging at a particular point on the surface.

Other atmospheric correction techniques are to be found in Slater et al.[371] and Holm et al.,[173] which utilize ground and aircraft (including sunphotometry) reflectance measurements to calculate aerosol path lengths; in Herman and Browning,[167] which utilize "raw" (i.e., satellite data prior to any onboard signal conditioning such as geometric registration) Thematic Mapper data to calculate a radiative transfer code; in Diner and Martonchik[84] and Kaufman and Sendra,[213] and many others. The atmospheric correction algorithms developed by Freemantle et al.[113] are used to provide in-flight adjustments to the high spectral resolution Compact Airborne Spectrographic Imager (CASI).

8.5 A THEMATIC MAPPER ILLUSTRATION

The Multispectral Scanners (MSS) originally launched aboard Landsat-1 (formerly ERTS-1) and duplicated on Landsats 2–5 have provided earth observations since 1972. The Landsat series of environmental satellites comprise sun-synchronous, near-polar, Earth orbiters. The MSS is a broadband four-channel radiometer that records (at $\sim$80 m $\times$ $\sim$60 m spatial resolution on a 16–18 day repetition cycle) reflected radiation in the wavelength intervals (0.5–0.6) μm; (0.6–0.7) μm; (0.7–0.8) μm; and (0.8–1.1) μm. These data have been extensively used for monitoring changes on the Earth's surface (primarily land features, explaining the name change from ERTS to Landsat). Since terrestrial change was the desired commodity in many of the satellite-oriented monitoring activities, the importance of absolute radiometric calibration of satellite multispectral scanning devices could be quasi-justifiably ignored. A major component of such terrestrial change studies involves the use of both supervised (based upon known or assumed spectral signatures of environmental parameters) and unsupervised (based upon no known or assumed spectral signatures) classification techniques in which similar multiband spectral signatures are clustered into synoptic patterns to facilitate terrain feature identification as well as the spatial and temporal variability of those features. Consequently, much of the Landsat work has, over the years, ignored the actual physical units of the recorded satellite data, utilizing instead the integer values of the digital data tapes without the need for a consideration of the calibration coefficients which would transform these integer values into radiance values. As an indication of this lack of radiometric definition, the parameter represented by such integer values stored at the satellite was extensively referred to in the literature by the somewhat enigmatic term *apparent radiance*. Since reliable values of subsurface volume reflectance are mandatory to the estimation of organic and inorganic concentrations in natural waters, these calibration coefficients cannot be ignored in water quality studies.

With the launch of Landsat-4 in 1982, Landsat Thematic Mappers (TM) have enabled the collection of environmental data in more spectral bands of higher spatial resolution and higher precision than are possible with the MSS. The Thematic Mapper is a multispectral scanning device that records (as integers that are proportional to) upwelling radiances in six reflective energy bands in addition to an emissive thermal band. The six reflective spectral bands have a spatial resolution of 30 m, while the thermal band has a spatial resolution of 120 m.

Subsequent to its upward transmission through the atmosphere, the spectral radiance exiting the air-water interface is registered from the optical sensors aboard the satellite and ultimately recorded as digital numbers N_j, where $1 \leq j \leq 6$ represents the jth spectral band of the TM. TM radiances L_j (in units of W m^{-2} sr^{-1}) are proportional to N_j, and, as discussed by Markham and Barker,[256,257] Price,[334] Slater et al.,[370] and others, can be expressed in terms of sensor calibration gains α_j (in units of W m^{-2}sr$^{-1}\mu$m^{-1}count^{-1}) and offsets β_j (in units of W m^{-2}sr$^{-1}\mu$m^{-1}), namely:

$$L_j = \alpha_j N_j + \beta_j \tag{8.29}$$

where j is the jth spectral band of the TM.

The spectral characteristics of the Landsat-5 TM sensors are recorded in Table 8.1. Values of the calibration coefficients α_j and β_j for the six reflective bands are taken from Price[334] and values of the extraterrestrial solar irradi-

TABLE 8.1

Spectral Characteristics of the Landsat-5 Thematic Mapper (TM)

Spectral Band	Center Wavelength (μm)	Band Width (μm)	Spectral Solar Irradiance E_{solar} (W m^{-2}μm^{-1})	Calibration Coefficients	
				α (W m^{-2}sr^{-1}μm^{-1}N^{-1})	β (W m^{-2}sr^{-1}μm^{-1})
TM1 (blue)	0.486	0.066	1957	0.062	−1.5
TM2 (green)	0.570	0.081	1829	1.17	−2.8
TM3 (red)	0.660	0.067	1557	0.806	−1.2
TM4 (near ir)	0.840	0.128	1047	0.815	−1.5
TM5 (near ir)	1.676	0.216	219.3	0.108	−0.37
TM6 (thermal)	11.450	2.100			
TM7 (near ir)	2.223	0.252	74.52	0.057	−0.15

Solar irradiance values taken from Markham and Barker.[258]
Calibration Coefficients taken from Price.[334]

ance, $E_{solar}(\lambda)$, for the reflective TM bands are taken from Markham and Barker.[258]

These satellite-observed values of L_j [transformed by equation (8.29) into the upwelling radiance spectra $L_u(h, \lambda, \theta_V, \phi_V)$ of Section 8.1] must then be converted into the subsurface volume reflectance spectra $R(0^-,\lambda)$. It is these volume reflectance spectra which must then be optimized, via the appropriate optical cross section spectra, to yield the co-existing concentrations of aquatic components. The optical cross section spectra we have been discussing throughout this book have been confined to the visible spectral wavelengths. Consequently, as seen from Table 8.1, only the spectral bands TM1, TM2, and TM3 can be utilized with the optical cross section spectra that would generally be available. In theory, three spectral bands should be sufficient for estimating three unknown concentrations (i.e., chlorophyll, suspended mineral, and DOC). However, it is the optical properties of the aquatic components which determine the rate of change of subsurface volume reflectance with changing concentrations of organic and inorganic aquatic components. As we have seen from equation (5.11) the rate of change of subsurface volume reflectance with wavelength (i.e., the slope, at a particular wavelength, of a measured or inferred volume reflectance) is a complex function of the absorption and scattering cross sections of these aquatic components. Reiterating from Section 5.8, not only are judiciously selected values of wavelengths required, but so also are precise values of the spectral slopes at each of those wavelengths. In a continuous subsurface volume reflectance spectrum, many of these slopes could be near-zero in magnitude. The broadbands of the TM could be characterized by the inclusion of slopes of both positive and negative values. While a broadband three-point upwelling spectrum is far from being an ideal candidate for multivariate "curve-fitting" optimization analysis, it is, nonetheless, a Landsat data-gathering limitation that must be reluctantly accommodated if an attempt is made to apply the satellite data to a study of inland and/or coastal water quality. Unfortunately, such restrictive spectral sensitivity is a limitation common to environmental satellites other than the Landsat series, and the accommodation of precise optical cross section spectra within the limitations of existing satellites is a requisite to the utilization of these satellites for monitoring the bioproductivity of non-Case I waters. Therefore, all of the considerably more highly resolved optical cross section spectra pertinent to an aquatic target (encompassing the spectral range included within TM1, TM2, and TM3) should be utilized in the multivariate optimization of the volume reflectance spectra inferred from the three-point upwelling radiance spectrum recorded by the Landsat TM (or any other environmental satellite sensor).

As detailed in Jerome et al.,[204] and illustrated in equation (5.3), the subsurface volume reflectance $R_1(0^-,0^\circ)$, integrated over the wavelength band of TM1, and represented by the radiance L_1 recorded in TM1, can be defined, for the condition of vertical incidence and for $0 \leq b_B/a \leq 0.25$ (where b_B is the backscattering coefficient and a is the absorption coefficient of the water column), by:

$$R_1(0^-, 0^\circ)$$

$$= 0.319 \sum_{\lambda=0.45\,\mu m}^{\lambda=0.52\,\mu m} \left[\frac{(b_B)_w(\lambda) + x_1(b_B)_{Chl}(\lambda) + x_2(b_B)_{SM}(\lambda)}{a_W(\lambda) + x_1 a_{Chl}(\lambda) + x_2 a_{SM}(\lambda) + x_3 a_{DOC}(\lambda)} \right] \quad (8.30)$$

where $(b_B)_w(\lambda)$, $(b_B)_{Chl}(\lambda)$, and $(b_B)_{SM}(\lambda)$ are the backscatter cross sections, at wavelength λ, for pure water, chlorophyll, and suspended mineral, respectively; $a_w(\lambda)$, $a_{Chl}(\lambda)$, $a_{SM}(\lambda)$, and $a_{DOC}(\lambda)$ are the scattering cross sections, at wavelength λ, for pure water, chlorophyll, suspended mineral, and dissolved organic carbon, respectively; and x_1, x_2, and x_3 are the co-existing concentrations of chlorophyll, suspended mineral, and dissolved organic carbon, respectively.

Equation (8.30) would be applicable for a radiance band that displays a flat wavelength response function, i.e., a band sensitivity that is independent of wavelength. Such, however, is not the case for the TM bands which display quasi-sinusoidal sensitivity functions over their wavelength ranges.[334] Consequently, equation (8.30) is more appropriately written as:

$$R_1(0^-, 0^\circ) = 0.319 \sum_{\lambda=0.45\,\mu m}^{\lambda=0.52\,\mu m} \left[\frac{x_i (b_B)_i(\lambda)}{x_i a_i(\lambda)} \right] S_1(\lambda) \quad (8.31)$$

for $\theta = 0^\circ$ and $0 \leq b_B/a \leq 0.25$,

where $\quad x_i$ = concentration of the ith aquatic component,

$(b_B)_i(\lambda)$ = backscatter cross section of the ith component at wavelength λ,

$a_i(\lambda)$ = absorption cross section of the ith component at wavelength λ, and

$S_1(\lambda)$ = sensitivity of TM1 at wavelength λ.

Similarly, the integrated subsurface volume reflectance $R_2(0^-,0°)$ represented by the radiance L_2 recorded by TM2 will be defined as:

$$R_2(0^-, 0°) = 0.319 \sum_{\lambda=0.52\,\mu m}^{\lambda=0.60\,\mu m} \left[\frac{x_i(b_B)_i(\lambda)}{x_i a_i(\lambda)} \right] S_2(\lambda) \tag{8.32}$$

where $S_2(\lambda)$ = sensitivity of TM2 at wavelength λ.

The integrated volume reflectance $R_3(0^-, 0°)$ represented by the radiance L_3 recorded by TM3 will be defined as:

$$R_3(0^-, 0°) = 0.319 \sum_{\lambda=0.63\,\mu m}^{\lambda=0.69\,\mu m} \left[\frac{x_i(b_B)_i(\lambda)}{x_i a_i(\lambda)} \right] S_3(\lambda) \tag{8.33}$$

where $S_3(\lambda)$ = sensitivity of TM3 at wavelength λ.

Following Jerome et al.,[204] incorporating the solar zenith angle dependence of volume reflectance, equations (8.31), (8.32), and (8.33) may be adapted to the form

$$R(0^-, \theta_W) = R(0^-, 0°)/\mu_o \tag{8.34}$$

where μ_o = cosine of the in-water refracted angle θ_w for the incident radiation distribution characterized by a solar zenith angle of θ.

As discussed earlier, the upwelling radiance L_{wj} from the surface of the water is related to the band radiance L_j received by the Thematic Mapper through an expression of the form:

$$L_{wj} = \frac{L_j - L_{Aj}}{T_j} \tag{8.35}$$

where $\quad L_j$ = radiance recorded in the jth band of the TM,

L_{Aj} = radiance recorded in the jth band of the TM attributable to atmospheric interactions with downwelling radiation, and

T_j = transmittance of the atmosphere at jth band wavelengths.

We have briefly discussed several atmospheric models that have been developed to estimate values of L_{Aj} and T_j, thus allowing conversions of upwelling radiance at the Landsat TM into estimates of upwelling radiance just above the air-water interface. As evident from equations (8.29) to (8.35), however, the harsh reality of water quality monitoring from Landsat (and, indeed, from all of the currently available satellite platforms) is the considerably less-than-ideal spectral resolution of the space-borne radiometers. The radiometric calibration factors such as α_j and β_j, the sensor band sensitivity functions $S_j(\lambda)$, and the extant, local atmospheric absorption and scattering vagaries, while problematical, can, at least to varying degrees of satisfaction, be incorporated into data analysis schemes. The broadband characteristics of the satellite sensors, however, place gargantuan obstacles before the aquatic data interpreter. As we have discussed in Chapter 5, for remote sensing to realize maximum potential in the determination of water quality parameters of inland and coastal waters, it is imperative that the pertinent optical cross section spectra (along with their seasonal variations) be determined for each major water body being remotely sensed. The minimum spectral resolution necessary for the computation of cross sections would appear to be ~20 nm, i.e., the equivalent of 15 channels between 410 nm and 690 nm (Bukata *et al.*[42]). As seen in the above Thematic Mapper example, such spectral resolution is resoundingly denied the Landsat platform. Equations (8.30) to (8.33) represent integrals of finely resolved spectral optical cross sections defining finely resolved subsurface volume reflectance spectra upon which optimization "curve-fitting" analyses must be performed to yield the co-existing concentrations of aquatic matter. The satellite records broadband radiance values (characterized by a non-constant wavelength sensitivity), each of which must be fine-tuned to incorporate the known cross section spectra. The limited number of broadband upwelling radiance values then must be further fine-tuned to estimate subsurface volume reflectance spectra of sufficient resolution to respond to optimization analyses.

Such extraction of spectral infrastructure from spectral broadbands is far from simple. These limitations in spectral sensitivity have produced an emphasis on the use of chromaticity coordinates (as outlined in Chapter 6 and discussed for the TM by Jaquet and Zand[188]) and the use of broadband ratioing (as will be discussed in the the next section of this chapter). There certainly are restricted sets of circumstances under which such ratioing and chromaticity analyses can be of consequence to the remote sensing of natural waters. However, to maximize the probability of using the already voluminous bank of existing satellite-acquired data in a meaningful evaluation of natural water resources, it logically follows that the totality of remotely acquired data should be incorporated into interpretative schemes (regardless of possible inherent shortcomings in their spectral resolution). In this regard, attempts to improve the spectral capability of planned satellite missions are being seriously addressed, although the future of such medium-to-high resolution satellite devices is still unclear. The most appropriate devices for water quality applications remain the high spectral resolution airborne spectrometers such as the Compact Airborne Spectrographic Imager (CASI), the Airborne Ocean Color Imager (AOCI), and possibly the Airborne Visible Infrared Imaging Spectrometer (AVIRIS). CASI was a modification of an earlier sensor, the Fluorescence Line Imager (FLI) designed for aquatic studies. AVIRIS, a prototype for future spacecraft sensors, was originally designed for mineral exploration and covered the spectral range (1200–2400) nm, but has been extended to include the (450–1200)-nm range. It is still not certain whether AVIRIS has sufficient sensitivity to be valuable for detailed studies of complex water bodies, although Carder et al.[57] have obtained very encouraging results in delineating dissolved and particulate constituents in the coastal waters of the Tampa Bay plume.

8.6 REMOTE SENSING OF WATER QUALITY PARAMETERS

Since the early days of the Landsat program, oceanic chlorophyll concentrations have been estimated from satellite altitudes with reasonable accuracy. The operative chlorophyll retrieval algorithms are generally empirical regressions based upon selected ratios of spectral radiances recorded by the satellite sensor (Gordon and Clark,[139] Gordon et al.,[141] Morel,[271] Smith and Wilson,[379] and others). Expressed as a function of upwelling radiance $L(\lambda)$, such single component retrieval algorithms assume the form:

$$\langle \text{CHl} \rangle = x \left[\frac{L(\lambda_1)}{L(\lambda_2)} \right]^y \tag{8.36}$$

where $\langle \text{Chl} \rangle$ is the estimated near-surface aquatic chlorophyll concentration, $L(\lambda_1)$ and $L(\lambda_2)$ are the upwelling radiances at wavelengths λ_1 and λ_2, respectively, and x and y are empirically determined constants. Such empirical regressions, however, completely ignore the scattering and absorption characteristics of the organic and inorganic matter indigenous to the aquatic region under observation. The successes of such local regressions in the mapping of mid-oceanic chlorophyll concentrations is a direct consequence of the fact that chlorophyll pigments are the principal colorant of Case I waters. The optical complexity of inland and coastal waters, however, prohibits the luxury of single-component extraction algorithms and underscores the need for multiple-component extraction methodology.

This need to estimate simultaneously the co-existing concentrations of organic and inorganic matter comprising optically complex waters is illustrated in Figure 8.3. Herein is depicted the situation encountered when attempting to extract a single aquatic component concentration (e.g., chloro-

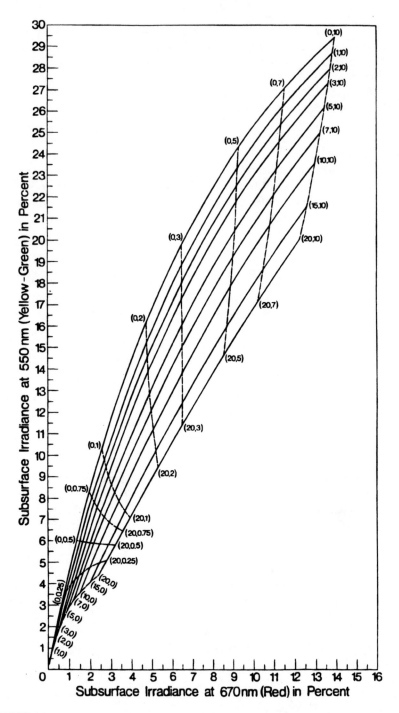

FIGURE 8.3

Subsurface irradiance reflectances (volume reflectances) that would be simultaneously observed in the Nimbus-7 CZCS bands at 550 nm (yellow-green) and 670 nm (red). Each point on the duoisoplethic curves is defined by the coordinates (Chl*a*, SM). (From Bukata, R. P., Bruton, J. E., Jerome, J. H., Jain, S. E., and Zwick, H. H., *Appl. Opt.*, 20, 1704–1714, 1981. With permission.)

phyll) from a remote observation over non-Case I waters. Figure 8.3 plots, in duo-isoplethic form, the anticipated subsurface volume reflectance that would be simultaneously observed in two of the Nimbus-7 Coastal Zone Color Scanner (CZCS) bands, namely, the yellow-green (550 nm) and the red (670 nm). The volume reflectance duoisopleths of Figure 8.3 were generated from optical cross section spectra appropriate to Lake Ontario waters.[38] Each point on the duoisopleth is defined by co-existing concentrations of chlorophyll a and suspended minerals, i.e., by the coordinates (Chl$a$, SM). Each point, therefore, represents the intersection of an isoorganic curve (constant chlorophyll concentration) and an isoinorganic curve (constant suspended mineral concentration). Values of Chl$a$ displayed in the figure range from 0 to 20 mg m^{-3} (1 mg m^{-3} = 1 μg/l) and values of SM range from 0 to 10 g m^{-3} (1 g m^{-3} = 1 mg/l), covering the water quality concentrations encountered in most Case I and non-Case I waters. The dissolved organic carbon (DOC) concentration is kept fixed at 2.16 g C m^{-3}, the average concentration in Lake Ontario during the period for which the optical cross section spectra were determined. Other values of DOC concentrations would, of course, result in other duoisoplethic diagrams. Comparable duoisoplethic diagrams utilizing (Chl$a$, DOC) or (SM, DOC) as the plotted points could be constructed not only for the CZCS bands displayed in Figure 8.3, but for any of the other bands of Nimbus-7 or other environmental satellites. The Case I single-component chlorophyll extraction algorithms of equation (8.36) result in isoinorganic curves (such as the curve in the lower portion of Figure 8.3) for a zero concentration of SM and plotted for the points labelled (0,0) to (20,0)] relating the radiance recorded in one satellite band to the radiance recorded in another band. The fact that several empirically determined Case I chlorophyll retrieval algorithms are currently in use is a consequence of natural water bodies being optically defined by myriad such interwoven isoplethic curves. Restricting remote estimation of chlorophyll concentrations to the limitations of Figure 8.3 can, therefore, dramatically downgrade the credibility of remote estimations of chlorophyll concentration.

This downgrade in credibilty is shown in Figure 8.4 taken from Bukata et al.[47] wherein chlorophyll concentrations directly measured in Ladoga Lake, Russia, are plotted against chlorophyll concentrations inferred from upwelling radiance spectra just above the air-water interface. These upwelling radiance spectra were utilized in conjunction with six different oceanic algorithms in common usage (the empirical regression relationships are included within Figure 8.4) to predict the chlorophyll concentrations. The lack of agreement, not only between measured and predicted chlorophyll concentrations for each Case I algorithm, but also amongst the predictions of the six Case I algorithms themselves, is clearly evident.

By contrast, Figure 8.5, also taken from Bukata et al.,[47] compares the measured chlorophyll concentrations with the predictions of chlorophyll concentrations using the multivariate otimization analyses (see Section 5.9)

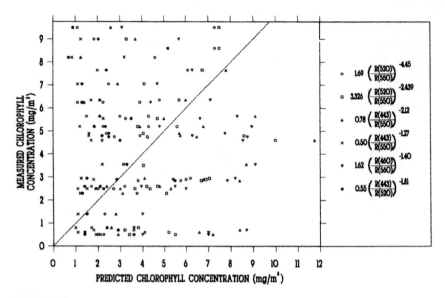

FIGURE 8.4

Directly sampled chlorophyll concentrations in Ladoga Lake plotted against chlorophyll concentrations predicted by six different contemporary chlorophyll-retrieval algorithms. (From Bukata, R. P., Jerome, J. H., Kondratyev, K., Ya., and Pozdnyakov, D. V., *J. Great Lakes Res.*, 17, 470–478, 1991. With permission.)

with the optical cross section spectra appropriate to Ladoga Lake given in Bukata *et al.*[46] The differences between Figures 8.4 and 8.5 are indeed striking. Similar comparisons exist between measured and predicted suspended mineral concentrations and between measured and predicted dissolved organic carbon concentrations.

Remote sensing is now routinely being used to monitor carbon budgets and bioproductivity of terrestrial and mid-oceanic ecosystems. Contrastingly, however, this technology has been virtually unused in freshwater studies despite the fact that it has long been recognized as the only feasible way of obtaining synoptic data from large lakes or temporal data from a large number of small lakes (such as, for example, would be scattered throughout a boreal forest biome). The fundamental obstacle to the remote sensing of inland and coastal waters is the typically complex admixtures of optically competitive organic and inorganic components residing therein. As seen from the above discussion, extracting the spectral signature of a single aquatic component from a background of co-existing and optically competitive aquatic components is not trivial. Thus, the optical complexity of non-Case I waters renders unsuitable the single-component retrieval algorithms developed for oceanic waters. Multiple-component retrieval algorithms which estimate co-existing concentrations of organic and inorganic aquatic compo-

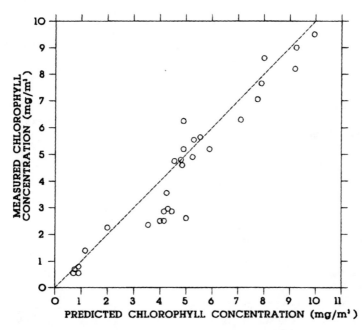

FIGURE 8.5
Directly sampled chlorophyll concentrations in Ladoga Lake plotted against chlorophyll concentrations predicted by use of multivariate optimization analyses and optical cross section spectra. (From Bukata, R. P., Jerome, J. H., Kondratyev, K., Ya., and Pozdnyakov, D. V., *J. Great Lakes Res.*, 17, 470–478, 1991. With permission.)

nents are, therefore, required for the remote sensing of inland and coastal waters. Obligatory to multiple-component monitoring are the optical cross section spectra of these aquatic components. Optical cross section spectra are functions of both space and time, and a paucity of regionally appropriate spectra currently exists. There is an increasing effort, on a global scale (largely due to major efforts directed toward the role of oceanic waters in climatic change distruptions to the carbon and water cycles), to obtain precise information on the absorption and scattering cross section spectra of chlorophyll-bearing biota. Unfortunately (due, perhaps, to the lack of a comparable concern regarding the role of lakes and coastal regions in climate change distruptions to the carbon and water cycles), there is not a parallel effort to obtain precise information on the absorption and scattering cross section spectra of the inorganic components of aquatic regimes. There are, however, indications that a slow (and hopefully sure) increased interest in the remote sensing of inland and coastal waters is emerging, and that this interest is coupled with the realization that bio-optical modelling of the type we have been discussing needs to play an integral role in remote sensing of non-Case

I waters. Jaquet *et al.*[189] discuss this remote sensing prospect as well as presenting *in situ* optical data obtained in Lakes Zug and Lucerne (central Switzerland) coincident with an AVIRIS overpass.

It could be mentioned that such a combined remote sensing and optical cross section project is currently being performed as part of the Canada/ United States Boreal Ecosystem-Atmosphere Study (BOREAS) at lake sites within the Prince Albert National Park in Saskatchewan, Canada, and involves participants from the National Water Research Centre (R. P. Bukata and J. H. Jerome), NASA/Ames Research Center (R. C. Wrigley and R.A. Armstrong), the Institute for Space and Terrestrial Science (J. H. Miller), the National Hydrology Research Institute (M. S. Evans), and the Centre in Mining and Mineral Exploration Research (E. A. Gallie). The aims of the project include utilizing the Compact Airborne Spectroscopic Imager, CASI (Borstad and Hill,[27] Anger *et al.*,[9] Harron *et al.*[162]), and the Airborne Ocean Color Imager, AOCI (Wrigley *et al.*[432]), as means of estimating near surface dissolved organic carbon concentrations and phytoplankton photosynthesis in boreal lakes. Results of this study should be available shortly.

The determination of optical cross section spectra requires both the presence and a large concentration range of each of the organic and inorganic aquatic colorants. The success of an integrated program of optical and water quality data collection, bio-optical modelling, and optimization analyses, therefore, can be site-limited. The perversity of natural waters in their failure to comply to these concentration range requirements oftimes precludes the *in situ* determination of optical cross sections, and laboratory techniques such as described in Kiefer and SooHoo,[217] Mitchell,[265] and Gallie[119] are required to obtain at least *some* of the pertinent cross section values.

8.7   A VERY BRIEF INVENTORY OF ENVIRONMENTAL SATELLITES

Throughout our discussions in this and previous chapters, we have periodically referred, in a somewhat cavalier manner, to such space vehicles as Landsat and Nimbus, to such space hardware as CZCS and TM, and to such airborne hardware as CASI and AVIRIS. It might be of value to present a brief overview of the existing satellites that have been and/or are being used as a source of aquatic color data (generally referred to in the scientific and technical literature as *ocean color data*). This section is not intended to be a surrogate for the volumes of detailed mission-oriented documentation that are readily available from NASA, NOAA, and various government, academic, and private sector space-related institutes. Rather, its intent is to provide in a single location mention of available sensors/satellites along with brief discussions of their *raison d'etres.*

Satellite oceanography is generally considered to have been inaugurated with the 1972 launch of the first in the Landsat series of a four-channel broadband Multispectral Scanner System (MSS) (500–600; 600–700; 700–800; 800–1100 nm). The MSS has a spatial resolution of 80 m covering a 100-nautical mile square area. In 1982 the Thematic Mapper (TM) launched aboard Landsat-4 succeeded the MSS with an improved spatial resolution of 30 m and a pre-selected set of more finely resolved spectral bands designed for monitoring terrestrial parameters (primarily vegetation). The seven selected TM bands (see Table 8.1) do not constitute a continuous spectrum, but rather comprise three visible (450–520; 520–600; 630–690 nm), three near-infrared (760–900; 1550–1750; 2080–2360 nm), and one thermal infrared (10400–12500 nm) wavelength bands. Unlike the 30-m resolution of the other TM channels, the TM thermal band has a spatial resolution of 120 m. The 16–18 day repetition cycles of the Landsat space vehicles, coupled with low radiometric sensor sensitivities, renders both the MSS and the TM less-than-ideally suited to the study of aquatic environments.

The 1986 launch of the Systeme Probatoire d'Observation de la Terre (SPOT) with three spectral bands (500–590; 610–680; 790–890 nm) and a spatial resolution of 20 m (with a 60-km swath), while providing an improvement on repetitive viewing, added yet another terrestrially oriented mission.

The Coastal Zone Color Scanner (CZCS) launched aboard Nimbus-7 was the first satellite sensor to specifically target aquatic environments. Four narrow visible spectral bands (20 nm bandwidths) were centered at 443 nm, 520 nm, 550 nm, and 670 nm to accommodate pigment absorption (443; 670 nm), pigment scattering (550 nm), and a spectral pivot point selected to characterize an equilibrium between the optical properties of chlorophyll and "pure" water for an aquatic system in which the only foreign matter is chlorophyll a (520 nm). [Recall the discussions in Section 5.3, wherein it was seen, from equation (5.8) and Figure 5.7, that for chlorophyll-bearing biota indigenous to the hypothetical water mass considered for illustrative purposes, this spectral pivot point was located at $\lambda = {\sim}497$ nm, i.e., $(\partial R / \partial C_{Chl})$ $= 0$ at $\lambda {\sim}497$ nm.] A 700–800-nm band was included to distinguish land and clouds from open water. The CZCS was more radiometrically sensitive than the MSS and the TM, but had a spatial resolution of ${\sim}1$ km and a swath width of 1600 km for overlapping day-to-day coverage. As illustrated in Platt and Sathyendranath,[317] satellites are essential to the determination and monitoring of marine primary production at the scale of regional ocean basins. In this regard (as will be discussed more fully in Chapter 9), it is necessary to, amongst other requirements, utilize satellite data to extract near-surface values of chlorophyll concentrations from which non-uniform depth profiles of chlorophyll concentrations can be inferred and aquatic primary production estimated. Although the Coastal Zone Color Scanner was functional only during the years 1978 to 1986, its data have been invaluable to obtaining near-surface aquatic chlorophyll concentration maps (see, for example, Hovis *et al.*,[175] Gordon *et al.*,[140] Bidigare *et al.*,[24] Morel and Berthon,[275] Collins *et al.*,[63] Carder *et al.*,[54] Balch *et al.*,[15] Esaias *et al.*,[99] Austin and Petzold,[13] Weaver and Wrigley,[422] among numerous others). The eight years of CZCS operations resulted in thousands of synoptic images, and their interpretation has resulted in numerous scientific articles and reports. A Goddard Space Flight Center atlas generated by Hovis *et al.*[176] contains representative samples of these images along with brief discussions of some of the significant results.

We have continuously emphasized throughout this book that, despite the problems and limitations associated with the remote sensing of Case I waters (We will detail the difficulties encountered in remotely estimating phytoplankton productivity in both Case I and non-Case I waters in Chapter 9), oceanic monitoring successes, such as those attributable to the CZCS, result from the freedom of mid-oceanic waters from optical impacts of terrestrial matter. As we have discussed and illustrated in Section 8.6, this freedom results in the frequent applicability of single-component concentration extraction algorithms to such waters. Thus, although by virtue of its name, the Coastal Zone Color Scanner implies an application to the water color/ water quality relationships of non-Case I waters, its lack of spectral resolution restricts its value (and the value of similar sensors) to Case I waters.

The CZCS approach to the estimation of aquatic primary production has been extended to the soon-to-be-launched privately owned SeaStar satellite (Lyon and Willard[252]). The SeaStar satellite(s), to be launched from airborne space vehicles Pegasus and/or Taurus, will contain a new-generation CZCS instrument, the Sea-viewing Wide Field-of-View Sensor (SeaWiFS). Additional bands have been added at 412 nm (to assist in the delineation of yellow substances through their blue wavelength absorption) at 490 nm (to increase sensitivity for chlorophyll concentration estimations), and at 765 nm and 865 nm (near-infrared bands to assist in the removal of atmospheric interventions). The aims of the SeaWiFS mission, in addition to ending the hiatus in the satellite mapping of oceanic chlorophyll and primary production resulting from the inoperable CZCS, include assessing the role of the ocean in the global carbon and other biogeochemical cycles (re-emphasizing the need we have mentioned in Section 1.6 to employ reliable techniques to infer biological parameters from physical measurements). To achieve these goals SeaStar will provide a repetition frequency of every second day, and will be directed toward evaluating the variability in the annual cycle of marine bioproductivity, the timing and spatial distribution of spring chlorophyll blooms, and the relationship between plankton and the carbon storage capacities of global waters. Whether or not such noble goals can be realized for coastal zone waters, let alone inland and estuarial waters, has yet to be determined. However, it is considerably less risky to be optimistic that the Case I water successes conceded to the CZCS operation can not only be

TABLE 8.2

Spectral Characteristics of the Coastal Zone Color Scanner (CZCS), Airborne Ocean Color Imager (AOCI), and Sea-viewing Wide Field-of-View Sensor (SeaWiFS)

Sensor	Band Number	Band Center (nm)	Band Width (nm)
CZCS	1	443	20
	2	520	20
	3	550	20
	4	670	20
	5	750	100
	6	11,500	2,000
AOCI	1	444	23
	2	490	20
	3	520	21
	4	565	20
	5	619	21
	6	665	21
	7	772	60
	8	862	60
	9	1,012	60
	10	10,395	3,900
SeaWiFS	1	412	20
	2	443	20
	3	490	20
	4	510	20
	5	555	20
	6	670	20
	7	765	40
	8	865	40

duplicated by SeaWiFS, but perhaps exceeded. SeaStar/SeaWiFS, like Nimbus-7/CZCS, has a spatial resolution of ~1 km.

In anticipation of the SeaWiFS launch NASA obtained a high altitude aircraft sensor, the Airborne Ocean Color Imager (AOCI) that simulated the spectral bands and radiometric sensitivity of SeaWiFS. Table 8.2 provides an intercomparison of the spectral characteristics of the CZCS, the AOCI, and SeaWiFS. It is seen that the AOCI lacks the SeaWiFS 412 nm band but adds a visible band at 619 nm (the original SeaWiFS design contained a 620 nm band that was deleted in favor of the "yellow-substance band" at 412 nm), a near-infrared band at 1012 nm (atmospheric correction over more-than-moderately turbid waters), and a thermal band. The AOCI has a spatial resolution of 50 m (swath width of 35 km) at an aircraft altitude of 20 km.

There are several post-SeaWiFS sensors in various stages of development. A Japanese Space Agency instrument called the Ocean Colour and Temperature Sensor (OCTS) with similar bands (plus thermal), radiometric sensitivity, and spatial resolution to SeaWiFS is scheduled for an early launch. The

Moderate Resolution Imaging Spectrometer-Tilt (MODIS-T), whose future with the restructured Earth Observing System program (Eos)[292] is currently uncertain, was envisioned as a finer spectral resolution SeaWiFS with 64 contiguous 10 nm bands in the (400–1040)-nm range, any 16 of which could be selected for transmission at any time. The tilt feature of MODIS-T is a mechanism designed to advance the sensor field-of-view look angle to avoid sun-glint from the air-water interface (a feature also present in the CZCS). The Moderate Resolution Imaging Spectrometer-Nadir (MODIS-N), also planned for the Eos, technically is not a spectrometer, but rather a potpourri of terrestrially and aquaticly oriented bandwidths selected for a variety of environmental targets and objectives. Its restriction to nadir viewing directions could result in a substantial sun-glint contamination problem for the acquired data. Similar to MODIS-T is an European Space Agency sensor, the Medium Resolution Imaging Spectrometer (MERIS) planned for launch on the first European Earth Observing System (EEos). It will be capable of transmitting any selected 15 of its 64 bands, and will have a slightly better spatial resolution (~0.5 km) than the other ocean color sensors.

Another possible Eos system with an unclear status is the High Resolution Imaging Spectrometer (HIRIS) comprising 224 contiguous 10 nm bands in the spectral region (450–2400) nm and a spatial resolution of 20 m. HIRIS, however, was conceived for terrestrial (primarily forest and agriculture) studies. HIRIS is designed to be a target-of-opportunity, rather than a continuously operational, instrument.

Mention should also be made of the polar orbiting series of meteorological satellites of the TIROS-N/NOAA A-J genre and generally included within the acronym POES (Polar Orbiting Environmental Satellites). The availability of low-cost personal computer systems and low-cost satellite tracking station systems has enabled research teams to acquire, and thus, interpret TIROS/NOAA satellite data on local-issue priority targets and environmental issues.[145,292] The sensor of note on POES is the Advanced Very High Resolution Radiometer (AVHRR). The acronym AVHRR, unfortunately, totally misrepresents the sensor since it is neither "advanced" nor of "very high resolution" (either spectrally or spatially). Technically, in fact, the instrument is not a particularly sophisticated radiometer, comprising as it does only four or five very broad bands, a visible red band (580 nm–680 nm), two near-infrared bands (725 nm–1100 nm, 3550 nm–3930 nm), and one or two thermal bands (10.50 μm–11.50 μm, or 10.30 μm–11.30 μm and 11.50 μm–12.50 μm). The spatial resolution of the AVHRR is a circular field-of-view of ~1.2 km diameter, a reasonably convenient pixel size for oceanic and coastal oceanic aquatic regions, but inconvenient for lakes much smaller than those comprising the Laurentian Great Lakes. Despite obvious limitations, the AVHRR data collection and application protocols have proven to be both popular and valuable in monitoring changes in broadband spectral signatures over both terrestrial and oceanic regimes. Much of this value is associated with (a) the 12-hour repitition frequencies of the NOAA satellites (a daytime and a nighttime overpass for mid-latitude locations), (b) extensive spatial coverage that can be obtained from a satellite tracking station (for example, a tracking station situated along the Canada/United States border near Winnipeg, Manitoba, can obtain synoptic views out to both the Pacific and the Atlantic coasts of North America. Twice daily, essentially instantaneous, spatial thermal contours of the entire Great Lakes and their confining basin can be readily obtained and recorded), and (c) the relatively low-cost general monitoring facility it provides.

The above paragraphs have presented a very brief overview of currently available satellite platforms/sensors that have been and/or are being used or considered for applications to aquatic color (perhaps more appropriately stated as applications to ocean color). If they are to be so considered, then the full spectral information contained within the recorded data (spectral resolution limitations notwithstanding) must somehow be utilized, if possible, in analyses and interpretation schemes. Reiterating our comments at the close of Section 8.5, however, a high spectral resolution is required if the quality and productivity of inland and coastal waters are to be incorporated into remote sensing protocols. This high spectral resolution is currently available on the Compact Airborne Spectrographic Imager (CASI) and the Airborne Visible Infrared Imager (AVIRIS). At this writing, most applications of AVIRIS have been in the absorption features of plant canopies, although CASI has been applied to both terrestrial and aquatic spectral targets.[28]

As yet, there are no CASI-type spectrometers aboard existing or scheduled environmental satellites. Therefore, if satellite monitoring of non-Case I waters is to be implemented, techniques and/or models must be developed to overcome the modest spectral sensitivities inherent to existing operational satellite radiometers. In addition to the need to employ realistic scattering and absorption parameters to remove atmospheric obfuscation from the modestly resolved radiance spectrum recorded at the satellite altitude, the need to reduce a considerably less modest set of optical cross section spectra to the broadband character of the satellite data stream (as discussed in Section 8.5) must also be addressed. To date, successes enjoyed by the oceanographic chlorophyll mapping activities have precluded serious attention to this requirement.

Chapter 9

PRIMARY PRODUCTION IN NATURAL WATER

9.1 PHOTOSYNTHESIS

Photosynthesis is a process resulting in the biological combination of chemical compounds in the presence of light. It commonly refers to the production of organic substances (primarily sugars) from carbon dioxide and the water residing within green plant cells, provided the cells are sufficiently irradiated to allow plant chlorophylls to act as an agent in transforming radiant energy into an organic compound. The natural and cultivated flora of a region provide the principal photosynthetic organisms to the terrestrial biome. The benthic and planktonic algae as well as higher forms of plant life and some species of bacteria provide the principal photosynthetic organisms to the aquatic biome. To capture radiant light and use the light energy to form organic chemical compounds, these photosynthetic organisms contain certain pigments within intricate subcellular plant structures known as *chloroplasts* or intricate bacterial substructures known as *chromatophores*.

The photosynthetic process of plants and bacteria have been extensively reviewed in books and journals (Duysens,[96] Murrell,[286] Thomas,[396] Goodwin,[128–130] Halldal,[156,157] among numerous others) and will not be meticulously re-investigated here. Rather, we shall state that photoreactions are a quantum dynamical effect described in terms of stimulated transitions among the vibrational and rotational states of matter such as the chlorophyll molecule. Short-lived electronically excited states (resulting in stimulated emission of photons due to the process of *fluorescence*) give way to long-lived chemically excited states resulting in reduction of carbon dioxide and oxidation of a hydrogen donor (generally water in algae or a sulfur compound such as thiosulfate).

Recall from the basic definition of photonic energy (equation 1.2) that shorter wavelength quanta contain more energy than longer wavelength quanta. For example, it would take about 1.53 times as many photons of 650 nm to possess the same energy content as photons of 425 nm. A single photon, irrespective of its energy, cannot excite more than one molecule. Under identical ambient conditions, therefore, a beam of red light of a specified total energy has the potential to excite more molecules than blue light of identical total energy. Thus, if the photosynthetic process is studied from the perspective of aquatic component pigment responses to impinging radiation, it is essential to consider not only the intensity of the radiation, but also its spectral distribution.

The PAR absorption cross section spectra of planktonic chlorophyllous pigments clearly are pertinent to the photosynthetic process. Also pertinent is knowledge as to which pigments are photosynthetically active and which are photosynthetically inactive as well as knowledge of the spectrally dependent effectiveness of the photosynthetically active pigments. This effectiveness, as discussed in Section 7.6, results in the photosynthetic usable radiation (PUR) being the pertinent optical factor of photosynthesis and is determined by the photosynthetic efficiency of the indigenous chlorophyllous pigmentation. *Photosynthetic efficiency* can be simply defined as the ratio of the energy stored by phytoplankton as chemical energy in the form of carbohydrates, to the total energy absorbed by its pigments. It has long been observed that the photosynthetic efficiency in algae has a maximum upper limit of ~25%, a figure that does not increase irrespective of incident photon intensity. Photosynthetic efficiency is measured by precise laboratory determinations of *action spectra* (organisms' wavelength-dependent response functions to incident irradiance spectra). These action spectra, which reflect the absorption spectrum of the specific pigment considered, will not be fully discussed here. Rather, the reader is referred to the pioneering works described in Haxo and Blinks,[165] Yocum and Blinks,[435] French et al.,[115] Halldal,[155,156] Duysens,[97] and others.

Once determined, an organism's action spectrum should enable the determination of both the identity and efficiency of pigments that are responsible for a particular photoreaction such as the rate of uptake of carbon dioxide in photosynthesis. All phytoplankton contain the photosynthetically active chlorophyllous pigment chlorophyll *a* (Chl*a*). The pigment chlorophyll *b* (Chl*b*) is very similar to chlorophyll *a*, the only difference being that its molecule contains an aldehyde (CHO) as opposed to a methyl (CH_3) radical. Chlorophyll *c* (Chl*c*) is a pigment present in some brown algae, diatoms, crysomonads, dinoflagellates, and cryptomonads, while chlorophyll *d* (Chl*d*) can be found in a few species of red algae. There are also rare occurrences of the accessory pigment chlorophyll *e* (Chl*e*). In addition to the main categories of chlorophyllous pigments, different forms of Chl*a* have been routinely observed in living plants from both absorption spectra measurements and photophysiological analyses (see Halldal,[154] French,[114] Butler[53]). These different chlorophyll *a* forms manifest as a variety of absorption peaks clustered in the red (and in some instances the near-infrared) regions of the spectrum. Other photosynthetically active plant pigments include the carotenoid group,

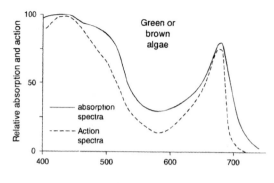

FIGURE 9.1
Hypothetical absorption and action spectra for a representative green or brown alga.

of which carotene (the red and orange isomeric hydrocarbon found in carrots) is the most well-known member and the phycobilin group of which phycoerythrin and phycocyanin are the principal members.

It is virtually impossible to classify natural waters into depth layers or spatial regions that are populated by algae of predictable chlorophyllous pigmentation. Mixtures of green, red, and brown algae may exist at all depths (red algae, while present in oceanic waters, are absent from inland waters). As we have seen in our discussions of the optical cross sections of chlorophyll a (without focus on the sub-classes of the pigment), the absorption cross section of aquatic chlorophyll a (the results of many workers) displays, in general, a peak at about 675 nm. The varying full width/half maximum resolution of this absorption peak is illustrative not only of the specific difficulty in resolving the sub-pigments of Chl$a$, but also of the general difficulty in resolving Chl$a$ from b, c, d, and e chlorophyllous pigments. This is one reason why we advocate that the desirability of an ideal water quality remote sensing concentration extraction methodology be tempered with a sensible restriction of that method to a manageable number of organic and inorganic variables. Despite its obvious shortcomings, the four-component optical model described earlier, wherein Chl$a$ is considered a surrogate for the organic component of natural waters, provides, at least for the present, a reasonable approach to the realities of remotely sensing optically complex natural waters.

As an illustrative representation of the terms discussed in this section, Figure 9.1 sketches a hypothetical absorption (solid line) and a hypothetical action (dotted line) spectrum for a typical green or brown alga that would be indigenous to natural waters. Figure 9.2 sketches a hypothetical absorption (solid line) and a hypothetical action spectrum (dotted line) for a typical diatom or red alga. Individual algal species will, of course, display distinct action spectra. However, in general, green and brown algae display action spectra that, while being spectrally distinct from their Chl$a$ absorption spectra, nevertheless basically adhere to the same spectral shape, namely a double peaked appearance with maxima in the blue and red wavelengths and a minimum in the green. Red algae display action spectra essentially inverse

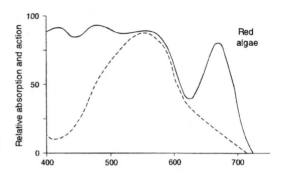

FIGURE 9.2
Hypothetical absorption and action spectra for a representative red alga.

to those of the green and brown algae, i.e., single maximum at green wavelengths and minima at the blue and red extremities. Also, red algae are seen to absorb light more effectively over a wider range of wavelengths than their green and brown counterparts.

Recall from Section 7.5 that as the irradiance levels of Great Lake-type (non-Case I) waters decrease with increasing depth, the radiation in the blue region of PAR becomes less significant and the radiation in the red region of PAR becomes more significant. It is interesting to compare the Lake Erie photic zone PAR spectra shown in Figure 7.15 with the action spectra of green and brown algae shown in Figure 9.1 and the action spectra of red algae shown in Figure 9.2. Such an intercomparison provides insight into the efficiency of indigenous chlorophyllous pigments within the aquatic region most important to the photosynthetic process. It is seen that the spectral shape of the red algae action spectra is in phase with the light availability spectrum, while the spectral shape of the green and brown algae action spectrum is in antiphase with the light availability spectrum. Photic zone primary production by red algae proceeds considerably more effectively than photic zone primary production by green or brown algae. Primary production by green or brown algae proceeds more effectively near the surface than it does at greater depths.

In addition to the photochemical processes discussed above, there are also enzymatical processes controlling phytoplankton photosynthesis. Enzymes are protein-like molecules which comprise a major component of the organic matter found in unicellular algae. Whereas photochemical processes involve the absorption of light quanta by photosynthetically active pigments, enzymatical processes involve the presence of enzymes within the algae that act as temperature-dependent catalysts in the chemical reactions. The net photosynthetic process can be limited by either of these two tandem processes. Excessively high photochemical rates (as compared to enzymatic rates) results in light limitation, while excessively high enzymatic rates (as compared to photochemical rates) results in an overabundance of organic matter being photosynthetically produced.

The biological response of natural waters to impinging light, therefore, is clearly dependent upon not only the light spectrum of the incident radiation but also the populations of indigenous algal species and the temperature gradients defining the water column.

9.2 PHYTOPLANKTON PHOTOSYNTHESIS (PRIMARY PRODUCTION)

The terms primary production and primary productivity have undergone subtle changes in usage since their origins in trophic dynamic theory (Lindeman[248]). *Primary production* is the chemical energy contained within an ecosystem as a direct result of photosynthesis. *Primary productivity* is the sum of all the photosynthetic rates within that ecosystem. For the assessment of aquatic primary production, therefore, the singlemost important parameter is the daily integral of photosynthesis occurring within the entire water column. Historically, this photosynthesis has been determined by *in situ* determinations of oxygen production or carbon (^{14}C) uptake at a variety of depths. Primary production is generally recorded in units of g Carbon m^{-2} yr^{-1}. Since phytoplankton are the principal photosynthesizers in Case I waters, it is not unreasonable to assert that phytoplankton photosynthesis and primary production are interchangeable nomenclature for mid-oceanic systems. Such is not the case, however, for non-Case I waters wherein terrestrially derived organics and photosynthesis by macrophytes and periphyton can dominate aquatic carbon budgets.

Determination of near-surface Chl*a* concentrations is a principal objective of the remote monitoring of inland and coastal waters. Since it is phytoplankton photosynthesis that can be best estimated from a knowledge of near-surface Chl*a* concentrations, it would be of consequence to briefly discuss, at this stage, an algorithm in general use to calculate *in situ* phytoplankton photosynthesis.

The relationships required to estimate aquatic photosynthesis (Smith,[372] Fee,[104-107] Vollenweider,[415] Patten,[304] Bannister,[16]) are:

1. Solar Photosynthetically Available Radiation (PAR) (400–700 nm) as a function of time, i.e., E_{PAR} vs t
2. Photosynthesis P as a function of PAR, i.e., P vs E_{PAR}
3. PAR as a function of aquatic depth, E_{PAR} vs z

Relationship 1 is obtained by direct measurement in air above the water body. Relationship 3 is determined from the attenuation coefficient for PAR, K_{PAR}, which, as discussed earlier, is a function of solar angle θ (i.e., time t).[105] Bukata *et al.*,[45] however, conclude that while the diurnal variation of K_{PAR}

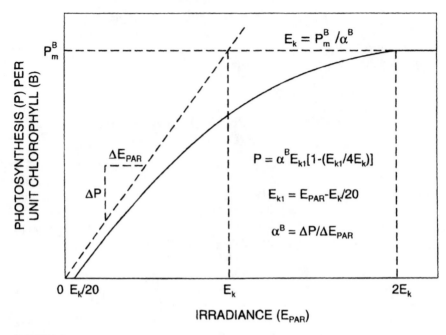

FIGURE 9.3

Photosynthesis per unit chlorophyll concentration displayed as a function of irradiance E_{PAR}.

has a significant impact on calculated daily irradiation (integration at a particular depth of subsurface irradiance over daylight hours), its impact on estimates of primary production is quite small. On-site values of $K_{PAR}(\theta)$ are obtained from depth profiles of the subsurface PAR irradiance. The essence of the primary production (phytoplankton photosynthesis) calculation is relationship 2, namely P as a function of E_{PAR}.

As detailed in Talling,[392] Steeman Nielsen,[385] Shearer *et al.*,[365] and Fee,[107] Figure 9.3 illustrates the manner in which relationship 2 may be calculated from the extant chlorophyll concentration B and two physiologically meaningful photosynthesis parameters (transfer coefficients) P_m^B and α^B. The chlorophyll concentration is designated as B in this discussion rather than as Chl$a$, x_{Chl}, or other nomenclature used in this book to be consistent with existing scientific literature on primary production. The superscript B in the transfer coefficients P_m^B and α^B does not represent an exponent. Rather (again not to be at variance with the published literature), it is used to indicate that both the transfer coefficients are normalized to the aquatic chlorophyll concentration (a surrogate for *B*iomass) at the specific time and depth of primary production determination. P_m^B is the maximum rate of photosynthesis per unit chlorophyll concentration, and α^B is the slope of the production per unit chlorophyll versus irradiance curve (translated to $E_{PAR} = 0$). This initial slope is a direct consequence of the photochemical aspect of phytoplanktonic photosynthesis (a function of the type and quantity of indigenous photosynthetic pigments). The near-horizontal asymptotic portion of the curve at high E_{PAR} values represents the maximum rate of the enzymatic aspect of the phytoplanktonic photosynthesis (a function of the type and quantity of cellular enzymes and the specific temperature at which the primary production is determined).

The irradiance E_{PAR} at the point in the photosynthesis per unit chlorophyll versus irradiance curve of Figure 9.3 that the initial and asymptotic horizontal slopes intersect is designated as E_k. This irradiance was originally designated as I_k by Talling,[392] a designation still in use today. However, we will use the irradiance nomenclature E for consistency. E_k thus represents a form of compromise between the photochemical and the enzymatic aspects of the photosynthesis process and is defined by the ratio of the transfer coefficients P_m^B and α^B.

Figure 9.3 describes the water column photosynthesis per unit chlorophyll (P/B). The expression for the P versus E_{PAR} curve for a water column containing a chlorophyll concentration B can then be mathematically defined[107] as:

$$P = 0 \quad \text{for } E_{PAR} < E_k/20 \tag{9.1}$$

$$P = BP_m^B \quad \text{for } E_{PAR} \geq 2E_k \tag{9.2}$$

$$P = B\alpha^B E_{k1}(1 - E_{k1}/4E_k) \quad \text{for } E_k/20 \leq E_{PAR} \leq 2E_k \tag{9.3}$$

where

$$E_k = P_m^B/\alpha^B \quad \text{and} \tag{9.4}$$

$$E_{k1} = E_{PAR} - E_k/20 \tag{9.5}$$

are empirically determined constants consistent with the work of Jassby and Platt[191] and Platt and Sathyendranath.[317] Being an empirical fit between the initial and horizontal P versus E_{PAR} slopes, equation (9.3) assumed a parabolic form. It is seen that photosynthesis is zero at low values of irradiance ($E_{PAR} < E_k/20$). The form of equation (9.3) would predict photosynthesis to be zero when E_{k1} has a value equivalent to $4E_k$. From equation (9.5) this corresponds to an E_{PAR} value of $4.05E_k$. The occurrence of such high irradiance inhibition of photosynthesis, however, is generally considered as improbable for *in situ* observations of photosynthesis (Marra,[260] Welschmeyer and Lorenzen[423]). Further, as argued by Fee,[106] even if such high irradiance inhibition did occur *in situ*, it would be of little quantitative significance. Consequently, equation (9.2) is taken as operative for $E_{PAR} \geq 2 E_k$.

The total daily phytoplankton photosynthesis (approximate primary production PP) throughout the water column would then be given as the integration of the P versus E_{PAR} curve over the depth of the photic zone and the length of the daylight period.

$$PP = \int_{t_d}^{t_s} \int_{z_{0.01}}^{z_{1.0}} P(z, t)dzdt \tag{9.6}$$

where $z_{1.0}$ = depth of the 100% irradiance level, i.e., just beneath the air-water interface

$z_{0.01}$ = depth of the 1% irradiance level, below which occurs no significant contribution to primary production

t_d = time of local dawn

t_s = time of local sunset.

Vollenweider[415] introduced a set of parameters to define the P vs E_{PAR} curve. The Fee[104] integration of the Vollenweider[415] model primary production model may be expressed as:

$$PP = P_{opt}\delta \int_{-\lambda/2}^{\lambda/2} \int_{0.01E(0^-,t)/E_k}^{E(0^-,t)/E_k} \frac{dydt}{K_{PAR}(t)\{(1 + y^2)[(1 + (ay)^2]^n\}^{0.5}} \tag{9.7}$$

where P_{opt} = optimum rate of photosynthesis per unit volume of water of chlorophyll concentration B,

a, n = parameters of the Vollenweider model,

δ = P_{max}/P_{opt} where P_{max} is the maximum rate of photosynthesis per unit volume (BP_m^B) when a or $n = 0$,

$E(0^-,t)$ = irradiance just below the air-water interface at time t,

y = $E(0^-,t)/E_k$ where E_k is the light saturation parameter [equation (9.4)] when a or $n = 0$,

λ = length of local day (sunrise to sunset), and

$K_{PAR}(t)$ = irradiance attenuation coefficient for PAR at time t.

In Fee's original integration of the Vollenweider model the irradiance attenuation coefficient was considered to be a constant determinable from Beer's Law and an irradiance depth profile. However, as discussed in Section 3.8 and illustrated in Figure 3.17, K_{PAR} displays a strong dependence on solar zenith angle. Thus, in equation (9.7) $K_{PAR}(t)$ is shown as a function of time.

In capsule, therefore, the procedure for estimating *in situ* phytoplankton photosynthesis is as follows:

1. Obtain direct measurements of the incident above water irradiance E_{PAR} as a function of time. This incident irradiance is a composite of direct solar irradiance and diffuse sky irradiance.

2. For a given instantaneous value of incident irradiance E_{PAR} determine the depth profile of PAR from either direct measurements of the subsurface irradiance levels or from predetermined values of the time-dependent attenuation coefficient $K_{PAR}(\theta)$.

3. Determine the local photosynthesis P versus E_{PAR} curve discussed above. This necessitates determinations of the primary production to obtain the transfer coefficients P_m^B and α^B. Water samples are generally collected for laboratory determinations of chlorophyll and organic carbon concentrations. Samples are laboratory incubated under a variety of PAR levels for a set incubation time, after which ^{14}C uptake is measured and photosynthesis rates calculated. Details of the techniques for such laboratory determinations of ^{14}C uptake and photosynthesis rates are given in Vollenweider,[414] Strickland and Parsons,[390] Stainton *et al.*,[384] Shearer *et al.*,[365] Fee *et al.*,[108] among others. The units of α^B are mg carbon (or oxygen) m² per mg Chl per einstein. The units of P_m^B are mg carbon (or oxygen) per mg Chl per hr.

4. Calculate the volumetric phytoplankton photosynthesis rate by integrating the local photosynthesis over the entire depth of the photic zone ($z_{1.0}$ to $z_{0.01}$).

5. Calculate the daily phytoplankton photosynthesis rate by integrating the volumetric phytoplankton photosynthesis over the daylight hours (t_d to t_s).

9.3 REMOTELY ESTIMATING PHYTOPLANKTON PHOTOSYNTHESIS

Ancillary *in situ* optical monitoring and direct sample collection can only complement the collection of remotely sensed data during intensive scientific field campaigns. To be cost effective, remote sensing of phytoplankton photosynthesis/primary production should be a reasonable surrogate for *in situ* monitoring, particularly for those aquatic regions whose dimension and/or geographic location prohibit direct access or continual sampling or both. Consistent with our discussions to this point, the procedure for remotely estimating phytoplankton photosynthesis would be as follows:

1. Obtain E_{PAR} as a function of time. If feasible, the aircraft or satellite could supply solar and/or sky irradiance data at the aircraft or satellite altitude. A suitable atmospheric intervention model could transform this irradiance to that which would be recorded as downwelling E_{PAR} at the air-water interface. It is also possible that some aquatic targets could be situated near stations that routinely monitor near-surface incident radiation. In many instances, however, it would be necessary to rely upon simulations of the solar and sky irradiances based on existing models and knowledge of the region being remotely sensed.

2. Transfer the downwelling irradiance field through the air-water interface to obtain the maximum value of E_{PAR} available for phytoplankton photosynthesis.

3. Obtain through on-site coordinated programs of direct optical monitoring and sample collection (either in concert with or prior to the remote sensing activity) the optical cross section spectra of the indigenous Chl*a*, SM, and DOC. If such direct determinations are not possible, optical cross section spectra appropriate to comparable waters *might* be of consequence.

4. Transfer the upwelling radiance spectrum recorded at the remote platform into the upwelling radiance spectrum that would be recorded at a remote platform located just above the water surface. It is essential to remember that the atmospheric correction algorithms applied to the radiance spectrum recorded over non-Case I waters must accommodate the convoluted signal emanating from such waters. Thus, some of the atmospheric models appropriately developed for use with oceanographic monitoring missions such as the CZCS and based upon the premise of negligible return from the 670-nm band are inappropriate for coastal and inland waters[430] due to the ubiquitous occurrence of nonzero subsurface volume reflectance at 670 nm in such waters.[37] Remote sensing of non-Case I waters requires either the use of atmospheric intervention models based upon the premise of negligible energy return in the near-infrared band[431] or upon atmosphere simulations such as the LOWTRAN 7 model,[232] which require input values of many atmospheric parameters. For most situations, however, obtaining a suite of atmospheric parameters is impractical, and models such as LOWTRAN contain a menu of pre-set possible atmospheric conditions to select for use in their calculations.

5. Transport the upwelling radiance spectrum immediately above the water surface through the air-water interface and convert this spectrum into the volume reflectance spectrum that would be observed immediately below the air-water interface.

6. As detailed in Chapter 8, use appropriate multivariate optimization techniques in conjunction with the optical cross section spectra to determine the co-existing concentrations of Chl*a*, SM, and DOC that were responsible for generating the subsurface volume reflectance spectrum estimated from the remotely recorded upwelling radiance spectrum.

7. Obtain the values of the pertinent primary production transfer coefficients P_m^B and α^B for the natural water body being studied. Ideally, this would involve direct water sampling and laboratory incubations followed by ^{14}C uptake and photosynthesis determinations. This, too, is most often an unaccessible avenue. Unfortunately, the value of possibly using optical cross section spectra of optically comparable water bodies as surrogates for inaccessible water bodies may not be reflected in such a use of possible surrogate transfer coefficients.

8. Use the primary production transfer coefficients along with the near-surface estimate of chlorophyll *a* concentration to estimate phytoplankton photosynthesis (approximate primary production).

The development of algorithms or techniques that will yield primary production (despite the inappropriateness of the nomenclature, we will refer

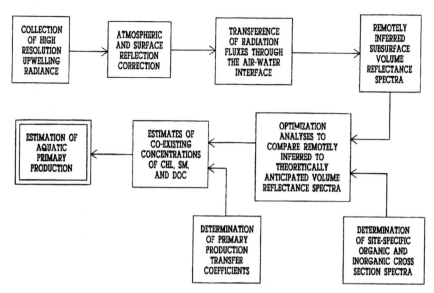

FIGURE 9.4
Flow diagram of activities essential to the remote estimation of aquatic primary production.

to phytoplankton photosynthesis as primary production in our discussions) as a function of near-surface chlorophyll *a* distribution is, of course, complicated by the non-uniform vertical profiles of chlorophyll-bearing biota defining natural waters.

The activities essential to the remote estimation of aquatic primary production are schematically illustrated in the flow diagram of Figure 9.4. In essence, two independent, albeit interrelated, activities proceed in tandem. Remote measurements of spectral radiance are performed over the aquatic target. *In situ* optical data and water samples are collected to obtain the organic and inorganic scattering and absorption cross section spectra, and depth profiles of PAR and primary production are used to determine the chlorophyll transfer coefficients P_m^B and α^B. Atmospheric correction, water quality, optimization, and primary production models are then utilized with the collected data and calculated model parameters in an effort to arrive at the remotely estimated magnitude of primary production/phytoplankton photosynthesis.

Gordon and McCluney[136] determined that for a sensor viewing the ocean at any wavelength λ, 90% of the recorded signal originates from within a so-called *penetration depth*, z_P, given by the inverse of the downwelling irradiance attenuation coefficient $K_d(\lambda)$. From equation (1.68), this corresponds to one attenuation length $\tau(\lambda)$, i.e., the depth, $z_{0.37}$, at which the downwelling irradiance $E_d(\lambda)$ falls to ~37% of its value just beneath the air-water interface. Thus, satellite observations over natural waters are restricted to the upper-

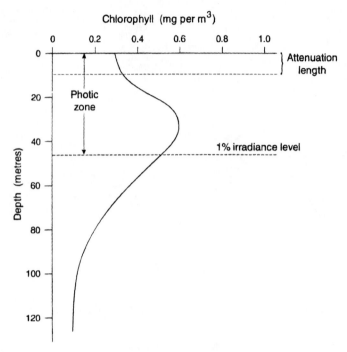

FIGURE 9.5
Hypothetical Case I water column (K_{PAR} = 0.1 m^{-1}) illustrating the relative locations of the photic zone, the chlorophyll depth profile, and the remote sensing penetration depth (the upper attenuation length).

most attenuation length ($0 \leq z \leq z_{0.37}$). The photic zone is considered to extend to the depth of the 1% irradiance level, i.e., to $z_{0.01}$. The photic zone, therefore, extends to 4.605 penetration depths. Classifying the biological nature of a water body on the basis of its surface layer is certainly non-optimal. Unfortunately, this shortcoming of satellite monitoring remains unresolvable.

Due to such physical and biological forcing functions as currents, density gradients, light fields, vertical migration, and growth cycles, the vertical profile of pigment concentrations in Case I waters is not uniform. The existence of a peak in the pigment profile, the *deep chlorophyll maximum (DCM)* is discussed by Cullen[70,71] and shown to be an extensive feature of oceans. This DCM is generally located within the photic zone and deeper than the penetration depth. Assuming a hypothetical value of mid-oceanic K_{PAR} of ~0.1 m^{-1}, the photic zone would extend to a physical depth of ~46 m. This hypothetical water column with its DCM is illustrated in Figure 9.5. Note that the chlorophyll concentration just beneath the air-water interface is illustrated as a value intermediate to the DCM and the concentration at great depths. Although not universal, such surficial chlorophyll concentration, as well as the general Gaussian shape of the chlorophyll-depth profile of Figure 9.5 is a prolific feature of Case I waters. Also note that the upper attenuation length (in this example the upper physical 10 m of ocean water) is the depth from which the remote estimate of chlorophyll is obtained. To account for such layering in the aquatic pigment profiles, techniques must be developed from which the primary production (phytoplankton photosynthesis) of the entire water column may be inferred from remotely sensed estimates of near-surface chlorophyll concentrations.

Consequences of such non-uniform pigment profiles to the remote sensing of ocean chlorophyll and biomass have been considered by Lewis *et al.*,[246] Platt *et al.*,[316] Platt and Sathyendranath,[317] and Sathyendranath and Platt.[349,350] The authors consider that pigment depth profiles, $B(z)$, such as illustrated in Figure 9.5 can be generalized, at time t, by a Gaussian curve superimposed upon a constant background, i.e.:

$$B(z) = B_0 + \frac{h}{\sigma\sqrt{(2\pi)}} \exp\left[-\frac{(z - z_m)^2}{2\sigma^2} \right] \qquad (9.8)$$

where
$\quad z_m$ = depth of the chlorophyll maximum
$\quad \sigma$ = standard deviation or spread of z_m
$\quad B_0$ = constant background chlorophyll concentration
$h/[\sigma\sqrt{(2\pi)}]$ = factor that determines the height of the chlorophyll maximum above B_0.

As the depth of the chlorophyll maximum, z_m, increases, the ability of the satellite to directly observe it decreases. The spread σ of this maximum would not significantly alter the radiance spectrum recorded at the satellite provided the chlorophyll peak remains well contained within the photic zone. For values of z_m closer to the surface, however, the value of σ can impact the reading at the satellite since as the spread becomes larger, more chlorophyll can become accessible to remote observation. For near-surface values of z_m, only the lower portion of the Gaussian distribution will lie outside the photic zone and a substantial percentage of the total pigment concentration might be found above the 37% irradiance level. In such cases, near-surface estimates of surface chlorophyll could provide more realistic estimates of the oceanic primary production, provided, of course that these near-surface values of z_m can be verified or confidently assumed. It must be remembered, however, that as the chlorophyllous pigment concentration increases, the value of K_{PAR} increases and the magnitude of the attenuation length diminishes.

Following Smith,[372] Platt *et al.*[319] express the depth dependence of primary production upon available subsurface irradiance at the remote sensing site by the relation:

$$P(z) = \frac{B(z)\alpha^B E(z)}{\{1 + [\alpha^B E(z)/P_m^B]^2\}^{1/2}} \qquad (9.9)$$

where $P(z)$ is the instantaneous primary production at depth z. α^B, P_m^B, and $E(z)$ are as previously defined, and $B(z)$ is the non-uniform chlorophyll depth profile of equation (9.8). Although not obligatory, it is not unreasonable to assume, at least for Case I waters, that the transfer coefficients α^B and P_m^B are depth invariant. Denman and Gargett[82] observed that the physical barrier imposed by a shallow thermocline resulted in disparate phytoplanktonic photoadaptation profiles within each of the layers (i.e., distinct slopes of the P vs. E_{PAR} curves). They also observed that the diurnal cycle of vertical motion of phytoplanktonic cells within the upper layer generally proceeds slowly enough to enable the cells to adapt to the high irradiances near the air-water interface, yet rapidly enough to redistribute the photosynthesis per unit chlorophyll roughly evenly throughout the epilimnion.

As written, equation (9.9) accommodates the assumption that the slope α^B is linear and that $E(z)$ is $E_{PAR}(z)$. For the general case of a known spectral depth distribution of PAR, $E(z,\lambda)$, and a known non-linear slope function $\alpha^B(z,\lambda)$, such wavelength dependency should also be accommodated. Further, the subsurface irradiance field, $E(z,\lambda)$, is a function of in-water refracted solar zenith angle θ_r and can be partitioned into a direct sunlight component, $E_d(z,\lambda,\theta_r)$, and a diffuse skylight component, $E_{sky}(z,\lambda)$. It is the daylight progressions of these functions E_d and E_{sky} that must be constructed from the solar constant $E_{solar}(\lambda)$, the atmospheric transparency coefficents $T_i(\lambda)$, the cloud cover, and the aquatic attenuation coefficients.

To incorporate a depth-dependent and non-linear slope function, a spectral and angular dependence of the subsurface PAR light field, and a solar and sky partitioning of $E(z,\lambda)$ into the photosynthesis-light consideration, equation (9.9) may be rewritten (Platt et al.[319]) as:

$$P(z) = \frac{B(z)\Lambda(z)}{\{1 + [\Lambda(z)/P_m^B]^2\}^{1/2}} \qquad (9.10)$$

where $\Lambda(z) = \sec\theta_r \int \alpha^B(z,\lambda)E_d(z,\lambda,\theta_r)d\lambda + 1.2 \int \alpha^B(z,\lambda)E_{sky}(z,\lambda)d\lambda$.

Integration of equation (9.9) or (9.10) over the photic zone for the daylight hours [equation (9.6)] then will yield the phytoplankton photosynthesis (approximate primary production) for the aquatic region being remotely monitored.

9.4 PHYTOPLANKTON PRIMARY PRODUCTION FROM ESTIMATES OF AQUATIC CHLOROPHYLL

Equations (9.8) and (9.10) address two of the principal concerns of the remote estimation of aquatic primary production, namely (a) a non-uniform distribution of aquatic chlorophyll concentration with depth and (b) a non-linear relationship between photosynthesis P and irradiance E_{PAR}. If B_{RS} is the instantaneous value of near-surface (upper attenuation length) chlorophyll concentration estimated from a remote platform, this B_{RS} must be converted into the instantaneous photic zone chorophyll depth profile $B(z)$ at the time of observation.

For Case I waters Morel[272] has shown that a statistically significant relationship exists between the total areal pigment content B_{TOT} (taken as the depth profile of Chl$a$ concentration uncorrected for phaeophytin contamination integrated over the photic zone, in mg m^{-2}) and the depth of the photic zone $z_{0.01}$ (depth in m of the 1% irradiance level):

$$B_{TOT} = 4910 z_{0.01}^{-1.34} \qquad (9.11)$$

A "mean" chlorophyllous pigment concentration B_m (in mg of Chl$a$ plus phaeophytin per m^3) would then be defined as B_{TOT} divided by $z_{0.01}$.

Morel[272] also reports that for Case I waters, the "mean" attenuation coefficient for downwelling PAR, $K_{PAR}(z_{1.0}$ to $z_{0.01})$, can be related to the "mean" chlorophyllous pigment concentration B_m by:

$$K_{PAR}(z_{1.0} \text{ to } z_{0.01}) = 0.121 B_m^{0.428} \qquad (9.12)$$

Equations (9.11) and (9.12), which may facilitate *in situ* determinations of oceanic primary production, are of reduced value for remote determinations, however, if the upper attenuation length of the ocean is not predictably representative of photic zone chloropyllous pigment. [Equations (9.11) and (9.12), of course, do not hold for inland and coastal waters since such waters are also under the influence of sediments and/or dissolved organic matter].

Gordon and Clarke[138] have introduced the concept of an *equivalent homogeneous ocean*, wherein the stratification of the optical properties of a layered ocean can be expressed in terms of a depth-weighted (over the uppermost attenuation length) volume reflectance $R_{dw}(\lambda)$, which is itself a function of a corresponding depth-weighted ratio of scattering and absorption coefficients, $[b(\lambda)/a(\lambda)]_{dw}$ which, as we have been discussing, is, in turn, a function of the co-existing organic and inorganic aquatic components. For Case I waters each of the optical properties $b(\lambda)$ and $a(\lambda)$ can be expanded in terms of the optical cross section spectra of pure seawater $[a_w(\lambda)$ and $b_w(\lambda)]$, phytoplankton $[a_{Chl}(\lambda)$ and $b_{Chl}(\lambda)]$, and yellow substance $[a_{YS}(\lambda)$ and $b_{YS}(\lambda)]$.

Consistent with the *equivalent homogeneous ocean concept*, Sathyendranath and Platt[350] show that:

$$[b(\lambda)/a(\lambda)]_{dw} = \int (b/a)(z)f(z)dz \bigg/ \int f(z)dz \qquad (9.13)$$

and that the signal from a layered chlorophyll distribution $B(z)$ recorded at a satellite at any wavelength would be equivalent to the signal recorded by that satellite over a homogeneous water column of satellite-weighted chlorophyll concentration $B_w(\lambda)$ given by:

$$B_w(\lambda) = \int B(z)f(z)dz \Big/ \int f(z)dz \qquad (9.14)$$

In both equations (9.13) and (9.14), the integration is performed over the penetration depth (i.e., from the air-water interface to the depth of the 37% irradiance level, $z_{0.37}$).

The weighting function $f(z)$ is given by:

$$f(z) = \exp\left[-\int_0^z 2K(\lambda, z')dz' \right] \qquad (9.15)$$

where $K(\lambda,z')$ is the downwelling irradiance attenuation coefficient at depth z' for a wavelength λ. [The inverse of $K(\lambda,z)$ defines the depth of the upper attenuation length, i.e., penetration depth].

As we have discussed in Chapter 8, the single component algorithms that have generated oceanic chlorophyll maps are a consequence of regressing chlorophyll concentrations with ratios of radiances recorded at different wavelengths, λ_1 and λ_2. Each of these radiances is a consequence of a wavelength-dependent depth-weighted volume reflectance [i.e., a consequence of the sets of absorption and scattering coefficients $a(\lambda_1)$ and $b(\lambda_1)$; $a(\lambda_2)$ and $b(\lambda_2)$]. Two values of wavelength-dependent satellite-weighted chlorophyll concentration (which may or may not be of significant variance) result, B_w (λ_1) and $B_w(\lambda_2)$.

Assisting to a significant degree in estimating optical behaviour of Case I waters is the co-varying nature of chlorophyll and yellow substance concentrations in many regions. Thus, if the shape of the pigment profile is either known or can be reasonably approximated, chlorophyll concentrations of the oceanic column can be confidently extracted from remote observations of the upwelling radiances. Sathyendranath and Platt[350] illustrate, however, that the assumption of a homogeneous ocean can lead to relative errors in excess of 100% in satellite estimations of both total pigment content and photic depths of Case I waters. This renders all the more understandable the discrepancies among the oceanic chlorophyll concentration extraction algorithms illustrated in Figure 8.4.

Recapping, the most significant factors governing the photosynthetic processes in natural waters include:

1. The spectral composition and intensity of the solar and sky photon flux incident upon the air-water interface,

2. The attenuation of this incident flux in its subsurface propagation,

3. The presence, relative populations, and volumetric distributions of chloro-phyll-bearing biota and the photosynthetically-active pigments con-tained therein.

The subsurface attenuation of incident radiation is directly controlled by the concentrations and inherent optical properties of the organic and inorganic matter comprising the water column. In deep-sea waters, phyto-plankton are both the principal photosynthesizers and the principal attenua-tors (other than the water itself) of subsurface light. The bio-optical properties of oceans are, therefore, tightly controlled by the indigenous pigmented algal cells (and to a subordinate degree by their derived detrital products). Thus, the above factors are interrelated in a more readily discernible manner in oceanic waters than they are in inland and coastal regimes.

9.5   CHLOROPHYLL AND BIOMASS OF OCEAN WATERS

The Coastal Zone Color Scanner (CZCS), launched aboard the Nimbus-7 satellite in 1978 and remaining operational until 1986, has enabled assess-ments of the oceanic distributions of chlorophyll concentrations and *phyto-plankton biomass* (Putnam,[338] Müller-Karger et al.[278]). To qualify as a unit of measurement for living matter, however, biomass requires a focus on a biological parameter. For particular studies, biomass has been oftimes related to such specific parameters as chlorophyll, carbon, or nitrogen. Consistent with Strickland,[389] Platt and Irwin,[315] and others, we will define phytoplank-ton biomass to be the organic carbon content of the phytoplankton popula-tion. Clearly, however, such a definition requires a relationship between the chlorophyll and the organic carbon content of algae, a relationship that can be highly variable in both space and time. This relationship becomes all the more essential if remote sensing is to be applied to the monitoring of aquatic biomass. Measuring the organic carbon content of natural phytoplankton in an environment free from the contaminative impacts of detritus and zooplankton, however, is virtually impossible, and so the rather unfortunate circumstance arises whereby chlorophyll and a derived or postulated carbon-to-chlorophyll ratio are required to remotely estimate aquatic biomass. There-fore, it is generally assumed that a remote estimate of near-surface chloro-phyll concentration constitutes a remote estimate of near-surface biomass.[138] For more detailed discussions see Eppley and Peterson,[98] Harris,[158] Bannister and Laws,[18] Putnam,[338] Platt and Sathyendranath,[318] and others.

9.6   CHLOROPHYLL AND BIOMASS OF INLAND AND COASTAL WATERS

The impact of land-derived nutrients on the primary production of once oligotrophic oceanic waters is becoming increasingly significant,[420] thereby focusing concern on semi-enclosed basins such as the Caribbean Sea, the Baltic Sea, the Mediterranean Sea, the Gulf of Mexico, the Gulf of Finland, the Gulf of St. Lawrence, and other areas. The biological productivity of inland waters is subject to similar focussed concerns. The presence of a variety of non-co-varying colored components in inland and coastal waters prohibits the use of single-component algorithms to extract the near-surface chlorophyll concentration from a remote measurement of the upwelling radiance spectrum. This necessitates the use of multi-component concentration algorithms of the type described in Chapter 8 to extract near-surface chlorophyll concentrations from their backgrounds of co-existing organic and inorganic matter. For coastal waters the problem often, but not always, reduces to resolving phytoplankton pigments from dissolved organic matter (DOM). Phytoplankton and DOM may or may not co-vary. Carder et al.[55,56] utilize specific inherent optical properties as a means of quantifying chlorophyll in the presence of DOM. For inland and coastal waters the problem invariably remains resolving phytoplankton pigments from DOM and suspended inorganic matter (SIM). The DOM and SIM, in addition to rendering single component concentration algorithms inappropriate, also invariably reduce the penetration depth z_P [due to the higher value of $K_d(\lambda)$], as well as impacting the depth profiles of chloropyllous pigments. Profiles containing one or more maxima are readily observed in inland waters of varying degrees of clarity. The assumption of a Gaussian distribution describing aquatic pigment depth profiles, therefore, is not always a valid one.

Near-surface chlorophyll concentrations (along with the concurrent concentrations of suspended solids and dissolved organic carbon) can be extracted from remote spectral observations if the regional aquatic component cross section spectra are either known or can confidently be assumed. If cautiously approached, the concept of an *equivalent homogeneous water body* could be equally as valid as the concept of an *equivalent homogeneous ocean*, and much of the ongoing work in estimating the primary production of oceans on the basis of their near-surface chlorophyll concentrations could be directly applied to estimating the primary production of inland and coastal waters on the basis of *their* near-surface chlorophyll concentrations.

As an attempt to overcome the perverse variabilities in the depth profiles of chlorophyll concentrations in inland and coastal waters, it might be worth considering an approach suggested for Case I waters by Morel and Berthon.[275] This approach considered that two extreme aquatic configurations could accommodate the chlorophyll concentration profiles of oceanic waters. The first extreme is taken to be stratified oligotrophic oceanic waters wherein nutrient depletion is observed in the near-surface layer. In this case the maximum chlorophyll concentration (DCM) would be downward-driven. The second extreme is taken to be well-mixed waters wherein sufficient irradiance levels have established eutrophic conditions. In this case the vertical distribution of chlorophyll, $B(z)$, throughout the photic zone would

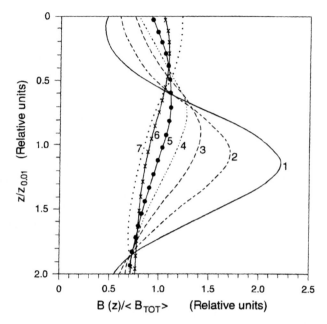

FIGURE 9.6
Seven generic dimensionless chlorophyll depth profiles for Case I waters. (Adapted from Morel, A. and Berthon, J.-F., *Limnol. Oceanogr.*, 34, 1545–1562, 1989).

be quite uniform. Understandably, a wide range of intermediate chlorophyll profiles within these extremes are encountered in natural waters. It is perhaps not unreasonable, however, to reduce this wide range of profiles to a limited number of generic chlorophyll depth distributions pertinent to specific types of water columns (and on through to specific water bodies) depending upon the trophic status of the water, the aquatic vegetation growth cycle, and the intensity and duration of the wind-induced vertical mixing. In general the average total chlorophyll content of the photic zone, $<B_{TOT}>$ can be given as:

$$<B_{TOT}> = (z_{0.01})^{-1} \int_0^{z_{0.01}} B(z)dz = (z_{0.01})^{-1}B_{TOT} \qquad (9.16)$$

where $z_{0.01}$ is the depth to the 1% subsurface irradiance level (i.e., the depth of the photic zone) and B_{TOT} is the total chlorophyll concentration residing in the photic zone.

Statistically analyzing over 4000 oceanic station data sets, Morel and Berthon illustrated the presence of seven generic classifications of oceanic waters. These generic profiles are schematically illustrated in Figure 9.6. The seven profiles are plotted in terms of a dimensionless chlorophyll concentration, $B(z)/<B_{TOT}>$, and a dimensionless depth $z/z_{0.01}$, i.e., the depth is reckoned as a fraction of the photic depth, and the chlorophyll concentration is reckoned as a fraction of the total photic zone chlorophyll. Curves 1 and 7 represent the chlorophyll depth profile limits of stratified oligotrophic and well-mixed eutrophic waters, respectively. As natural waters depart from oligotrophy and approach eutrophy, Curves 2 through 6 illustrate a progressive collapse of the range of chlorophyll values structuring the vertical concentration profiles along with a corresponding upward migration of the DCM.

Curves such as illustrated in Figure 9.6 illustrate that for some inland and perhaps more coastal waters whose $B(z)$ depth profiles display a single maximum (i.e., waters whose photic zone depth can be reliably related to total photic zone chlorophyll concentration), equations (9.8), (9.9), and (9.10) can be applied to primary production estimates, provided, of course, that the seven $B(z)$ distributions represented in the Morel and Berthon work can be supplied with appropriate values of $\alpha^B(z,\lambda)$, P_m^B, σ, h, z_m, and θ_r. Values of $E_d(z,\lambda,\theta_r)$ and $E_{sky}(z,\lambda)$ could be either determined from numerical simulation models or derived from direct actinometric measurements obtained from ground-based or coastal zone network stations.

Unfortunately, in non-Case I waters, the levels of light energy available for photosynthesis and the depth of the photic zone are not totally controlled by the chlorophyll content, B_{TOT}. They are also controlled by the presence of other optically responsive aquatic components that do not necessarily co-vary with B_{TOT}. Uncontrolled and unpredictable interplay of these optically competitive components generally renders the use of equations (9.8), (9.9). and (9.10) considerably less attractive.

For inland and coastal waters displaying vertical inhomogeneities in their optical properties the approach to the remote estimation of phytoplankton photosynthesis (approximate primary production PP) should be based upon integration of a P versus E_{PAR} curve that accommodates a non-linear $\alpha^B(z,\lambda)$ and responds to a chlorophyll depth distribution that does not necessarily display a single maximum.

The irradiance attenuation coefficient for PAR, $K_{PAR}(z,\lambda,\theta_r)$, is controlled by the time-dependent concentrations of aquatic chlorophyll, suspended sediments, and dissolved organic matter. The depth-dependency of $K_{PAR}(z,\lambda,\theta_r)$ emphasizes the importance of the vertical distribution of aquatic chlorophyll concentration for equation (9.10) to provide a realistic estimate of aquatic primary production.

It might not be unreasonable to suggest that a thorough statistical analysis of the variety of $B(z)$ profiles encountered in the photic zones of inland and coastal waters might also reveal a limited number of generic dimensionless chlorophyll profiles of the type determined for oceanic waters by Morel and Berthon. These could then be utilized for the photic zone depth integration of equation (9.10). The multivariate optimization technique discussed in Chapters 5 and 8 enables remote estimates of the co-existing concentrations of the three major optically responsive aquatic components (chlorophyll, suspended sediments, and dissolved organic carbon). Thus, it would be possible to assess the values of K_{PAR} and $z_{0.01}$ in terms of attenuation by not only chlorophyll, but by its optical competitors as well. This assessment could be accomplished by using judiciously selected optical cross section spectra $[a_i(\lambda), b_i(\lambda) B_i(\lambda)]$ obtained either from *a priori* knowledge of the water body or from previously determined local spectra. On the basis of these non-Case I dimensionless $B(z)$ profiles ($z/z_{0.01}$ versus $B(z)/<B_{TOT}>$) and the remotely estimated near-surface chlorophyll concentration B_w, the chlorophyll distribution over the photic zone might be estimable.

The vertical profiles of suspended sediments and dissolved organic matter in these waters remain problematical. In relatively clear Case II waters the values of K_{PAR} generally are in the range 0.1 m^{-1} to 0.5 m^{-1} (see the Great Lakes examples in Chapter 7) resulting in penetration depths z_p of 10 m to 2 m being commonly observed in open waters. The corresponding photic zones in these waters would extend to depths $z_{0.01}$ of about 46 m to about 9 m. Thus, for relatively clear open inland and coastal waters, a situation exists that is comparable to that for oceanic waters, namely, the photic zone generally has a greater vertical extension than the near-surface layer being remotely-sensed. For relatively turbid Case II waters (again see Chapter 7) the values of K_{PAR} generally are in the range 0.5 m^{-1} to 1.2 m^{-1}, resulting in penetration depths in the range 2 m^{-1} to 0.8 m^{-1} and photic depths in the range 9 m to 4 m. Again the photic zone depth exceeds the depth of water being remotely sensed. However, for the conditions of thermal lake stratification (well-defined epilimnion and hypolimnion), the photic zone often coincides approximately with the epilimnion. The epilimnion is the lake layer which most often displays vertical mixing with regard to its suspended mineral and dissolved organic matter content. Thus, although the vertical profiles of chlorophyll concentrations in non-Case I waters may be most irregular, a vertical homogeneity of SM and DOM concentrations in these waters may very often be a not-unreasonable assumption (Petrova[307]). The vertical profiles of SM and DOM are not unlike the dimensionless chlorophyll profiles for well-mixed eutrophic ocean waters (Curve 7 of Figure 9.6). This epilimnetic vertical mixing (and the resulting vertical invariance that may be ascribed to the SM and DOM concentration profiles) emphasizes the value of simultaneously extracting penetration depth concentrations of suspended sediments and dissolved organic carbon (along with the chlorophyll concentration) from the upwelling radiance spectrum.

9.7 PROTOCOL FOR REMOTELY ESTIMATING PRIMARY PRODUCTION IN NON-CASE I WATERS

On the basis of the discussions in the previous sections, a remote assessment of the diurnal phytoplankton primary production for inland and coastal

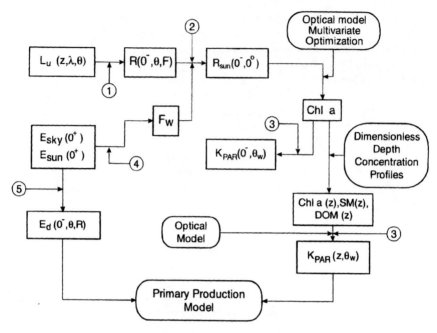

FIGURE 9.7

Generalized flow diagram outlining parameters and sequential operations involved in remotely estimating primary production in non-Case I waters. The mathematical formulism is described in the text.

waters could be performed provided the following information were available:

1. *A priori* knowledge of the (possibly generic) dimensionless chlorophyll profiles,

2. Diurnal records (or model estimations) of direct and diffuse solar irradiance (the global radiation) at the air-water interface, and

3. Remotely sensed estimates of the co-existing concentrations of chlorophyll, suspended minerals (or sediments), and dissolved organic matter. This requires the use of multivariate optimization analyses with appropriate optical cross section spectra.

Figure 9.7 is a generalized flow diagram outlining the parameters and sequential operations that we suggest in attempting a remote estimation of the diurnal phytoplankton primary production in optically competitive non-Case I waters. Operations commence at the upper left-hand corner with the recording of an upwelling radiance spectrum $L_u(\lambda, \theta)$ at the remote platform. To facilitate the interpretation of Figure 9.7, numerals are inserted throughout the diagram that represent the mathematical formulisms required to move along the flow diagram. These sequential mathematical formulisms and their purposes are listed below. For simplicity, wavelength dependence will not be explicit but implied. In describing the protocol of Figure 9.7, we will reiterate material that we have covered in previous Chapters.

1. Reiterating from Section 8.1, the recorded upwelling radiance spectrum, $L_u(z,\lambda,\theta)$, just above the air-water interface ($z = 0^+$) is converted into the subsurface volume reflectance spectrum $R(z,\lambda,\theta)$ just below the air-water interface ($z = 0^-$).

$$R(0^-, \theta) = \frac{Q[L_u(0^+, \theta) - 0.0212\alpha E_{sky} - f_2 E_{sun}]}{0.544(0.944 E_{sky} + f_4 E_{sun})} \tag{9.17}$$

where Q = ratio of the upwelling irradiance below the water surface to the upwelling nadir radiance below the water surface,

α = spectrally dependent ratio of the downwelling zenith sky radiance to the downwelling sky irradiance,

f_4 = reflection of the solar irradiance due to surface waves,

f_2 = reflection of the solar irradiance into the field of view of the sensor due to surface waves,

E_{sun} = spectrally dependent downwelling irradiance from the sun, and

E_{sky} = spectrally dependent downwelling irradiance from the sky.

2. The subsurface volume reflectance just beneath the air-water interface $R(0^-)$ will, of course, be remotely estimated under variable conditions of solar zenith angle θ and relative fractions of diffuse and direct downwelling global radiation. Thus, it is necessary to normalize the subsurface volume reflectance to a standard solar zenith angle and a standard downwelling radiation. Again reiterating from Section 8.1, the estimated subsurface volume reflectance just beneath the air-water interface can be compartmentalized into components responsive to the direct and diffuse components of the global radiation.

$$R(0^-, \theta, F) = (1 - F)R_{sun}(0^-, \theta) + FR_{sky}(0^-) \tag{9.18}$$

where F = the fraction of the total downwelling radiation that is diffuse. The fraction that is direct would then be given by $(1 - F)$.

$R_{sun}(0^-,\theta)$ = the volume reflectance just beneath the air-water interface resulting from the direct fraction of the global radiation. $R_{sun}(0^-,\theta)$ is dependent upon θ.

$R_{sky}(0^-)$ = the volume reflectance just beneath the air-water interface resulting from the diffuse fraction of the global radiation. $R_{sky}(0^-)$ is independent of θ.

Results of Monte Carlo simulations[44] relating the effects on $R_{sun}(0^-,\theta)$ and $R_{sky}(0^-)$ of varying θ have shown that the subsurface volume reflectance, $R_{sun}(0^-,0^\circ)$, observable for a vertically overhead sun ($\theta = 0^\circ$), may be related to the remotely estimated subsurface volume reflectance $R(0^-,\theta,F)$ by:

$$R_{sun}(0^-, 0^\circ) = R(0^-, \theta, F)/[1.165F + (1 - F)(\cos\theta_r)^{-1}] \quad (9.19)$$

where θ_r is the in-water refraction angle associated with a solar zenith angle of θ.

It is this normalized value $R_{sun}(0^-,0^\circ)$ of the remotely estimated subsurface volume reflectance that may be used to provide comparisons of water masses remotely sensed at varying times of day (solar zenith angle θ) and varying conditions of global radiation (diffuse fraction F). From such internally comparable sets of normalized subsurface volume reflectance values and appropriate optical cross section spectra, the near-surface (penetration depth) chlorophyll a concentrations (along with co-existing concentrations of SM and DOC) may be estimated (Chl$a$ in Figure 9.7).

3. Reiterating from Section 3.8, the expressions derived by Kirk[227] can then be used to obtain the attenuation coefficients for direct incident solar radiation, $K_{direct}(\lambda, \theta_r)$, and for diffuse incident sky radiation, $K_{sky}(\lambda)$, namely:

$$K_{direct}(\lambda, \theta_r) = (\cos\theta_r)^{-1}[a(\lambda)^2 + (0.473\cos\theta_r \quad (9.20)$$
$$- 0.218)a(\lambda)b(\lambda)]^{1/2}$$

and

$$K_{sky}(\lambda) = 1.168[a(\lambda) + 0.168a(\lambda)b(\lambda)]^{1/2} \quad (9.21)$$

where $a(\lambda)$ and $b(\lambda)$ are the bulk absorption and bulk scattering coefficients for the aquatic medium ($a = \Sigma a_i$ and $b = \Sigma b_i$, with i representing H_2O, Chl$a$, SM, and DOM), and θ_r is the refracted solar zenith angle as measured immediately below the air-water interface.

For an incident PAR radiation comprised of a diffuse sky fraction F, the combined irradiance attenuation coefficient K_{PAR} is given as:

$$K_{PAR}(\lambda, \theta_r) = FK_{sky}(\lambda, \theta_r) + (1 - F)K_{direct}(\lambda, \theta_r) \quad (9.22)$$

where $F = E_{sky}(\lambda)/[E_{sun}(\lambda) + E_{sky}(\lambda)] = E_{sky}/E_{PAR}$.

4. The values of F are obtained from the measured or modelled values of E_{sky} and E_{sun}, namely, $F = E_{sky}/(E_{sky} + E_{sun}) = E_{sky}/E_{PAR}$. To accommodate for the value of F (the diffuse fraction of the above surface downwelling radiation) being transferred through the air-water interface, a subsurface value F_w (the diffuse fraction of the downwelling irradiance just beneath

the air-water interface) should be employed in the determination of the attenuation coefficient $K_{PAR}(\theta_r)$ (equation 9.22), i.e.,

$$K_{PAR}(\lambda, \theta_r) = F_w K_{sky}(\lambda, \theta_r) + (1 - F_w)K_{direct}(\lambda, \theta_r) \qquad (9.23)$$

F is readily transformed into F_w via the Fresnel reflection coefficients, i.e.,

$$F_w = F(1 - \rho_{sky})/[F(1 - \rho_{sky}) + (1 - F)(1 - \rho_{direct}(\theta)] \qquad (9.24)$$

where ρ_{sky} = Fresnel reflectivity of sky irradiance from a flat air-water interface and

$\rho_{direct}(\theta)$ = Fresnel reflectivity of solar irradiance directly propagating from the zenith angle θ.

5. For a calm water surface, the downwelling subsurface irradiance $E_d(0^-,\theta,R)$ can be obtained either directly from on-site measurements or indirectly estimated from above-surface values of downwelling irradiance $E_d(0^+,\theta,R)$ through the relationship:

$$E_d(0^-, \theta, R) = \left[1 + \frac{R(\theta)\rho_u(\theta)}{1 - R(\theta)\rho_u(\theta)}\right][1 - \rho_d(\theta)]E_d(0^+, \theta, R) \qquad (9.25)$$

where $\rho_d(\theta)$ = coefficient of reflection of above-surface downwelling irradiance for solar zenith angle θ and

$\rho_u(\theta)$ = coefficient of reflection of subsurface upwelling irradiance for solar zenith angle θ.

Jerome et al.[204] have shown that

$$\rho_u(\theta) = 0.271 + 0.249\mu_0 \qquad (9.26)$$

for direct incident solar radiation, and

$$\rho_u = 0.561 \qquad (9.27)$$

for a cardioidal diffuse incident radiation distribution. μ_0 is defined, from Snell's Law, as $\cos[\sin^{-1}(n \sin\theta)]$, n being the relative refractive index of the water. Values of $\rho_d(\theta)$ may be calculated from Fresnel's equation, while ρ_d for a diffuse cardioidal distribution can be taken as 0.066 (Jerlov[198]).

6. The primary production model to which the $K_{PAR}(z,\theta_r)$, $E_d(0^-,\theta,R)$, and remotely estimated chlorophyll concentration are to be used as inputs must be selected. The Vollenweider-Fee primary production model [equation (9.7)] considers a smooth P versus E_{PAR} curve and has been demonstrated to work

well for *in situ* primary production studies of inland waters. The Platt *et al.*
model [equation (9.10)] allows for a non-linear slope of the P versus E_{PAR}
function and uses a Gaussian distribution to define the subsurface ocean
chlorophyll depth profile $B(z)$. This approach has proven successful in the
remote sensing of primary production in Case I waters. Both models require
sets of input parameters. Assuming selection of the Platt *et al.* primary
production model, values of the transfer coefficients $\alpha^B(z,\lambda)$ and P_m^B must be
obtained, as well as expressions for non-Case I pigment depth profiles $B(z)$
that may or may not follow the Gaussian distribution of equation (9.8). The
relationships among $B(z)$ and the concurrent depth profiles of suspended
particulates and dissolved organic matter in inland and coastal waters should
also be investigated. Thus, one area of currently required research is to
establish such depth profiles and test the validity of possibly cataloguing
generic chlorophyll profiles for specific inland and coastal regimes.

Recapping in dialogue, the suggested series of operations (Figure 9.7)
for remotely estimating the primary production in inland and coastal (as
well as oceanic since Case I waters can be regarded as a subset of non-Case
I waters) may be outlined as follows:

1. Obtain an upwelling radiance spectrum of high spectral resolution, $L_u(\lambda,\theta)$,
   above the water body being studied or monitored. Also directly measure
   or otherwise obtain/estimate the global radiation at the air-water interface,
   that is measure, model, or estimate E_{sun}, E_{sky}, and F.

2. Convert the atmospherically adjusted $L_u(\lambda,\theta)$ spectrum into the subsurface
   volume reflectance spectrum $R(\lambda,\theta)$ just beneath the air-water interface.

3. Use the multivariate optimization technique in conjunction with *a priori*
   or directly determined pertinent scattering and absorption cross section
   spectra to deconvolve the subsurface volume reflectance spectrum into co-
   existing concentrations of near-surface chlorophyll, suspended sediments,
   and dissolved organic carbon (or additional components if the bio-optical
   model discussed throughout this book can be expanded).

4. From the global radiation and the Fresnel reflectivities, determine the
   downwelling irradiance just beneath the air-water interface.

5. Determine the attenuation coefficients for the direct, diffuse, and total
   subsurface PAR irradiance, i.e., determine $K_{direct}(\lambda,\theta_r)$, $K_{sky}(\lambda,\theta_r)$, and
   $K_{PAR}(\theta_r)$. The subsurface diffuse fraction F_w is required for K_{PAR}.

6. From *a priori* knowledge of the area, from direct measurements in the area,
   from model outputs, or from mathematical simulations, obtain reliable
   estimates of the chlorophyll, suspended mineral, and dissolved organic
   carbon dimensionless depth profiles $z/z_{0.01}$ versus $B(z)/<B_{TOT}>$, $SM(z)/$
   $<SM_{TOT}>$, and $DOC(z)/<DOC_{TOT}>$, respectively.

7. Use the Platt *et al.* model [equations (9.8) through (9.10)] or comparable
   primary production model to determine phytoplankton primary
   production.

9.8   AVAILABLE LIGHT FOR PHOTOCHEMICAL AND PHOTOBIOLOGICAL ACTIVITY

To realistically evaluate photochemical and photobiological reactions occurring in natural water bodies, one must possess precise information regarding the actual energy density to which the chemical or biological component is exposed. This energy density is crucial to any estimate of the rates at which these reactions will proceed.

Global radiation, comprising a direct solar and a diffuse sky irradiance downwelling on the air-water interface, possesses inherent geometric properties related to the arrival directions of the impinging radiation. Subsurface downwelling irradiances determined from such global radiation values will contain corresponding directional biases. A photosynthesizing phytoplankton cell suspended in water receives light energy not only from the downwelling irradiance [the E_{PAR} entries of the equations of the previous Sections, E_{PAR} being the PAR wavelength band of $E_d(\lambda)$] but also from the upwelling irradiance to which it is subjected. Consequently, it is the *scalar irradiance*, $E_o(z)$, at a particular wavelength λ, rather than the vector irradiance, $E_d(z)$, at that wavelength which is more representative of the cell's light regime. The scalar irradiance is the total energy per unit area arriving a point from all directions when all directions are equally weighted. Thus, $E_o(z)$ is the integration of the radiance distribution $L(\theta,\phi)$ at a point over 4π space. When divided by the speed of light in water, E_o becomes a measure of the energy density at a given point in the aquatic medium. As discussed in Section 1.3, despite its omnidirectional nature, a quasi-directionality is conveniently ascribed to scalar irradiance, giving rise to downwelling scalar irradiance E_{od} and upwelling scalar irradiance E_{ou} as counterparts to downwelling vector irradiance E_d and upwelling vector irradiance E_u. From Preisendorfer,[328] the scalar irradiance is related to the upwelling and downwelling irradiances $E_u(z)$ and $E_d(z)$ and the total absorption coefficient $a(z)$ by the relationship:

$$E_o(z) = -\frac{1}{a(z)} \frac{d}{dz} [E_d(z) - E_u(z)] \tag{9.28}$$

From curve-fitting to Monte Carlo computer simulations for a homogeneous water column, Jerome et al.[204] determined the ratio E_o/E_d as a function of volume relectance $R(\theta_r = 0°)$ and θ_r (the in-water refraction angle corresponding to a given collimated incident beam from a solar zenith angle θ) for various depths z within the water column. They showed that at all depths throughout the photic zone and for all solar zenith angles the ratio of scalar irradiance to downwelling irradiance (E_o/E_d) was numerically >1 and steadily increased with increasing subsurface volume reflectance. At the lower limit of the photic zone $(z_{0.01})$, this ratio may reach or even exceed 2. This is illustrated in Figure 9.8 taken from Jerome et al.[204] (their Figure 6). Therein

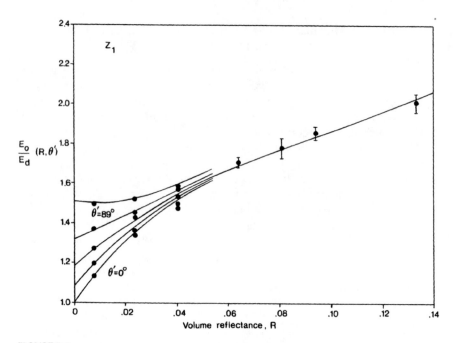

FIGURE 9.8

Monte Carlo determinations of E_o/E_d at photic depth ($z_{0.01}$) as a function of volume reflectance R for solar zenith angles θ from $0°$ to $89°$. (Adapted from Jerome, J. H., Bukata, R. P., and Bruton, J. E., *Appl. Opt.*, 27, 4012–4018, 1998.)

are plotted the E_o/E_d ratio at photic depth $z_{0.01}$ as a function of $R(\theta_r = 0°)$ for solar zenith angles θ from $0°$ to $89°$. Figure 9.8 quantitatively substantiates the importance of scalar irradiance, $E_o(z)$ (i.e., actual energy density to which a chemical or biological aquatic component is exposed), in assessing photochemical and/or photobiological reactions. This logically implies the necessity of a relevant modification to the Vollenweider-Fee productivity model [equations (9.1)–(9.5)].

Recall that irradiation is time-integrated irradiance. Jerome *et al.*[205] derived one set of multiplicative factors, $F_\theta(0^-,R)$, to convert a daily irradiation just above the air-water interface into a daily scalar irradiation just below the air-water interface, and a second set of multiplicative factors, $F_{z\theta}(R)$, to convert a depth-averaged ("effective") daily irradiation for a specific photic zone into a depth-averaged ("effective") daily scalar irradiation for that specific photic zone. These multiplicative factors would then be defined as:

$$F_\theta(0^-, R) = \frac{\int E_o(0^-, \theta, R)d\theta}{\int E_d(0^+, \theta, R)d\theta} \qquad (9.29)$$

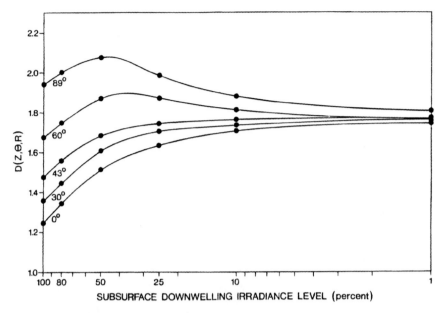

FIGURE 9.9
Depth dependencies of the ratio of downwelling irradiance to scalar irradiance [$D(z,\theta,R)$] for
volume reflectance $R(0°) = 0.08$. (From Jerome, J. H., Bukata, R. P., and Bruton, J. E., *J. Great
Lakes Res.*, 16, 436–443, 1990. With permission.)

and

$$F_{z\theta}(R) = \frac{\displaystyle\iint E_o(z, \theta, R)dzd\theta}{\displaystyle\iint E_d(z, \theta, R)dzd\theta} \tag{9.30}$$

where 0^- and 0^+ respectively refer to values of z immediately below and
immediately above the air-water interface, and θ is the solar zenith angle.
Equation (9.29) indicates an integration over time (i.e., θ) and equation (9.30)
indicates an integration over both time and depth. The multiplicative factors
can be determined as a function of the volume reflectance of the natural
water mass for incident global radiation comprised of direct solar and diffuse
sky radiation distributions for specific latitudes and specific seasons, thereby
generating site-specific nomograms that enable the transition of $E_d(0^+)$ to
$E_o(0^+)$ for a wide range of natural water types.

Figure 9.9 is taken from Jerome et al.[205] (their Figure 2) and illustrates
the depth dependencies throughout the photic zone of the ratio $E_o(z,\theta,R)/$
$E_d(z,\theta,R)$, designated in the figure as $D(z,\theta,R)$, for a volume relectance R
of 0.08 and a variety of solar zenith angles θ. Of course, different depth

dependencies would result from the use of other volume reflectance values. The multiplicative factor $F_{z\theta}(R)$ was then determined as the depth-averaged integral of the appropriate $D(z,\theta,R)$ curve multiplied by the corresponding value of $E_d(z,\theta,R)$ for the volume reflectance and solar zenith angle under consideration. That is:

$$F_{z\theta}(R) = \frac{\displaystyle\int_{\theta_1}^{\theta_2}\int_{z_{0.01}}^{z_{1.0}} D(z,\,\theta,\,R)E_d(z,\,\theta,\,R)dzd\theta}{\displaystyle\int_{\theta_1}^{\theta_2}\int_{z_{0.01}}^{z_{1.0}} E_d(z,\,\theta,\,R)dzd\theta} \qquad (9.31)$$

Notice that the numerator in equation (9.31) reduces to $E_o(z,\theta,R)$, thereby conforming to the definition of $F_{z\theta}(R)$ given in equation (9.30). The depth limits of integration of equation (9.31) are the air-water interface $z_{1.0}$ and the depth of the photic zone $z_{0.01}$. The time limits of integration define the length of the local daylight hours, namely the solar zenith angle θ_1 defining local sunrise and the solar zenith angle θ_2 defining local sunset. An expression for $E_d(0^+,\theta,R)$ may be obtained either from direct measurement or from global radiation models such as discussed by Kondratyev,[235] Davies et al.,[79] Frouin et al.,[116] and others.

Equation (9.31) essentially weights the local value of $D(z,\theta,R)$ according to the corresponding local value of $E_d(z,\theta,R)$. Thus, $F_{z\theta}(R)$ is the ratio of an "effective" scalar irradiance $E_o(z,\theta,R)$ to an "effective" downwelling irradiance $E_d(z,\theta,R)$.

For any layer of a water body the "effective" downwelling irradiance may be calculated as the ratio

$$\int_{z_1}^{z_2} E_d(0^-,\,\theta,\,R)\exp(-Kz)dz \bigg/ \int_{z_1}^{z_2} dz$$

where z_1 and z_2 are the depth limits of the water layer being considered and K is the irradiance attenuation coefficient appropriate to that layer. For a photic zone defined by $z_{1.0}$ and $z_{0.01}$, the "effective" downwelling irradiance is $\sim 0.215E_d(0^-,\theta,R)$.

Although the photic zone is generally and conveniently taken to be contained within the 100% and 1% subsurface irradiance levels, these limits can be argued to be somewhat arbitrary. The multiplicative factor $F_\theta(0^-,R)$ does not involve a depth integration. The multiplicative factor $F_{z\theta}(R)$ does. However, as seen from equation (9.31) and the illustrative $D(z,\theta,R)$ depth profiles of Figure 9.9, the effect of alternate irradiance level definitions of the photic zone will produce a minimal impact on the depth integration of $D(z,\theta,R)$ values weighted by rapidly attenuating $E_d(z,\theta,R)$ values. Jerome et *al.*[205] report that extending the integration limit from the 1% irradiance level to the 0% irradiance level produces an infinitesimal change in the value of $F_{z\theta}(z,\theta,R)$, while reducing the integration limit from the 1% to the 10% irradiance level results in a 1.5% change in the value of $F_{z\theta}(z,\theta,R)$. Reducing the integration limit to $z_{0.30}$ results in a 10% change in the value of $F_{z\theta}(z,\theta,R)$.

Values of $F_{z\theta}(R)$ can, of course be determined for any incident global radiation distribution, day of the year and location on the Earth, and water body (subsurface volume reflectance). First the appropriate $D(z,\theta,R)$ function is weighted according to $E_d(z,\theta,R)$ and integrated over the depth of the photic zone. The resulting $F_{z\theta}(R)$ function is then integrated over the local daylight hours. Figure 9.10, also taken from Jerome *et al.*[205] (their Figure 6), illustrates $F_{z\theta}(R)$ as a function of $R(0°)$ for the spring and fall equinoxes and the winter and summer solstices at latitudes between 40°N and 50°N. Illustrated are the situations for direct incident radiation[235] and for a cardioidal diffuse incident radiation, the latter being independent of season. Figure 9.10 illustrates the multiplicative factors that will convert the "effective" daily irradiation that would be collected *in situ* into the "effective" daily irradiation that would more appropriately describe the photosynthetic activity occurring within the water column. It can be seen that the values of $F_{z\theta}(R)$ can be quite substantial. Even for water bodies that display volume reflectance values R of zero, the "effective" daily scalar irradiation within the photic zone is 15%–45% higher than the "effective" daily irradiation for that photic zone. This figure increases to 85%–110% for $R = 0.14$, a volume reflectance value commonly encountered in non-Case I waters.

However, the multiplicative factor that is more readily obtainable and more readily applicable and, therefore, perhaps more germane to remote sensing is $F_\theta(0^-,R)$, which restricts its focus to the air-water interface. This factor may be used to convert values of above-surface daily irradiation into values of daily scalar irradiation just below the surface (equation (9.29). Values of above-surface irradiation may be obtained either from direct measurements or from solar irradiance models such as presented by Zepp and Cline.[436] The only hydro-optical parameter required is $R(0°)$, the subsurface volume reflectance for a solar zenith angle of 0°.

Both photochemical and photobiological processes are strongly wavelength dependent. Subsurface volume reflectance and impinging global radiation are also strongly wavelength dependent. The discussions in this section have not explicitly included wavelength as an independent variable. The emphasis on primary production in this chapter renders the PAR wavelength band as implicit. Further, these discussions tacitly assume that the $R(0°)$ and the global radiation possess spectral consistency and do not violate the spectral requirements of the implied aquatic chemical and biological processes.

Values of $F_\theta(0^-,R)$ can be determined for any global radiation distribution and water type at any geographic location for any time of year. Similar to

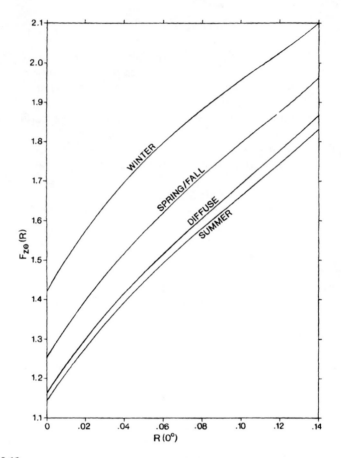

FIGURE 9.10

$F_{z\theta}(R)$ as a function of $R(0°)$ for the equinoxes and solstices ior direct and diffuse incident radiation distributions. These are the multiplicative factors to convert "effective" daily irradiation into "effective" daily scalar irradiation. (From Jerome, J. H., Bukata, R. P., and Bruton, J. E., *J. Great Lakes Res.*, 16, 436–443, 1990. With permission.)

Figure 9.10, Figure 9.11, also taken from Jerome *et al.*[205] (their Figure 5), plots $F_\theta(0^-,R)$ as a function of $R(0°)$ for the spring and fall equinoxes and the winter and summer solstices for both direct and diffuse (cardioidal) global radiation distributions again for latitudes between 40°N and 50°N. These factors enable the conversion of the above-surface daily irradiation into the daily scalar irradiation just below the air-water interface.

In capsule, the scalar irradiance, $E_o(z)$, at a given depth z can, under certain sets of aquatic conditions, be greater than twice the downwelling irradiance, $E_d(z)$ at that depth. Although not discussed here, analyses by Madronich[254] have shown that scalar irradiance of *actinic flux* (defined as the

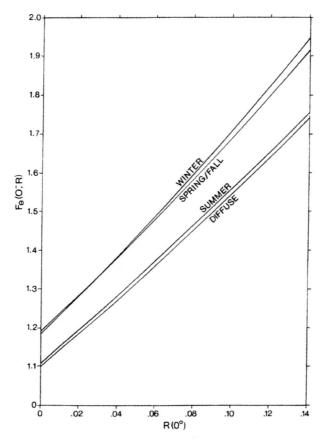

FIGURE 9.11

$F_\theta(0^-,R)$ as a function of $R(0°)$ for the equinoxes and solstices for direct and diffuse radiation distributions. These are the multiplicative factors to convert incident above-surface daily irradiation into daily scalar irradiation just below the air-water interface. (From Jerome, J. H., Bukata, R. P., and Bruton, J. E., *J. Great Lakes Res.*, 16, 436–443, 1990. With permission.)

short wavelength region of the radiation spectrum, e.g., violet and ultraviolet, responsible for chemical changes) in the atmosphere can also be much greater than the downwelling irradiance for conditions of extensive cloud cover and/or high ground albedos. For a latitude range appropriate to the Laurentian Great Lakes, and subsurface volume reflectance values in the range 0.0 to 0.14, the multiplicative factors $F_\theta(0^-,R)$ and $F_{z\theta}(R)$ vary between 1.1 and 2.1.

To properly incorporate the scalar irradiance into the estimate of aquatic primary production, the use of conversion factors such as the multiplicative factors $F_\theta(0^-,R)$ and/or $F_{z\theta}(R)$ may become mandatory. If so, these multiplicative factors would have to be determined for specific geographic locations and specific times of year, in addition to specific water masses (i.e., specific volume reflectances R).

9.9   AN EXAMPLE OF PRIMARY PRODUCTION IN LADOGA LAKE

We will now illustrate, as an example of some of the operational protocol discussed in this chapter, unpublished results from a field campaign on Ladoga Lake, Russia. Although no remote sensing *per se* was performed during this campaign, the optical measurements performed above and below the air-water interface are pertinent to the preceding discussions.

Daily spectral measurements of the global PAR incident radiation $E_{PAR}(0^+)$ were made in late June 1989 aboard a small research vessel in the northwestern coastal zone of Ladoga Lake. Concurrent with these actinometric observations were *in situ* measurements of volume reflectance spectra just below the surface and measurements of the vertical distribution of phytoplankton chlorophyll concentration throughout the photic zone. Figure 9.12 illustrates a daily profile of E_{PAR}. Figure 9.13 illustrates the chlorophyll depth concentration profile recorded that same day. A sample directly recorded subsurface volume reflectance spectrum, $R(0^-)$, is illustrated in Figure 9.14. The subsurface data displayed in Figures 9.13 and 9.14 were collected under a cloudless sky and an almost perfectly calm air-water interface at local noon ($\theta = 35°$). The diurnal value of the global PAR incident radiation (Figure 9.12) was calculated from continuously recorded data to be 37 einsteins per m². The subsurface volume reflectance spectrum displayed a peak value of 0.040 at $\lambda = 580$ nm.

Use of the optical cross section spectra for Ladoga Lake discussed in Chapter 8 in conjunction with the bio-optical model and multivariate optimization technique resulted in the retrieval from the volume reflectance spectrum of 7.1 mg per m³ of Chl$a$, 0.8 g per m³ of SM, and 8 g carbon per m³ of DOM as estimates of the co-existing near-surface aquatic concentrations. The value of $K_{PAR}(0^-)$ deduced on the basis of the bio-optical model and the retrieved concentration values of Chl$a$, SM, and DOM was 1.8 m^{-1}, a value gratifyingly close to the $K_{PAR}(0^-)$ value of 1.7 m^{-1} resulting from the direct vertical profiling of the upwelling and downwelling irradiance fluxes. This $K_{PAR}(0^-)$ corresponds to a penetration depth z_p (the depth of the 37% subsurface irradiance level) of ~0.6 m.

The following input values of the Vollenweider-Fee model parameters of equation (9.7) [see Vollenweider[415] and Fee[104] for more detailed definitions of these parameters] were selected as appropriate to Ladoga Lake: $a = 1.0$; $n = 0.5$; $v = 0.01$; $\delta = 0.26$; $\lambda = 12$ hours; $P_{opt} = 80$ mg C per m³; $E_k = 800$ μeinsteins per m² per sec. It was assumed that the depth profiles of SM and DOM remained invariant from their estimated near-surface values through-

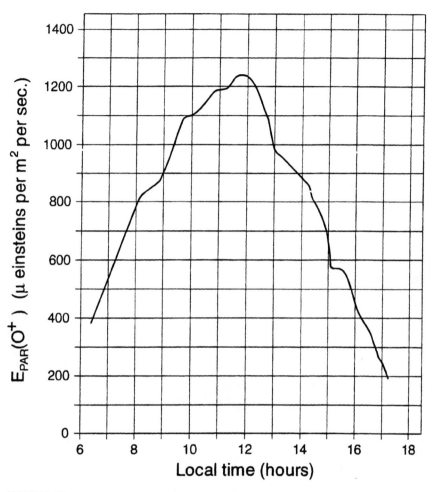

FIGURE 9.12
Daily profile of E_{PAR} over Ladoga Lake, June 1989.

out the photic zone. The chlorophyll depth profile was taken from Figure 9.13. Integration of the Vollenweider-Fee parametric equation over t and z yielded a daily phytoplankton photosynthesis (approximate primary production) PP value of 346 mg C per m^2 per day. This estimate was in good agreement with direct measurements of primary production in Ladoga Lake performed during the same season [values of PP in the range (320–380) mg C per m^2 per day were reported by Petrova[307]]. The PP value determined from the volume reflectance estimates was not corrected for scalar irradiance since the direct measurements reported in Petrova were values for down-welling irradiance.

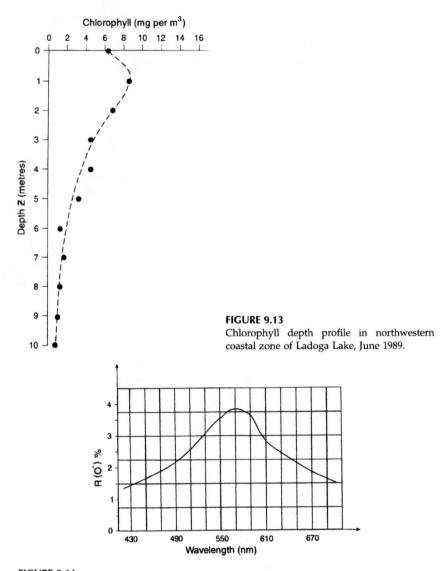

FIGURE 9.13
Chlorophyll depth profile in northwestern coastal zone of Ladoga Lake, June 1989.

FIGURE 9.14
Directly recorded volume reflectance spectrum in northwestern coastal zone of Ladoga Lake, June 1989.

Admittedly, the above example was an *in situ* as opposed to a remote sensing exercise. Nonetheless, it does provide evidence to support the operational protocol presented in the flow diagram of Figure 9.7. Remotely sensed upwelling radiance spectra over optically complex waters can yield reliable estimates of the subsurface volume reflectance spectra. The availability of optical cross section spectra for indigenous aquatic constituents can enable the deconvolving of this volume reflectance spectrum into the organic and inorganic matter responsible for that volume reflectance. The assumption of a thoroughly homogeneous SM and DOM concentration in the photic zone of a stratified lake proved to be non-problematical. The shortcoming of the investigation lay in the use of a directly measured chlorophyll depth profile. It is evident that more efforts need to be directed toward the establishment (should such establishment indeed be possible) of a set of generic dimensionless chlorophyll profiles that could be used for inland and coastal water bodies.

Chapter 10

AQUATIC REMOTE SENSING IN REGIONAL AND GLOBAL ENVIRONMENTAL MONITORING

10.1 ECOSYSTEMS AT RISK

It has become indisputably evident that the unsustainable combination of an ever-escalating global population and the incessant injection, over several decades, of chemical impurities into the atmospheric, terrestrial, and aquatic components of the Earth's network of ecosystems has generated an inter-related myriad of environmental stresses and ecosystem perturbations that can never be completely revoked. There is growing fear, for example, that the Earth will experience a regionally dependent and globally extensive climate change over the upcoming century, a consequence of, among other determinants, increased concentrations of atmospheric "greenhouse" gases. Among the most prominent of "greenhouse" gases are water vapor (H_2O), carbon dioxide (CO_2), methane (CH_4), nitrous oxide (N_2O), ozone (O_3), and halocarbons such as chlorofluorocarbons (CFCs).

While the physical and chemical processes leading to the satellite-observed reductions in concentrations of stratospheric ozone are well understood, the biogeochemical impacts of the anticipated geographically dependent enhancements in ground-level UV-A and UV-B radiation upon the Earth's ecosystems are far from being as well understood. In fact, perhaps due to the somewhat understandable complacency resulting from centuries of a relatively undisturbed stratospheric ozone layer, the biogeochemical effects of *ambient* levels of ground-level ultraviolet radiation are themselves not fully understood, particularly with regard to the mutative and/or adaptive responses of flora and fauna to such damaging radiation. Further, the

295

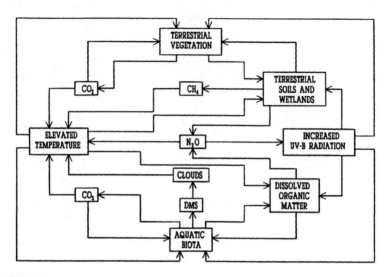

FIGURE 10.1
Conceptual interlinking of the carbon cycle with the global climate system. The top half represents terrestrial and wetland ecosystems, while the bottom half represents aquatic ecosystems.

possible latitude variability of these responses could result in shifting populations that would upset ecosystem stabilities currently in place.

It is generally acknowledged by aquatic ecologists that the carbon cycle is paramount to the fundamental chemical framework that not only determines the structure and dynamics of the aquatic food web, but also supplies, along with other elements such as phosphorus, a standard upon which to legislate controls to regulate and/or restore water quality and/or aquatic ecosystem population dynamics.

The conceptual interlinking of the carbon cycle with the global climate system is simplistically sketched in Figure 10.1. Herein two aspects of global change (greenhouse gas emission resulting in elevated near-surface temperatures and ozone depletion resulting in latitude-dependent elevated concentrations of ground-level ultraviolet radiation) are linked through the carbon cycle to both the terrestrial and aquatic ecosystems. The top half of Figure 10.1 represents terrestrial (and wetland) ecosystems, while the bottom half represents aquatic ecosystems. Illustrated in Figure 10.1 are such interactions and feedback loops as:

1. Changes in CO_2 emissions from organic matter in terrestrial soils, wetlands, and marine, inland, and coastal waters caused by alterations in decomposition as a consequence of atmospheric changes.

2. Anticipated changes in primary productivity in terrestrial and aquatic ecosystems due to combined effects of temperature change and enhanced UV-B radiation.

3. Increased emmisions of methane gas from wetlands subjected to increased warming.

4. Disruptions to the nitrogen cycle and possible increases in N_2O emissions due to low altitude atmospheric warming.

5. Aquatic biota responses (adaptive and arguably proactive) to enhanced UV-B fluxes manifesting as modified dimethylsulfide (DMS) emissions from aquatic ecosystems.

6. Changes in cloud albedo as a consequence of the modified DMS emissions.

7. Reinforced effects of global warming as a consequence of possible increases in CO_2, N_2O, and DMS degassing.

What is noticeably absent from the simplistic Figure 10.1 are the direct socio-economic effects of climate change on human activity and life-style and the health hazards of enhanced ground-level UV-B radiation such as skin cancer, eye damage, and immunosuppression. This chapter, however, will be restricted to ecosystems and, whether the ecosystems be terrestrial or aquatic, the fundamental questions that must be asked are rather obvious. Unfortunately, the answers are not as obvious, and the search for these answers forms the basis for research priorities of the upcoming decades. A *partial* list of such questions would include (in framing these questions the word *climate* will be considered in its all-encompassing definition of prevailing conditions surrounding a point or a region and, therefore, will include radiation fields):

1. What are the temperature tolerances of indigenous plant and animal populations to the stress of climate change?

2. What are the protective pigment tolerances of plant and aquatic biota to the stress of climate change?

3. What are the impacts of climate change on terrestrial and aquatic nutrients?

4. What are the impacts of climate change on terrestrial and aquatic food chains? What impacts are directly encountered by the members of the food chain? What impacts are reverberations throughout the food chain as a consequence of non-identical impacts to lesser-order or higher-order members of the indigenous food chain?

5. What are the impacts of induced migrations in the locations of ecozones (for example, in the Canadian Arctic, these ecozones would include permafrost regions, ice-fields, boreal forests, tundra, sub-Arctic woodlands, temperate forests, and grasslands)?

6. What is the influence of climate change on biochemical processes? In particular, what is the impact on carbon production, transport, and fate? What are the possible disruptions to the dynamic equilibrium of the carbon cycle (in aquatic ecosystems this equilibrium is the balance amongst primary production, sedimentation, groundwater transport, air-water interchange, and consumption by herbivores and heterotrophic bacteria)?

Particular attention is generally directed towards dissolved organic carbon (DOC), although dissolved inorganic carbon (DIC) cannot be ignored.

7. What are the impacts of climate change on the hydrological cycle, including soil moisture storage, river ice transport, evaporation, run-off, snow-melt, and fluctuations in precipitation patterns and water levels?

8. What baseline data and/or information is required to identify or establish trends that can be indisputably associated with climate change? What reliable direct monitoring is required, and of what parameters? (What information is required concerning species composition and biodiversity? What climate change parameters need be monitored, and by what means? What life-quality parameters need be monitored, and what suitable biological indicators need be selected? What time scales are required to establish trends, and what reliable historical data exist? What fluctuations in indigenous flora and fauna data and/or weather records represent natural evolutions of an undisturbed climate? What fluctuatuions represent anthropogenic impacts?)

There are, of course, strains upon the stability of terrestrial and aquatic ecosystems that are consequences of chemical intrusions into an ecosystem either directly through toxic contaminant loadings and spills and the use of pesticides and herbicides or indirectly through the long-range transport of waterborne and/or airborne pollutants. There are also ecosystem impacts resulting from the depletion of non-renewable resources and/or insufficient regeneration of renewable resources. Here, too, while consequences of these strains to ecosystem stability and biodiversity are anticipated, the details of the magnitudes of these consequences, along with the probable degree of environmental regeneration, are far from certain.

In general, addressing the impacts of atmospheric change to human and ecosystem well-being is, quite logically, proceeding along three concurrent avenues:

1. Limitation

Historic multinational agreements have been signed to prevent further atmospheric devastation. The Framework Convention on Climate Change was signed by 154 countries at the 1992 United Nations Conference on Environment and Development in Rio de Janiero, Brazil. For example, Canada has agreed to stabilize its "greenhouse" gas emissions to 1990 levels by the year 2000.

Over 110 countries have signed the 1987 Montreal Protocol (tightened shortly thereafter by the London Protocol), which called for the phasing out of virtually all ozone-attacking reagents (halons and chlorofluorocarbons) by the year 2000.

While such agreements are most essential and certainly welcome, they can only arrest the acceleration of the problem, they cannot rectify the existing problem.

2. Mitigation

Mitigation (i.e., reducing the severity of the impact while not necessarily eliminating the impact) is the area that requires much research (medical, physical, biochemical, social, and economic) since mitigation is dependent upon the answers to the above and other questions concerning the nature and importance of the interactive feedback loops linking climate change to the Earth's ecosystems.

3. Adaptation

There is every likelihood that some of the impacts to the Earth's biosphere will either never be mitigated or that the mitigation will be a long time coming. Adaptation is what needs to occur while mitigation measures are being sought, developed, or implemented. Adaptation measures to atmospheric change are evident throughout the world and include, but are not limited to, such activities as:

- Planting of tree and crop species that are more tolerant of a variable and changing climate
- Development of seed species and/or seed coatings that are rugged and weather-resistant
- Moving of farming, fishing, and/or industrial sites to areas that are less sensitive to climate change, as well as reducing the legal extraction from the environment at such sites
- Defining and protecting endangered species
- Developing alternates for products of species or regions at risk, including recycling activities
- Using sunscreen lotions, sunglasses, and protective apparel
- Developing alternate energy sources and reducing the use of fossil fuels
- Incorporating climate change considerations into construction design and using building materials that reflect those considerations
- Improving the efficiency of irrigation, damming, and waterflow-diverting practices
- Protecting vulnerable shorelines from erosion, and restricting the use of such shorelines for housing or recreational activities

and many other adaptive measures and strategies. The difficulty with adaptation strategies, however, is that, understandably, they tend to be driven by human agendas (i.e., by the health risk, threat, financial cost, and inconvenience to society). Ecosystems, therefore, must fall within the domain of the ecologist to ensure their intermediate adaptation and ultimate mitigation.

10.2 SATELLITES AND GLOBAL CHANGE

Anthropogenically induced climate changes have necessitated global participation in solar/terrestrial research programs of unprecedented complexity. Coordinated long-term multidisciplinary climatology programs such as the IGBP: International Geosphere-Biosphere Program (National Research Council[295,296]), the GCIH: Global Change; Impact on Habitability (NASA[287,288], the JGOFS: Joint Global Ocean Flux Study (U.S.JGOFS[406]), and MPE: Mission to Planet Earth (NASA[290,291]) have incorporated the remote monitoring of terrestrial and oceanic processes from satellite altitudes into their operational philosophies. The anticipated integral role of satellite monitoring in the upcoming decades of multinational climate change impacts studies further accentuates the need to convert space-acquired data into reliable estimates of environmentally significant parameters.

The goals of the multinational multidisciplinary global change programs are, as to be expected, all-encompassing and very ambitious. Of particular concern to both the IGBP and the GCIH are the biogeochemical cycles of carbon, nitrogen, sulfur, phosphorus, and water, as well as such ecosystem health determinants as solar radiation, soil fertility, water quality, and air quality. With regard to remote sensing, the IGBP strategy divides environmental issues along the disciplinary categories agriculture, forestry, geology, hydrology, and oceanography. No direct limnological category was included, although coastal regions were incorporated into oceanography through such topics as near-shore processes, coastal tides and currents, toxic contaminants, and bathymetry.

The Earth's oceans, being simultaneous sources and sinks of carbon, as well as reservoirs of plant and animal life, are principal concerns of the multinational climatology programs. Due to large oceanic expanses, spatial inaccessibilities, surveillance logistics, and synoptic coverage requirements, satellites have become the major vehicles for monitoring the productivity patterns within Case I waters (satellites serve a comparable function for productivity patterns within terrestrial forests and cultivated croplands). As discussed earlier, the principal application of remotely sensed upwelling radiance spectra over Case I waters (as well as over terrestrial vegetation) has been to map chlorophyll pigment concentrations using either the ratios of spectral radiance in selected wavelength regions of the visible spectrum or ratios of radiances recorded in the visible red and near-infrared regions of the spectrum. Also as we have seen, such simplistic ratioing of radiances at dissimilar wavelengths will not serve to extract chlorophyll concentrations from the concentrations of co-existing organic and inorganic matter residing in non-Case I waters.

Coastal waters (including continental shelves, continent-bound seas, and large estuarine systems) account for only ~10% of the surface area of the global oceans. However, they presently contribute ~40% of the total global aquatic primary production (Wollast,[429] Schlesinger[356]). This disproportionality is largely a consequence of anthropogenic nutrient loadings (Ryther and Dunstan,[346] Nixon[299]). It is reasonable to assume that this disproportionality of coastal to oceanic (i.e., non-Case I to Case I) primary production will continue to increase as population growth, urban development, and both industrial and agricultural discharges into natural waters continue to increase. Due to the (possibly escalating) prominence of non-Case I waters in the overall global issues of carbon exchange and bioproductivity, we have argued[47] that changes in bioproductivity of inland and coastal waters should be incorporated into studies of impacts of regional climate change in the same manner that changes in bioproductivity of oceanic waters have been incorporated into studies of impacts of global climate change. In order to do so, bio-optical models such as described in this book, based upon the absorption and scattering cross section spectra appropriate to indigenous aquatic organic and inorganic matter, are required.

The collaborative Canada/USA remote sensing and aquatic optics project, mentioned in Section 8.6 and being performed within the framework of the international BOREAS program, is designed to address the regional issues of lake bioproductivity and carbon exchange. The boreal forest biome currently sequesters globally significant quantities of carbon. Anthropogenic perturbations to the carbon cycle could disrupt the processes that enable the biome to store carbon, creating the possibility that the biome could change from being a net sink of carbon *from* the atmosphere to being a net source of carbon *to* the atmosphere. Such a positive feedback loop in the carbon cycle could result in ecological catastrophe. There is, therefore, an urgent need to quantify all important sources and sinks of carbon within the boreal forest biome in order to enable future changes to be predicted, understood, and managed. Lakes are a very prominent feature of the boreal biome. Thus, they constitute a potentially important element within the carbon budget of the boreal forest biome. Boreal lakes have been shown to be highly sensitive to changes in the physical climate system (Schindler et al.[354]). For over two decades, a number of physical, chemical, and biological variables in the aquatic basins of the Experimental Lakes Area (ELA) in northwestern Ontario have been measured under a variety of climatic extremes. Schindler et al.[355] report that DOC concentrations in many of the lakes have declined over the period, illustrating the sensitivity of the carbon cycle to changes in climate and reinforcing the need for better understanding of the role(s) of DOC in lacustrine processes in order to predict impacts of global change on freshwater systems. Further, such lacustrine processes must be integrated with corresponding carbon fixation and transformation processes occurring within the terrestrial ecosystems. Unfortunately, at this time, there is no way to definitively assess the relative importance of these various sources and sinks of carbon.

Since the multivariate optimization bio-optical model approach discussed in this book can enable remote estimates of the near-surface DOC
concentrations over the large number of small lakes (or, in some situations,
small number of larger lakes) scattered throughout the boreal biome, remote
sensing over inland waters affords the potential of supplying information
on one more piece of the carbon budget puzzle.

10.3   SATELLITES AND HYDROLOGY

As discussed in Section 2.3, the attenuation of UV light by dissolved
organic matter in freshwaters plays an important role in protecting aquatic
organisms from direct UV exposure. From Figure 2.3 it was seen that non-
Case I waters either of depths less than ~ 1 metre or containing DOC
concentrations less than (1–2) mg C/l could be specifically vulnerable to
the direct impacts of enhanced ground-level UV-B radiation. This would
normally limit aquatic concerns to arctic and alpine lakes, boreal lakes with
upland catchments, ponds and shallow lakes and streams, and wetlands
and their associated spawning sites and habitats. Near-surface "greenhouse"
warming, however, may be more important than stratospheric ozone thinning in increasing exposure times of aquatic organisms to UV-B (Williamson
et al.[427]) due to disruptions in water transparency, thermocline depth, and
nutrient-driven predation.

Thus, climate change impacts will be strongly influenced by basin
hydrology. Hydrology also plays a determinant role in the DOC concentrations of inland and coastal waters. Aquatic DOC originates largely in wet
soils and wetlands, and is generally transported into the water column
and bottom sediments by surface runoff and/or groundwater leaching. As
streamflow decreases, the sources of DOC become less effective, and the
corresponding increases in lake residence times result in accentuated DOC
degradation. Thus, decreases in streamflow can manifest as decreases in
aquatic DOC. Similarly, increases in streamflow can manifest as increases in
aquatic DOC. Also, under warmer, drier climates, concentrations of DOC in
inland waters can decline due to the combined effects of decreased carbon
sources, possible increased degassing, and increased exposure to ultraviolet
light and bacterial action.

It is well known that the dynamic nature of a region's aquatic environment is a convoluted consequence of physical, chemical, biological, socio-
economic, and anthropogenic parameters that are inescapably inter-linked.
These ecosystemic complexities notwithstanding, however, it has long been
accepted that the regional precipitation at time t, $P(t)$, may be expressed in
terms of five variables, namely:

$$P(t) = Q(t) + E_T(t) + \Delta S(t) + \Delta W_G(t) + \Delta W_S(t) \qquad (10.1)$$

where $Q(t)$ = runoff at time t,
        $E_T(t)$ = evapotranspiration at time t,
        $\Delta S(t)$ = change in regional water storage (lakes, rivers, snow, ice, vegetation, etc.) at time t,
        $\Delta W_G(t)$ = change in groundwater at time t, and
        $\Delta W_S(t)$ = change in soil moisture at time t.

In order to at least partially incorporate the impact of environmental perturbations into equation (10.1), it would be ideal if the term $P(t)$ could refer to the total precipitation input to the region of interest. As such, it would not necessarily define pure rainwater and/or pure snow. Rather, it would include *all* downwelling aquatic material at time t, including atmospheric particles, acid rain, greenhouse gases, carbon concentrations, and other components. The terms $E_T(t)$, $\Delta S(t)$, $\Delta W_G(t)$, and $\Delta W_S(t)$ would be accordingly defined. Such an ideal situation, however, has yet to be attained.

Differentiating equation (10.1) with respect to time yields the governing hydrological equation:

$$\frac{dP(t)}{dt} = \frac{dQ(t)}{dt} + \frac{dE_T(t)}{dt} + \frac{dS(t)}{dt} + \frac{dW_G(t)}{dt} + \frac{dW_s(T)}{dt} \qquad (10.2)$$

where we have replaced $\Delta S(t)$, $\Delta W_G(t)$, and $\Delta W_S(t)$ by $S(t)$, $W_G(t)$, and $W_S(t)$, respectively, with no loss of mathematical generality.

The hydrological processes defined in equations (10.1) and (10.2) are schematically illustrated in the generalized hydrological cycle diagram of Figure 10.2 for surface, subsurface, and above-surface water. It must be emphasized that while equations (10.1) and (10.2) conveniently reduce $P(t)$ to five terms, each of these terms requires directly measured values, theoretically inferred values, or mathematical definitions involving other variables. In addition to the solar radiation spectrum at the Earth's orbit, climatological parameters that are considered to impact the hydrological cycle include wind, atmospheric pressure, temperature, humidity, and cloud cover (i.e, parameters that are normally considered to define *weather*).

Table 10.1 reviews recommended hydrological and climatic variables, along with surrogate and other dependent variables, which could be required to obtain the basic variables of equations (10.1) and (10.2). Many of these variables have been reported in NASA documents as targeted parameters for the Earth Observing System Eos (NASA[290,291]). Many emerge from recommendations given in such sources as the Canadian Global Energy and Water Cycle Experiment (GEWEX) Workshop held in Saskatoon, Saskatchewan in 1989, Goodison *et al.*,[127] NASA Advisory Council,[293,294] and standard hydrology textbooks. Also included within Table 10.1 are suggestions (as put forward by Whiting and Bukata[424]) as to possible minimal restrictions on spatial resolution, accuracy, and frequency of measurement of the hydrological

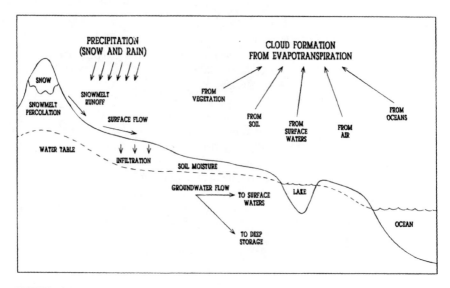

FIGURE 10.2
Generalized hydrological cycle diagram.

parameters. (We fully realize the subjectiveness of these suggestions, as well as the site-specific character of such suggestions.)

Alongside hydrological variables in Table 10.2 are some of the satellite remote sensing platforms and sensors that are or will be available for measuring or inferring those variables as part of the coordinated programs involving satellite monitoring. Remote sensing devices record the upwelling electromagnetic radiation (reflected or emitted) from some combination of the organic and inorganic constituents of the biosphere. No direct parameteric information *per se* is recorded at the aircraft/satellite altitude. Rather, the spectroradiometric data bits electronically stored at the remote platform must be converted, through appropriate multidisciplinary modeling activities, into *inferred* estimates of the physical, chemical, or biological variable being sought. Thus, the ability to extract hydrological and/or climate change variables from remotely-sensed upwelling spectral signatures, like the ability to extract co-existing concentrations of aquatic matter from upwelling spectral signatures, is dependent upon the models and algorithms that exist or are being developed to convert upwelling spectral radiances into numerical estimates of the desired variable.

Inarguably, what environmental satellites do, they do very well. Satellites offer an enormous potential for improving the temporal and spatial coverage of hydrometeorological data to meet the real-time needs of user agencies. They are essential for monitoring regions of a country's terrain that are inaccessible to conventional ground-based monitoring protocols. They pro-

TABLE 10.1

Hydrological and Climatic Variables Essential to Regional Climate Change

Variable	Spatial Resolution	Accuracy	Frequency
Precipitation	1 km	10%	1 per day
Precip. rate	< 100 km(h) × 5 km(v) >	?	?
Snow cover	1 km	10%	1 per day
Cloud cover	1 km	5%	4 per day
Cloud water content	50 km	0.1 kg m^{-2}	4 per day
Cloud top height	1 km(v)	0.5 km	4 per day
Columnar water vapor	100 km (h) × 100 km(v)	0.002 ppm	2 per day
pressure			
CO_2	1 km	10%	1 per day
CO	1 km	10%	1 per day
N_2O	1 km	10%	1 per day
CH_4	1 km	10%	1 per day
CFC	1 km	10%	1 per day
Aerosols	(10 × 10 × 1) km	5%	1 per day
Runoff	100 km	10%	1 per day
Lake area	30 m	2%	1 per day
Height (sfc)	100 m	5 m	1 per 10 yrs
Soil area	30 m	10%	1 per yr
Wind speed	50 km^2	1 m s^{-1}	1 per day
Temp. (lake sfc)	30 m	0.5°	2 per day
Area (biomass index)	30 km	10%	1 per mon
Biomass density	1 km	5%	1 per wk
Production rate	30 m	20%	1 per 3 days
Vegetation condition	30 m	15%	1 per 3 days
Land use	30 m	10%	1 per mon
Area (snow, ice)	1 km	10%	1 per wk
Temp. (snow sfc)	1 km	1°K	1 per 3 days
Albedo (snow)	?	?	?
Ice height	30 km	1 km	1 per 5 yrs
Volume (lake, river)	?	?	?
Groundwater area	30 km	5%	1 per 2 mon
(discharge, recharge)			
Vegetation vigor	30 m	10%	1 per mon
Irrigation practices	30 m	10%	1 per yr
Soil water content	1 km	10%	1 per 2 mon
Surface characteristics	30 km	20%	1 per 50 yrs
Terrain slopes	30 km	20%	once
Frozen ground boundaries	1 km	10%	1 per day
Temperature (troposphere)	100 km(h) × 5 km(v)	1°K	1 per day
Temp. (sfc)	1 km	1°K	2 per day
Temperature (cloud emission)	1 km	1°K	4 per day
Cloud albedo	50 km	5%	4 per day
Surface wind	50 km^2	1 ms^{-1}	1 per day

Abbreviations: sfc, surface; h, horizontal; v, vertical.

vide a singularly valuable snapshot in time of the spatial arena in which a myriad of physical, chemical, and biological forces interact to generate both obvious and subtle fluctuations within the Earth's ecosystems. They are excellent for monitoring *change* (as opposed to absolute numerical values) in both space and time [the very essence of the hydrology cycle mathematical formulism of equation (10.2)], and they are invaluable to such geometric applications as the delineation of landforms, coastlines, land-use boundaries, vegetation growth, lake area, ice-floes, extent of snow-cover, and general seasonal progressions.

It is evident from the entries in Table 10.2 that the estimation of hydrological variables from space requires recording of spectral returns that substantially transcend the visible wavelength interval to which we have restricted our discussions throughout this book. Consequently, we will not expand upon the various sensors and techniques that accommodate these wavelength intervals. Nor will we dwell upon opinions as to the capability of the satellite platforms/sensors to supply reliable data on the required hydrological variables. It would appear, however, that if the minimal suggested restrictions on spatial resolution, accuracy, and frequency of observations as listed in Table 10.1 can be deemed credible, it is not unreasonable to assume that, in addition to orbital constraints, sensor design limitations, signal obfuscations due to the intervening atmosphere, and shortcomings in currently available extraction algorithms and models, the inability to comply to the suggested minimal measurement requirements could handicap the role of environmental satellites in evaluating changes to the hydrology cycle. A dispassionate critique of existing satellite sensors and the role of remote sensing in the monitoring of hydrological parameters is, therefore, mandatory (A preliminary critique is given in Bukata et al.[50]). Also mandatory is an assessment of existing multispectral extraction algorithms and methodologies. This would include sensitivity analyses directed towards assessing regional climate model tolerances to the inherent inaccuracies in the remote sensing algorithms, as well as tolerances to the logistical limitations of satellite monitoring protocols. The results of such sensitivity analyses can ultimately dictate which variables should comprise obligatory ground-based collection from network stations, and what the quantified consequences would actually be if remote estimates of such variables were to be used as inputs to either regional or global climate change studies. Sensitivity analyses could be performed on the governing equations themselves.

10.4 CLIMATE CHANGE

It is obvious that, prior to assessing the impacts of climatic change on regional or global aquatic systems, it is essential to recognize that climate change is in fact occurring. Further, it is essential to recognize whether

TABLE 10.2

Techniques and Remote Sensing Platforms for Monitoring Hydrological Variables

Variable	Technique	Platform
Aerosols	lidar (surface)	—
Albedo (snow)	VIS, near-infrared	POES-AVHRR, Landsat
	microwave	SMMR, SPOT, Nimbus
	SAR	Radarsat
(cloud)	VIS, infrared	POES-AVHRR
	microwave	GOES-VAS, DMSP
Area (lake)	near-infrared	Landsat-TM4
(soil)	thermal infrared	HIRS, MSU, POES-AVHRR
	microwave	GOES-VAS
(snow)	VIS, near-infrared	POES-AVHRR
	microwave	GOES-SSM/I, DMSP
(groundwater)	thermal infrared	Landsat-TM6
	microwave	GOES-SSM/I, DMSP, AMSR
Biomass index	VIS, near-infrared	POES-AVHRR
	SAR	Radarsat
Biomass (density and production)	VIS, near-infrared	BIOME (Eos)
Cloud (type, height)	VIS	POES-AVHRR, GOES-VISSR
CO_2	direct monitoring	—
CO	infrared spectroscopy	ISAMS, UARS
CH_4	infrared spectroscopy	Space Shuttle-ATMOS
Frozen ground	direct observation	—
Lake height	ground-based COHAP	—
Lake ice	radar sounding	airborne
Landuse	VIS, infrared	POES-AVHRR
	microwave	GOES-SSM/I, DMSP
N_2O	infrared emmision	UARS
	infrared interferometry	Space Shuttle-ATMOS
Precipitation	infrared	GOES-J
	microwave	GOES-SMM/I, DMSP
	radar	TREM
Precipitation rate	radar	TREM
Runoff	thermal infrared	GOES-VISSR
	microwave	DMSP
Snow water equivalent	microwave	Nimbus-7-SMMR, GOES-SMM/I
Soil water content	microwave	GOES-SMM/I, DMSP, AMSU
Surface features	microwave	AMSU, DMSP
Temperature (cloud)	infrared	POES-TOVS
	microwave	GOES-VAS
(lake)	thermal infrared	POES-AVHRR, SPOT, Landsat-TM6
(ground)	thermal infrared	POES-AVHRR, HIRS
	microwave	GOES-VAS
(troposphere)	infrared	POES-TOVS
	microwave	GOES-VAS, AMSU
Terrain slope	DEM	SPOT

TABLE 10.2

Continued

Variable	Technique	Platform
Vegetation status	VIS, infrared	POES-AVHRR
Water vapor	microwave	GOES-SMM/I, DMSP,
		AMSU
	infrared	GOES-VAS
Windspeed	scatterometer	GOES-VAS

Abbreviations: VIS, visible; SAR, Synthetic Aperture Radar; AMSU, Advanced Microwave Sounding Unit; AMOS, Atmospheric Trace Molecules Observed by Spectroscopy; BIOME, Biological Imaging and Observational (Eos) Mission to Earth; COHMAP, Cooperative Halocene Mapping Project; DEM, Digital Elevation Model; DMSP, Defence Meteorological Satellite Program (Special Sensor, Microwave Imagery, Microwave Temperature Sounder); GOES, Geostationary Operational Environmental Satellite; GOES-SSM/I, Special Sensor for Microwave/Imaging; GOES-SSM/T, Special Sensor for Microwave/Temperature; GOES-VISSR, Visible Infrared Spin-Scan Radiometer; GOES-VAS, VISSR Atmospheric Sounder; HIRS, High Resolution Infrared Radiation Sounder; ISAMS, Improved Stratospheric and Mesospheric Sounder; Landsat, Land Satellite; MSU, Microwave Sounding Unit; POES, Polar-Orbiting Environmental Satellite; POES-AVHRR, Advanced Very High Resolution Radiometer; POES-TOVS, Tiros Operational Vertical Sounder; Radarsat, Radar Satellite; SMMR, Scanning Multichannel Microwave Radiometer; SPOT, Systeme Probatoire d'Observation de la Terre; TM, Thematic Mapper; TREM, Tropical Rainfall Explorer Mission; UARS, Upper Atmosphere Research Satellite.

climate change progressions are part of natural long-term and short-term variablities or are anthropogenically induced disruptions to such previously undisturbed temporal variations. Although the realization of such necessities is indeed obvious, distinctions amongst such climatic variabilities, unfortunately, are not.

While the term *climate* is intuitively understood to represent a general convolution of weather-related environmental conditions, a specific mathematical expression for *climate* is an elusive commodity. The term *climate change* is intuitively understood to represent the time-derivative of this elusive commodity. Adding further complexity to such a time-derivative is the above-mentioned need to distinguish natural climate change progressions (diurnal and seasonal variations, impacts of both predictable and non-predicable natural phenomena, etc.) from climatic progressions resulting from such perturbing influences as ozone depletion, near-surface temperature elevations, chemical changes to the carbon and nitrogen cycles, etc. These difficulties notwithstanding, however, it would indeed be convenient to present a definition of climatic change in terms of principal variables.

In a mathematical formulism similar to that of the hydrological equations (10.1) and (10.2), we suggest that such a time-dependent equation for a regional climate C_R might be functionally defined in terms of four basic variables, namely:

$$C_R = f(T, P, V, R) \qquad (10.3)$$

where T = near-surface air temperature
  P = precipitation
  V = near-surface wind velocity, and
  R = a term representing the "end-point" or "storage" status of aquatic matter within the region or basin under consideration (i.e., the state within the time-frame of the analyses beyond which no further transport or transformation will occur).

Then, regional climate change, dC_R/dt, would be given by:

$$\frac{dC_R}{dt} = \frac{\partial C_R}{\partial T}\frac{dT}{dt} + \frac{\partial C_R}{\partial P}\frac{dP}{dt} + \frac{\partial C_R}{\partial V}\frac{dV}{dt} + \frac{\partial C_R}{\partial R}\frac{dR}{dt} \qquad (10.4)$$

Notice that, while equation (10.4) defines a change in regional climate, it does not define C_R as a distinct single entity with an associated parametric unit. dC_R/dt is, rather, an index of changing conditions within a regional climate and thus may be more appropriately defined as an *index of regional climate change.*

The time derivatives of equation (10.4) are relatively well-defined and estimable. dT/dt may be obtained from temperature records. dP/dt is as defined by equation (10.2). dV/dt is the acceleration of near-surface wind. Both dP/dt and dV/dt can also be obtained or determined from data records. The "end-point" or "storage" removal term R is more controversially defined. Since equation (10.4) defines regional climate change (as opposed to global climate change), it is not restrained by the conservation laws that dictate the behavior of global climate change. The regional climate, being a non-closed system, may be characterized by aquatic matter which may accumulate or diminish from the region, depending upon the conditions existing at a particular time. R, then, may be confidently taken to represent the difference between output and input aquatic matter, i.e., R is essentially an increasing or decreasing residual for the basin under scrutiny, That is,

$$R(t) = \text{Input} - \text{Output as a function of time.}$$

From equation (10.1), $R(t)$ may be written as:

$$R(t) = P(t) - E_T(t) - Q(t)$$

$$= W_G(t) + W_S(t) + S(t)$$

$$= \text{groundwater} + \text{soil moisture} + \text{surface water storage.} \qquad (10.5)$$

Thus, $R(t)$ is the sum of all surface and subsurface water present in the basin at a particular time t. dR/dt is the time rate of change of the entire non-atmospheric water content of the basin. At any time t the basin may be far removed from total saturation. Thus, $R(t)$ may be regarded as a percentage basin saturation term, i.e., a non-atmospheric equivalent of relative humidity for a particular basin. dR/dt may then be considered as a hydrological change in regional non-atmospheric relative humidity. This is a direct analogy to global aquatic fate wherein aquatic transports occur from groundwater to rivers to oceans to great ocean depths (the ultimate "endpoint"). The regional counterpart of great ocean depths is subsurface water. Further, it is interesting to note that precipitation $P(t)$, which is the governing parameter for the hydrological cycle, is, quite appropriately, a variable in the regional climate expression.

While the total time derivatives of equation (10.4) do not present insurmountable problems to either scientific philosophy or direct numerical estimations, the partial derivative terms of equation (10.4) do accentuate the inherent difficulties in successfully relating regional climate change to even a reduced number of hydrologically significant environmental variables. The index of regional climate change dC_R/dt is dependent upon the climatic impact of near-surface air temperature, precipitation, surface windspeed, and subsurface residual storage capacity, each considered individually in the absence of the other hydrological driving forces. Setting up the required controlled research experiments or establishing verifiable environmental models to obtain such partial derivatives as functions of both space and time is certainly not a simple task. Nevertheless, such determinations of multivariate dependencies must be made if impacts of regional climate change are not only to be understood and predicted, but to be responsibly acted upon.

Each of the partial derivative climatic dependencies ($\partial C_R/\partial T$, $\partial C_R/\partial P$, $\partial C_R/\partial V$, and $\partial C_R/\partial R$), is itself an *index*, and thus also become *regional climate indices* (say, α_T, α_P, α_V, α_R) that will vary from aquatic basin to aquatic basin.

It is, of course, possible to assess mathematically the roles of the regional climate indices α_i and try to derive the interdependencies among the climate change variables considered herein. Such analyses should, of course, be performed. We will, however, not continue such analyses here. We shall simply say that the formulism of the climatic change expression and its parallelism in both form and parametric dependence to the governing hydrology equation suggest that the role of satellite sensing in acquiring data pertinent to climate change parallels the role of satellite sensing in acquiring data pertinent to hydrology. Thus, to obtain information relevant to monitoring parameters of climatic change, wavelength regions of the electromagnetic spectrum that transcend the optical range discussed throughout this book are required.

10.5   EFFECTS OF ENHANCED UV-B ON PHYTOPLANKTON PHOTOSYNTHESIS

Severe springtime stratospheric ozone depletion has been occurring over the Antarctic since the mid-1980s. As much as 10% of the southern hemisphere now lies under this seasonably depleted layer. Ozone concentrations 40% lower than 1960s values commonly occur. A combination of a warmer Arctic stratosphere, the predominantly Arctic land mass as opposed to the predominantly Antarctic water mass, the nature of polar stratospheric clouds, and the uneven geography of the Arctic inhibits the development of an Arctic ozone "hole" of the severity of the Antarctic ozone "hole." Nonetheless, Arctic springtime ozone losses of (5–10)% are common. The average global reduction in stratospheric ozone is currently estimated to be at least 3%. In addition to Antarctica, ozone depletion is now affecting much of North America, Russia, Europe, Australia, New Zealand, Southern Africa, and southern South America. Corresponding to the seasonal ozone depletion, associated enhanced ground-level UV-B fluxes have been observed on a short-term and regional but not on a long-term and global basis.

Since the maximum ozone reduction is directly above the Antarctic waters, and since UV-B is known to harm biological processes, much effort has been directed toward the biological aspects of UV-B penetration into oceanic waters (see reviews by Smith and Baker,[377] Häder and Worrest,[151] United Nations Environment Program,[404] Vincent and Roy,[412] Cullen and Neale,[74] and others). Inherent to the focus is the quantification of the net impact of UV-B on phytoplankton in a natural marine environment, in particular as to the induced changes within the photosynthetic responses. In general, a photoinhibition of phytoplankton photosynthesis is logically expected to result from an increase in subsurface UV-B radiation. Smith et al.[378] and Smith and Baker[376] established that such an inhibition of near-surface primary production due to subsurface UV-B radiation does indeed occur. Smith et al.[380] estimate (6–12)% reduction in primary production in the Antarctic marginal ice zone, along with a shift toward larger phytoplankton species. The *possibility* (not a *certainty*) exists, therefore, that such a perturbation to indigenous phytoplankton populations could result in a serious food-web problem to marine mammals.

Biological effects of absorbed ultraviolet radiation are functions of wavelength (weighting factors that define the *action spectra* or biological effectiveness as referred to in Section 2.3). Further, photoinhibition is also a function of the spectrally variant ratio of UV to PAR. This ratio, being dependent upon the irradiance attenuation coefficients of both UV, $K_{UV}(\lambda)$, and PAR, $K_{PAR}(\lambda)$, is a function of depth and a function of water type. To address this problem, Cullen et al.[72] and Cullen and Neale[73] presented an analytical model of photosynthesis (based upon controlled laboratory exposures of phytoplankton suspensions to a broad range of E_{PAR} and E_{UV}/E_{PAR} values) invoking biological weighting functions to incorporate both PAR and UV into the photosynthetic process. Defining E_{inh} as the biologically weighted dose rate of damage resulting from both the photosynthesis-inducing E_{PAR} and the photosynthesis-inhibiting E_{UV}, Cullen *et al.* defined the impacted rate of photosynthesis, P_{imp}, in terms of the rate of photosynthesis in the absence of photoinhibition, P, as

$$P_{imp} = P/(1 + E_{inh}) \tag{10.6}$$

where

$$E_{inh} = \varepsilon_{PAR} E_{PAR} + \sum_{280\ nm}^{400\ nm} \varepsilon(\lambda) E_{UV}(\lambda) \Delta\lambda \tag{10.7}$$

where ε_{PAR} = the relative biological efficiency for damage to photosynthesis (inhibition) due to E_{PAR} (i.e., in the absence of UV, ε_{PAR} = 0), and

$\varepsilon(\lambda)$ = the wavelength-dependent biological efficiency for damage to photosynthesis (inhibition) due to E_{UV} (i.e., in the absence of UV, $\varepsilon(\lambda)$ = 0).

The biological weighting functions (BWF) or action spectra ε_{PAR} and $\varepsilon(\lambda)$ are species-dependent and, therefore, vary in both time and space. The relationships between photoinhibition and decreased growth rates have not been totally determined and precise action spectra for aquatic biota remain a challenge. Thus, long-term impacts of fluctuating UV-A and UV-B fluxes remain difficult to predict. However, the fact that the action spectra are species-dependent suggests that changes in relative plankton populations are very likely in Case I waters located beneath stressed regions of the stratospheric ozone layer.

Lacking BWFs for indigenous chlorophyll-bearing biota, Smith and Baker[374] used the action spectrum for DNA (Setlow[363]) along with the spectral distribution of irradiance in a variety of marine waters and illustrated that (a) the shape of the DNA action spectrum $\varepsilon_{DNA}(\lambda)$ varied only slightly with depth, (b) the biologically effective radiation in the wavelength interval (290–340) nm [the product of the spectral irradiance at depth z, $E(\lambda)$, and the DNA action spectrum $\varepsilon_{DNA}(\lambda)$] diminished exponentially with depth, and (c) the associated irradiance attenuation coefficient for radiation damaging to DNA is approximately given by the irradiance attenuation coefficient for downwelling irradiance at 305 nm, i.e., $K_d(305)$.

As we have seen, remote sensing is a near-surface activity that must contend with volumetric information culled from the upper penetration depth (the inverse of the downwelling irradiance attenuation coefficient K_d). In Chapter 9 we have discussed the difficulties and hazards involved in estimating phytoplankton photosynthesis for the entire euphotic zone from remotely sensed estimates of near-surface concentrations of chlorophyll. The variation in depth of the deep chlorophyll maximum (DCM) and its location relative to the euphotic zone remain problematical in Case I waters wherein the depths of the 1% irradiance level for ultraviolet radiation can be several tens of meters. The variation of chlorophyll concentration with depth must then be combined with the variation of ultraviolet radiation with depth as well as with the biological responses of the indigenous chlorophyll-bearing biota [the BWF or action spectra ε_{PAR} and $\varepsilon(\lambda)$ of equation (10.7)].

Freshwaters are characterized by much higher absorption coefficients for ultraviolet radiation than are marine waters.[360] The pelagic ecosystems of freshwaters, in general, therefore, would appear to be less directly impacted by ultraviolet radiation than are the pelagic ecosystems of marine waters. However, as alluded to in Section 10.3, the impact of ultraviolet radiation on pelagic lifeforms is dependent not only upon the optical attenuation of downwelling light, but also upon the hydrodynamical processes occurring within the confining basin as well as the interplay of the carbon cycle feedback loops linking increases in ground-level ultraviolet radiation with increases in near-surface air temperatures. As evidenced by the above-referenced works of Cullen and his collaborators, Smith and his collaborators, and others, the action spectra appropriate to pelagic ecosystems are required to quantify the impacts of not only enhanced levels but also ambient levels of ultraviolet radiation (UV-A as well as UV-B) on natural waters. Further, it is the total (integrated over time) dosage of ultraviolet radiation (as opposed to the instantaneous dose) that determines the impact of UV light upon aquatic lifeforms. Thus, the time that aquatic organisms spend in the near-surface "danger zone" of pelagic freshwaters (the combined consequences of passive vertical circulation and enforced migrations created by disruptions to quiescent population dynamics within the food web) is a determining factor in UV damage to aquatic ecosystems. If, as is the case for a substantial number of freshwater systems, the pelagic waters contain dissolved organic matter concentrations in excess of (1–2) mg C/l, the planktonic communities do not require great depths to escape the "UV danger zone." The depth to which plankton can be transported due to passive circulation, however, is dependent upon the turbulent mixed layer in the near-surface waters, and the thickness of this turbulent layer is a function of basin climatology as well as the optical properties of the epilimnetic waters. For inland and coastal waters, the higher the concentrations of dissolved organic matter and/or suspended inorganic matter, the shallower will be the mixed layer as thermal stratification occurs. Thus, in turbid waters, passive vertical circulation does not carry phytoplankton to large depths. The protective turbidity within the water coupled with the enhanced planktonic mobility resulting from even very moderate surficial winds can certainly compensate for the shallowing of the upper mixed layer. Under a totally calm air-water interface, however, the steep vertical temperature gradient and the corresponding steep density gradient can greatly impede the downward movement of plankton trapped near the surface, and the damaging effects of UV radiation cannot be avoided for the duration of calmness. Such persistently calm conditions have been observed to result in the photoinhibition of plankton photosynthesis in lakes with near-surface thermoclines (Vincent et al.,[411] Neale and Richerson[297]). Ultraviolet radiation can liberate a free electron from dissolved organic matter. This free electron can combine with O_2 to form the superoxide anion O_2^- which then reacts with free protons to form hydrogen peroxide H_2O_2. Cooper and Lean[66] and Lean et al.[243] illustrate that the depth profile of hydrogen peroxide in natural water can provide a powerful tracer for vertical mixing events, and thus has the potential to estimate the diurnal exposure times of phytoplankton within the "UV danger zone." Hence, even though the pelagic ecosystems of freshwaters *appear* to generally be less directly impacted by UV radiation than are marine ecosystems, some combinations of shallow thermocline, persistently calm conditions, and organisms being mixed into the "UV danger zone" can create situations in which UV impacts to pelagic freshwater ecosystems can be comparable to or even exceed UV impacts to pelagic marine ecosystems.

The shallow PAR penetration depths coupled with the considerably shallower "ultraviolet photic zone" generally characterizing non-Case I waters may, under specific aquatic conditions, enhance the possible applications of remote sensing in evaluating impacts to aquatic ecosystems of fluctuating ultraviolet light regimes. If the obstacles to the remote monitoring of limnological and coastal primary productivity (as discussed in Chapter 9) can be overcome or circumvented, and such near-surface primary production is observed for persistently calm waters, this primary production could be considered to represent perturbations to the quiescent planktonic dynamics in response to ultraviolet radiation and/or climate change. Hence, a clever analysis scheme could benefit from those occasions for which measurements of impacts to near-surface phytoplankton photosynthesis can be considered as an appropriate (although not a mathematically equivalent) surrogate for the impacts to the phytoplankton photosynthesis of the total water column.

It should be noted that the discussion to this point has focussed upon modulated phytoplankton photosynthesis as a consequence of direct damage by ultraviolet radiation to the chlorophyll-bearing biota. Certainly, such direct damage will occur, but the pernicious impacts of ultraviolet radiation are not restricted to the lowest stature members of the food web. Damage to freshwater and marine invertebrates has been documented by numerous researchers for several years. However, the impacts of such damage at higher trophic levels (e.g., fish stock and fish yield) as well as the impacts at the lower trophic levels due to direct radiation damage at the higher trophic levels is still highly speculative. Bothwell *et al.*[29,30] using re-directed stream flow from the South Thompson River in British Columbia, Canada observed that an increased algal biomass can occur under direct exposure to ambient ultraviolet radiation in very shallow water even though the production per unit biomass was reduced. This was due to differential sensitivity to UV between algae and herbivores with the result that over a period of time there was little change in standing crop at lower trophic levels, but a reduced energy flow to higher trophic levels. It is, therefore, possible that even *ambient* ultraviolet radiation under certain conditions (e.g., quasi-controlled, exposed shallow water ecosystems) could produce a reduced yield in fish and/or bird species with no obvious change being observed in the algal populations or in the directly measured or remotely estimated phytoplankton photosynthesis. Such a possibility certainly warrants consideration in atmospheric change impacts studies on aquatic ecosystems that employ monitoring of aquatic bioproductivity as a principal component.

10.6 GLOBAL WATER BODIES AND THE GAIA HYPOTHESIS

It is now widely recognized that the Earth's climate and physical-chemical composition is unique (at least within our solar system) in its favourable disposition towards life. It is also widely recognized that relatively small changes in the dynamically balanced state that exists among such climatic and physical-chemical components could render these life-support capabilities completely unsuitable for contemporary biota. Less widely recognized, but apparently growing in support, is the theory first propounded by Lovelock[250] and Lovelock and Margulis[251] that the biosphere serves as more than merely a passive accommodation for environmental change. According to these workers, the biosphere (or more specifically, individual lifeforms) acts in a proactive manner so as to control the local or global climate and thereby optimize the environment's capacity for sustaining life. Such a biospheric "life force" and the theory that purports its existence bears the name Gaia, after the Greek deification of Mother Earth. In a chapter devoted to the uncertainties of climate change (ozone depletion, global warming, disruptions to the carbon, nitrogen, and hydrological cycles, threats to biodiversity) in a book devoted to aquatic optics, it seems somehow remiss not to include some dispassionate musings on climate change, aquatic optics, and Gaia.

The Gaia Hypothesis, of course, may forever remain directly unverifiable. Gaia's ascribed properties and powers are in direct conflict with those of the physical and chemical parameters that have traditionally become considered as climate regulators. Without intentionally fuelling such controversies (see Schneider[357]), suffice to say that the strongest (albeit circumstantial) evidence for support of the Gaia Hypothesis is that throughout the continuum of time generally ascribed to the presence of life on Earth (over 3×10^9 years), the chemical and physical properties of the Earth's environment have never been at variance with those optimal conditions required for the support of life. Some lifeforms have ceased to exist and new lifeforms have evolved, but the environmental conditions required to sustain life on Earth have themselves been sustained.

Since life abounds within global water bodies, aquatic resources form an integral component of Gaia speculations. Workers such as Twomey,[399] Shaw,[364] Twomey et al.,[400] and Meszaros[263] have extended the Gaia Hypothesis to atmospheric aerosols and suggest that the aerosols produced by the atmospheric oxidation of sulfur gases from biota may directly impact climate. More specifically, Charlson et al.[61] speculate that the dimethylsulfide emitted by oceanic phytoplankton directly serves as cloud condensation nuclei which modulate cloud development and, thereby, planetary albedo and the Earth's radiation budget. Charlson and his collaborators thus suggest that the oceanic phytoplankton act as biological regulators of the Earth's climate, a suggestion in agreement with the speculations of Shaw and Meszaros. Global climate control by oceanic phytoplankton, however, is challenged by Schwartz,[359] who indicates that anthropogenic sulfur dioxide, which exceeds marine emissions of dimethysulfide, shows no counterpart influence in either the present cloud component of planetary albedo or the temperature records of the past century. Sathyendranath et al.[351] using remotely sensed ocean color data, have indicated that phytoplankton distribution exerts a controlling influence on the seasonal evolution of sea surface temperature. Biological control of ocean/temperature interactions is also reported by Falkowski et al.[102] wherein satellite-acquired data suggest that variability in mid-oceanic albedo can be accommodated by natural marine and atmospheric processes that may have remained relatively constant since the onset of the industrial revolution. Environmental adaptability of blue-green algae was illustrated by Paerl et al.,[302] who reported that enhanced cellular carotenoid synthesis provide an effective protection against enhanced ultraviolet radiation fluxes.

Irrespective, however, of the controversial role of Gaia in ocean/atmosphere dynamics, it is reasonable to consider a feedback loop as existing between aquatic phytoplankton and downwelling solar irradiance above the air-water interface but below the clouds. It is also reasonable to consider directly measurable chlorophyll concentrations as a surrogate for phytoplankton populations and, therefore, to suggest that a feedback loop exists between chlorophyll concentrations and downwelling irradiance. Whether the phytoplankton passively respond to or proactively dictate (or both) specific variations in the downwelling solar irradiance, it is further reasonable to assume that phytoplanktonic adaptation (either passive or proactive) will manifest as physical changes (size, shape, structure, index of refraction) or even as changes to the indigenous species of chlorophyll-bearing biota. These changes will generate changes in the optical properties of the algal cells and, therefore, in the absorption and scattering optical cross section spectra of chlorophyll at that location, consistent with the observations of Paerl and his collaborators.

As discussed earlier, the optical cross sections of indigenous chlorophyll concentrations are subject to variabilities that are directly associated with species, seasonal growth cycles, and cell light history. They may also be a function of the very climate change they are and will continue to be used to monitor and investigate. This latter function is currently an unknown factor, the complexity of which could be related to the controversial role of Gaia. However, even without the added complexity of Gaia and the suggested feedback loop between chlorophyll concentrations and downwelling irradiance, the need to determine accurate values of optical cross section spectra is illustrated by Andre and Morel[8] who simulate the impact of varying atmospheric ozone on the remote estimates of marine phytoplankton using bio-optical algorithms developed for the Nimbus-7 CZCS. The authors report that the use of mean climatology values of ozone and pressure (rather than the actual values of ozone and pressure at the time of the remote measurements) can lead to inaccuracies in the estimates of phytoplanktonic pigment concentrations of a factor of two. It is reasonable to assume that such inaccuracies for the more optically complex inland and coastal aquatic regimes would be even larger. Since chlorophyll concentration predictions respond to changes in such climatic factors as ozone and pressure, the need to obtain reliable chlorophyll cross section spectra becomes all the more imperative.

We will consider one final musing on Gaia. It is in regard to the spectral cross section variability of chlorophyll-bearing biota that the thought-provoking, fascinating, and, indeed, appealing Gaia Hypothesis is considered. Scientific objectivity compels us, in the absence of definitive evidence, to neither support nor condemn the hypothesis. Human honesty and curiosity, however, compel us to admit unashamedly to being most intrigued by the concept of Gaia (Societal forces versus Nature may not be a fair fight, but could it be a fight nonetheless?). Cause and effect are often very difficult to segregate. Feedback loops abound within environmental ecosystems. Despite the ethereal qualities ascribed to Gaia, Gaia is, after all, a feedback loop. Phytoplankton change their physiology in response to their ambient light regime. This could result in changes to the absorption and scattering properties (optical cross section spectra) of the phytoplankton, which in turn would modify the ambient light environment, and so on. Gaia could, therefore, be considered as interchangeable with the naturally responsive components of a successfully evolved ecosystem. With or without invoking Gaia, the presence of phytoplankton/radiation field feedback loops such as suggested herein could profoundly affect the energy transfer processes within the water column, which will, in turn, affect the upwelling radiance spectrum recorded at a remote platform.

10.7   REMOTE SENSING AND WATER QUALITY

In this chapter we have briefly touched upon foreseeable roles of remote sensing in the upcoming decades of multinational and multidisciplinary studies of the Earth's inter-related ecosystems. Particular attention was directed towards impacts on aquatic ecosystems resulting from ozone depletion and disruptions to the hydrologic cycle. Implicit to such considerations is the quality of the water itself.

Water can no longer be considered an unlimited resource, even in those geographic regions containing large volumes of water. Natural water in general, and inland waters and coastal regions in particular, provide locale and sources of material for a variety of human activities (drinking, irrigation, agriculture, industry, recreation, tourism, waste disposal, etc.). The sustainable stabilities of many aquatic resources have been disrupted (in developing as well as developed countries) due to the use and abuse of these resources. The quality of a water body is determined not only by human activities, but also by the properties of its confining basin or watershed (climate, topography, geology, soils, vegetative cover, drainage) and by the properties of the water body itself (morphometry, currents, depth, mixing layers). *Water quality* is a descriptive term that refers to the state or condition of a water body in relation to a set of criteria established for its designated use. Water quality descriptors are sets of physical, chemical, and/or biological characteristics of a water body that serve as measures of the water body's compliance to that set of criteria. Gale[117] proposed a set of water quality descriptors from which Jaquet *et al.*[190] extracted those descriptors that they felt had the potential to be estimated by remote sensing. Included within that list of potentially estimable water quality descriptors were:

1. Directly measurable physical descriptors such as temperature, color, transparency, and turbidity,
2. Estimable (through bio-optical modeling) physical descriptors such as concentrations of suspended minerals and suspended solids,
3. Estimable (through bio-optical modeling) chemical descriptors such as dissolved organic carbon (DOC) and other dissolved organic matter (DOM),
4. Estimable (through algorithms or photogrammetry) hydrobiological descriptors such as water levels and aquatic vegetation species, and
5. Estimable (through bio-optical modelling) hydrobiological descriptors such as chlorophyll *a*.

Conventional water quality monitoring networks are designed to directly acquire data on a judiciously selected number of parameters that can be used to evaluate the quality of regional water for domestic use. Such monitoring networks rely upon *in situ* parameter monitoring as well as the collection of water samples for on-site evaluation or for transport to laboratories for further analyses. As such, monitoring stations can, within logistical and/or financial restrictions, consider collecting data on not only the abbreviated list of water quality descriptors given above, but on the full range of physical, chemical, microbiological, and even radioactivity descriptors suggested by Gale and others. While the temporal dimension of the water quality issue can usually be so monitored with sufficient accuracy, the limited number of stations associated with the water quality networks generally prohibits the spatial dimension of the water quality issue to be monitored with comparable accuracy. Remote sensing is very often considered as a means of addressing this shortcoming of monitoring networks by obtaining synoptic measurements of the optical returns (reflected or emitted) from the targeted water bodies. This optical return must then be transformed, through appropriate models, algorithms, or methodologies, into estimates of water quality parameters.

As considered in this book, water quality as defined from an optical/remote sensing perspective is based upon the specific inherent optical properties (absorption and scattering cross section spectra) of the organic and inorganic components of the water body being remotely monitored. Due to the low transparency of water at wavelengths longer than ~740 nm, the shorter (~390 nm–740 nm) wavelength region is generally used for remote sensing of water quality. Hence, strong emphasis is placed upon evaluating water quality in terms of water color. We have discussed at great lengths the impacts on water color of the co-existing and non-co-varying concentrations of chlorophyll, suspended sediments, and dissolved organic matter. For remote sensing concerned with the visible wavelength region of the spectrum, therefore, water quality must be defined in terms of the near-surface combinations of these aquatic components at the time of the remote observation.

Satellites containing instrument packages dedicated to monitoring the water quality of inland and coastal regions are currently not available, although the Landsat Thematic Mapper[289] does possess some applicability. Unfortunately, satellites dedicated to monitoring the water quality of non-Case I waters are not scheduled for the immediate future, even though the current emphasis on the SeaWiFS, HIRIS, and MERIS is comforting evidence that the need to monitor inland and coastal water quality is being addressed.

As a composite of discussions presented by Jaquet et al.,[190] Schneider et al.,[357] and Bukata and Jerome,[49] Table 10.3 lists some currently available environmental platforms (satellite and airborne) that are assessed according to their capabilities to provide water quality information (pertinent information on land and water). Entries designated by closed circles represent those water quality descriptors that can be *quantitatively* determined. Entries designated by open circles represent those water quality descriptors that can be be *qualitatively* determined. E suggests determinations that are relatively

TABLE 10.3

Capability of Environmental Platforms to Monitor Water Quality

Remote Sensor (Spatial Resolution)	Applications								
	Water							Land	
	Temp	Turb	Chl	DOC	SM	Oil	AquVeg	RipVeg	Land cover
Meteosat (kms)	oE								
NOAA-AVHRR (1 km)	●E	oH				oH		oH	
Landsat MSS (80 m)		oE				oH	oH		oE
Landsat TM (30 m)	●H	oE	●H	●H	●H	oE	oE	oE	oE
MOS-1 MESSR (50 m)		oH				oE	oH		oE
SPOT XS (20 m)		oE				oE	oE		oH
ERS-1	●E					oH			oH
High Spect. Res. (meters)		●E	●E	●E	●E	oE	oE	oE	oE
Low Spect. Res. (meters)		oE	oH			oE	oE	oE	oE
Thermal Scan.	●E								

easy to obtain. H suggests determinations that are relatively hard to obtain and require appropriate bio-optical models to extract the descriptor from the upwelling radiance spectrum. The applications for observations over water are listed as temperature (Temp), turbidity (Turb), chlorophyll concentration (Chl), dissolved organic carbon concentration (DOC), suspended mineral concentration (SM), oil, and aquatic vegetation (AquVeg). The applications for observations over land are listed as riparian vegetation (RipVeg) and land cover.

The first seven rows in Table 10.3 consider satellite platforms, while last three rows in Table 10.3 consider airborne sensors (high spectral resolution radiometers, low spectral resolution radiometers, and thermal scanners). Modest spatial and/or modest spectral sensitivity is used to reinforce our contention that water quality monitoring over non-Case I waters is beyond the capability of existing satellite platforms. The Landsat Thematic Mapper, however, used in conjunction with conventional *in situ* monitoring techniques and bio-optical models such as discussed throughout this book, does present the possibility for expanding a number of precise but spatially limited direct water quality measurements to a much larger spatial area. Upcoming remote sensing programs such as the American SeaWiFS, the European MERIS, and the Japanese ADEOS, which have incorporated coastal regions into their targeted priorities, will be able to provide satellite observations on less modest spectral and spatial scales. However, these platforms will retain limitations to their use as operational water quality monitors over inland and coastal regimes.

For simultaneous monitoring of the co-existing water quality descriptors Chl, SM, and DOC, both the high spectral and high spatial resolutions absent from satellite sensors are required. These resolutions are certainly inherent to airborne spectroradiometric imagers such as the CASI instrument, and we feel there is little doubt that such a device can be of tremendous benefit to the monitoring of both terrestrial and optically complex non-Case I aquatic ecosystems. Such high spectral and spatial resolution remote monitoring of inland and coastal water quality must proceed with ancillary *in situ* optical measurements to allow for periodic determinations of the absorption and scattering cross section spectra of the organic and inorganic matter being monitored.

The limited number of CASI-type instruments resident and in use across the globe is indicative of another obstacle that must be overcome if the remote monitoring of non-Case I water quality is to be incorporated into programs of regional effects of climate change and ozone depletion. The cost per line mile of remote sensing overflights is a luxury that very few research budgets can afford, particularly in the current climate of ever-diminishing research fundings. Even the possession of a spectrally sensitive airborne instrument does not necessarily guarantee affordability of or access to a dedicated aircraft. The luxury of remote sensing as a research tool is further emphasized by the unfortunate fact that, in the over two decades since the launch of Landsat-1, advances in engineering instrumentation and data collection, registration, and archiving have progressed at a rate that has far eclipsed the rate at which advances in analyses and interpretation of the remotely acquired data have progressed. This perceived tardiness of remote sensing enthusiasts in fully realizing the potential that has been continuously ascribed to remote monitoring, coupled with the very often prohibitive costs of the technology, has resulted in a skepticism that cannot be ignored. We have discussed some of the difficulties encountered in extracting meaningful environmental information from spectral data collected over optically complex water bodies. Such difficulties must be resolved if remote sensing is to be considered a viable supplement to the direct monitoring of environmental parameters.

REFERENCES

1. Adamenko, V. N., Kondratyev, K. Ya., Pozdnyakov, D. V., and Chekhin, L. P., *Radiative Regime and Optical Properties of Lakes* (Russian), Gidrometeoizdat Publishing Co., Leningrad, 300 pp, 1991.
2. Ahern, F. J., Goodenough, D. G., Jain, S. C., Rao, V. R., and Rochon, G., *Use of Clear Lakes as Standard Reflectors for Atmospheric Measurements*, Proc. 11th Int. Symp. Rem. Sens. Environ., Ann Arbor, MI, 583–594, 1977.
3. Ahern, F. J., Brown, R. J., Cihlar, J., Gautier, R., Murray, J., Neville, R. A., and Teillet, P. M., Radiometric correction of visible and infrared remote sensing data at the Canada Centre for Remote Sensing, *Int. J. Rem. Sens.*, 8, 1349–1367, 1987.
4. Ahn, Y.-H., Bricaud, A., and Morel, A., Light backscattering efficiency and related properties of some phytoplankters, *Deep-Sea Res.*, 39, 1835–1855, 1992.
5. Alekin, O. A., *Chemistry of the Ocean* (Russian), Gidrometeoizdat Publishers, Leningrad, 250 pp, 1966.
6. Alföldi, T. T. and Munday, J. C., Jr., Water quality analysis by digital chromaticity mapping of LANDSAT data, *Can. J. Rem. Sens.*, 4, 108–126, 1978.
7. Anderson, G. P., Clough, S. A., Kneizys, F. X., Chetwynd, J. H., and Shettle, E. P., *AFGL Atmospheric Constituent Profiles (0–120 km)*, Air Force Geophysics Lab., Hanscom AFB, Research Report No. AFGL-TR-86-0110, 1986.
8. Andre, J. M. and Morel, A., Simulated effects of barometric pressure and ozone content upon the estimate of marine phytoplankton from space, *J. Geophys. Res.*, 94, 1029–1037, 1989.
9. Anger, C. D., Babey, S. K., and Adamson, R. J., A new approach to imaging spectroscopy, *SPIE Imaging Spectroscopy of the Terrestrial Environment*, 1298, 72–86, 1990.
10. Ångström, A., The parameters of atmospheric turbidity, *Tellus*, 16, 56–71, 1963.
11. Austin, R. W., The remote sensing of spectral radiance from below the ocean surface, in *Optical Aspects of Oceanography*, Jerlov, N. G. and Steeman Neilsen, E., Eds., Academic Press, London, 317–344, 1974.
12. Austin, R. W. and Petzold, T. J., Considerations in the design and evaluation of oceanographic transmissometers, in *Light in the Sea*, Tyler, J. E., Ed., Dowden, Hutchison and Ross, Stroudsbury, 104–120, 1977.
13. Austin, R. W. and Petzold, T. J., The determination of the diffuse attenuation coefficient of sea water using the Coastal Zone Color Scanner, in *Oceanography from Space*, Gower, J. F. R., Ed., Plenum Press, New York, 239–256, 1981.
14. Baker, K. S. and Smith, R. C., Quasi-inherent characteristics of the diffuse attenuation coefficient for irradiance, *Proc. Soc. Photo-opt. Instr. Eng.*, 208, 60–63, 1979.
15. Balch, W., Evans, R., Brown, J., Feldman, G., McClain, C., and Esaias, W., The remote sensing of ocean primary productivity: Use of a new data compilation to test satellite algorithms, *J. Geophys. Res.*, 97, 2279–2293, 1992.
16. Bannister, T. T., Production equations in terms of chlorophyll concentration, quantum yield, and upper limit to production, *Limnol. Oceanogr.*, 19, 1–12, 1974.

17. Bannister, T. T., Empirical equations relating scalar irradiance to a, b/a, and solar zenith angle, *Limnol. Oceanogr.*, 35, 173–177, 1990.

18. Bannister, T. T. and Laws, E. A., Modelling phytoplankton carbon metabolism, in *Primary Productivity in the Sea*, Falkowski, P. G., Ed., Plenum Press, New York, 243–256, 1980.

19. Barica, J., Poulton, D. J., Kohli, B., and Charlton, M. N., Water exchange between Lake Ontario and Hamilton Harbour: Water quality implications, *Water Pollution Res. J. Can.*, 24, 213–226, 1988.

20. Bauer, D. and Morel, A., Étude aux petits angles de l'inicatrice de diffusion de la lumière par les eaux de mer, *Ann. Géophys.*, 23, 109–123, 1967.

21. Beardsley, G. F. and Zaneveld, J. R., Theoretical dependence of the near-asymptotic apparent optical properties on the inherent optical properties of sea water, *J. Opt. Soc. Amer.*, 59, 373–377, 1969.

22. Beeton, A. M., Relationship between Secchi disc readings and light penetration in Lake Huron, *Trans. Amer. Fish Soc.*, 87, 73–79, 1957.

23. Bethoux, I. P. and Ivanoff, A., Mesure de l'éclairement énergétique sous-marin, *Cah. Oceanogr.*, 22, 483–491, 1970.

24. Bidigare, R. R., Smith, R. C., Baker, K. S., and Marra, J., Oceanic primary production estimates from measurements of spectral irradiance and nutrient concentrations. *Global Biogeochem. Cycles*, 1, 171–186, 1987.

25. Bobba, A. G., Bukata, R. P., and Jerome, J. H., Digitally processed satellite data as a tool in detecting potential groundwater flow systems, *J. Hydrol.*, 131, 25–62, 1992.

26. Borstad, A., Brown, R. M., and Gower, J. F. R., Airborne remote sensing of sea surface chlorophyll and temperature along the outer British Columbia coast, *Proc. 6th Canadian Sym. Rem. Sens.*, 541–549, 1980.

27. Borstad, G. A. and Hill, D. A., Using visible range imaging spectrometers to map ocean phenomena, *Proc. Advanced Optical Instrumentation for Remote Sensing of Earth's Surface from Space, SPIE*, 1129, 130–136, 1989.

28. Borstad, G. E., Hill, D. A., and Kerr, R. C., Use of the Compact Airborne Spectrographic Imager (CASI); Laboratory examples, *Proc. 12th Can. Symp. Rem. Sens.*, IGARSS, Digest, 2081–2084, 1989.

29. Bothwell, M. L., Sherbot, D., Roberge, A. C., and Daley, R. J., Influence of natural ultraviolet radiation on lotic periphytic diatom community growth, biomass accrual, and species composition: Short-term versus long-term effects, *J. Phycol.*, 29, 24–35, 1993.

30. Bothwell, M. L., Sherbot, D. M., and Pollack, C. M., Ecosystem response to solar ultraviolet-B radiation: Influence on trophic-level interactions, *Science*, 265, 97–100, 1994.

31. Bricaud, A., Morel, A., and Prieur, L., Absorption by dissolved organic matter of the sea (yellow substance) in the UV and visible domains, *Limnol. and Oceanogr.*, 26, 43–53, 1981.

32. Bricaud, A. and Stramski, D., Spectral absorption coefficients of living and nonalgal biogenous matter: A comparison between the Peru upwelling and the Sargasso Sea, *Limnol. Oceanogr.*, 35, 562–585, 1990.

33. Brichambaut, C. P., *Rayonnement Solaire et Echanges Radiatifs Naturels*, Gauthier-Villars Publishers, Paris, 320 pp, 1966.

34. Bukata, R. P. and McColl, W. D., The utilization of sun-glint in a study of lake dynamics, in *Remote Sensing and Water Resources Management*, Thomson, K. P. B., Lane, R. K., and Csallany, S. C., Eds., American Water Resources Association, 351–367, 1973.

35. Bukata, R. P., Bobba, A. G., Bruton, J. E., and Jerome, J. H., The application of apparent radiance data to the determination of groundwater flow pathways from satellite altitudes, *Can. J. Spectrosc.*, 2, 48–58, 1978.

36. Bukata, R. P., Jerome, J. H., Bruton, J. E., and Jain, S. C., Determination of inherent optical properties of Lake Ontario coastal waters, *Appl. Opt.*, 18, 3926–3932, 1979.

37. Bukata, R. P., Jerome, J. H., Bruton, J. E., and Jain, S. C., Nonzero subsurface irradiance reflectance at 670 nm from Lake Ontario water masses, *Appl. Opt.*, 19, 2487–2488, 1980.

38. Bukata, R. P., Jerome, J. H., Bruton, J. E., Jain, S. C., and Zwick, H. H., Optical water quality model of Lake Ontario. 1. Determination of the optical cross sections of organic and inorganic particulates in Lake Ontario, *Appl. Opt.*, 20, 1696–1703, 1981.

39. Bukata, R. P., Jerome, J. H., Bruton, J. E., Jain, S. C., and Zwick, H. H., Optical water quality model of Lake Ontario, 2. Determination of chlorophyll *a* and suspended mineral concentrations of natural waters from submersible and low altitude optical sensors, *Appl. Opt.*, 20, 1704–1714, 1981.

40. Bukata, R. P., Bruton, J. E., and Jerome, J. H., Use of chromaticity in remote measurements of water quality, *Rem. Sens. Environ.*, 13, 161–177, 1983.

41. Bukata, R. P., Jerome, J. H., and Bruton, J. E., *Direct Optical Measurements of the Laurentian Great Lakes, Part II: An Optical Atlas*, NWRI Report No. 85-25, Aquatic Physics and Systems Division, National Water Research Institute, Burlington, Ont., 1985.

42. Bukata, R. P., Bruton, J. E., and Jerome, J. H., *Application of Direct Measurements of Optical Parameters to the Estimation of Lake Water Quality Indicators*, Environment Canada Inland Waters Directorate Scientific Series No. 140, 35 pp, 1985.

43. Bukata, R. P., Jerome, J. H., and Bruton, J. E., Relationships among Secchi disk depth, beam attenuation coefficient, and irradiance attenuation coefficient for Great Lakes waters, *J. Great Lakes Res.*, 14, 347–355, 1988.

44. Bukata, R. P., Jerome, J. H., and Bruton, J. E., Particulate concentrations in Lake St. Clair as recorded by a shipborne multispectral optical monitoring system, *Rem. Sens. Environ.*, 25, 201–229, 1988.

45. Bukata, R. P., Jerome, J. H., and Bruton, J. E., Determination of irradiation and primary production using a time-dependent attenuation coefficient, *J. Great Lakes Res.*, 15, 327–338, 1989.

46. Bukata, R. P., Jerome, J. H., Kondratyev, K. Ya., and Pozdnyakov, D. V., Estimation of organic and inorganic matter in inland waters: Optical cross sections of Lakes Ontario and Ladoga, *J. Great Lakes Res.*, 17, 461–469, 1991.

47. Bukata, R. P., Jerome, J. H., Kondratyev, K. Ya., and Pozdnyakov, D. V., Satellite monitoring of optically-active components of inland waters: An essential input to regional climate change impact studies, *J. Great Lakes Res.*, 17, 470–478, 1991.

48. Bukata, R. P., Bruton, J. E., and Jerome, J. H., Utilizing vegetation vigor as an aid to assessing groundwater flow pathways from space (Russian), *Studies of the Earth from Space*, 2, 107–118, 1991.

49. Bukata, R. P. and Jerome, J. H., Satellite remote sensing of water quality, in *Advances in Water Quality Monitoring*, Report of a World Meteorological Organization Regional Workshop, 7–11 March, Vienna, Austria, Tech. Rep. in Hydrology and Water Resources No. 42, 225–238, 1994.

50. Bukata, R. P., Jerome, J. H., and Whiting, J. M., Satellite remote sensing of water quantity, in *Advances in Water Quality Monitoring*, Report of a World Meteorological Organization Regional Workshop, 7–11 March, Vienna, Austria, Tech. Rep. in Hydrology and Water Resources No. 42, 239–248, 1994.

51. Bunsen, R., Blaue farbe des wassers und eises, *Jahresber. Fortschr. Chem.*, 1847, 1236, 1847.

52. Burt, W. V., Albedo over wind-roughened water, *J. Meteorol.*, 11, 283–290, 1954.

53. Butler, W. L., Spectral characteristics of chlorophyll in green plants, in *The Chlorophylls*, Vernon, L. P. and Seely, G. R., Eds., Academic Press, New York, 343–379, 1966.

54. Carder, K. L., Steward, R. G., Paul, J. H., and Vargo, G. A., Relationships between chlorophyll and ocean color constituents as they affect remote sensing reflectance models, *Limnol. Oceanogr.*, 31, 403–413, 1986.

55. Carder, K. L., Steward, R. G., Harvey, G. R., and Ortner, P. B., Marine humic and fulvis acids: Their effects on remote sensing of chlorophyll *a*, *Limnol. Oceanogr.*, 34, 68–81, 1989.

56. Carder, K. L., Hawes, S. K., Baker, K. A., Smith, R. C., Steward, R. G., and Mitchell, B.

G., Reflectance model for quantifying chlorophyll *a* in the presence of productivity degradation products, *J. Geophys. Res.*, 96, 20,599–20,611, 1991.

57. Carder, K. L., Steward, R. G., Chen, R. F., Hawes, S., and Lee, Z., AVIRIS calibration and application in coastal oceanic environments: Tracers of soluble and particulate constituents of the Tampa Bay coastal plume, *Photogramm. Eng. Rem. Sens.*, 59, 339–344, 1993.

58. Chambers, R. L. and Eadie, B. J., Nepheloid and suspended particulate matter in southeastern Lake Michigan, *J. Sedimentology*, 28, 438–447, 1981.

59. Chandrasekhar, S., *Radiative Transfer*, Oxford University Press, London, 393 pp, 1950.

60. Chapra, S. C. and Dobson, H. F. H., Quantification of the lake trophic typologies of Naumann (surface quality) and Thienemann (oxygen) with special reference to the Great Lakes, *J. Great Lakes Res.*, 7, 182–193, 1981.

61. Charlson, R. J., Lovelock, J. E., Andrae, M. O., and Warren, S. G., Oceanic phytoplankton, atmospheric sulphur, cloud albedo, and climate, *Nature*, 326, 655–661, 1987.

62. Charlton, M. N., Milne, J. E., Booth, W. G., and Chiocchio, F., Lake Erie offshore in 1990: Restoration and resilience in the central basin, *J. Great Lakes Res.*, 19, 291–309, 1993.

63. Collins, D. J., Kiefer, D. A., SooHoo, J. B., Stallings, C., and Yang, W. L., A model for the use of satellite remote sensing for the measurement of primary production in the ocean, in *Ocean Optics VIII, SPIE*, Orlando, FL, 335–348, 1986.

64. Commission Internationale de l'Éclairage (CIE), *Vocabulaire Internationale de l'Éclairage*, C. I. E. Pub. No. 1, 2nd ed., Paris, 136 pages, 1957.

65. Committee on Colorimetry, *The Science of Color*, Opt. Soc. Am., 6th ed., Washington, D.C., 385 pp, 1963.

66. Cooper, W. J. and Lean, D. R. S., Hydrogen peroxide dynamics in marine and freshwater systems, in *Encyclopedia of Earth System Science*, Vol. 2, Nierenberg, W. A., Ed., Academic Press, 527–535, 1992.

67. Cox, C. and Munk, W., Measurement of the roughness of the sea surface from photographs of the sun's glitter, *J. Opt. Soc. Am.*, 44, 838–850, 1954.

68. Cox, C. and Munk, W., Some problems in optical oceanography, *J. Mar. Res.*, 14, 63–78, 1955.

69. Cox, C. and Munk, W., *Slopes of the Sea Surface Deduced from Photographs of Sun Glitter*, Bulletin Scripps Inst. Oceanogr., Univ. Calif., 6, 401–488, 1956.

70. Cullen, J. J., *Chlorophyll Maximum Layers of the Southern California Bight and Possible Mechanisms of their Formation and Maintenance*, Ph.D. thesis, University of California, San Diego, 138 pp, 1980.

71. Cullen, J. J., The deep chlorophyll maximum: Comparing vertical profiles of chlorophyll *a*, *Can. J. Fish. Aquat. Sci.*, 39, 791–803, 1982.

72. Cullen, J. J., Neale, P. J., and Lesser, M. P., Biological weighting function for the inhibition of phytoplankton photosynthesis by ultraviolet radiation, *Science*, 258, 646–650, 1992.

73. Cullen, J. J. and Neale, P. J., Quantifying the effects of ultraviolet radiation on aquatic photosynthesis, in *Photosynthetic Responses to the Environment*, Yamamoto, H. and Smith, C., Eds., American Society of Plant Physiology, Rockville, Md., 46–50, 1993.

74. Cullen, J. J. and Neale, P. J., Ultraviolet radiation, ozone depletion, and marine photosynthesis, *Photosyn. Res.*, 39, 303–320, 1994.

75. Dave, J. V., *Development of Programs for Computing Characteristics of Ultraviolet Radiation, Technical Report, Scalar Case, Programs I-IV*, FCS-72-0009/0013, IBM Fed. Syst. Div., Gaithersburg, MD, 1972.

76. Dave, J. V., Influence of illumination and viewing geometry on the "tasseled-cap" transformation of LANDSAT MSS data, *Rem. Sens. Environ.*, 11, 37–55, 1981.

77. Dave, J. V., Halpern, P., and Braslau, N., *Spectral Distribution of the Direct and Diffuse Solar Energy Received at Sea-Level of a Model Atmosphere*, IBM Palo Alto Scientific Center Report No. G320–3332, 1975.

78. Dave, J. V. and Halpern, P., Effect of changes in ozone amount on the ultraviolet radiation received at sea level of a model atmosphere, *Atmos. Environ.*, 10, 547–555, 1976.

79. Davies, J. A., Schertzer, W. A., and Numez, M., Estimating global solar radiation, *Boundary-Layer Meteorol.*, 9, 33–52, 1975.

80. Davies-Colley, R. J. and Vant, W. N., Absorption of light by yellow substance in freshwater lakes, *Limnol. and Oceanogr.*, 32, 416–425, 1987.

81. Dekker, A. G., *Detection of Optical Water Quality Parameters for Eutrophic Waters by High Resolution Remote Sensing*, doctoral thesis, University of Amsterdam, 222 pp, 1993.

82. Denman, K. L. and Gargett, A. E., Multiple thermoclines are barriers to vertical exchange in the subarctic Pacific during SUPER, May 1984, *J. Mar. Res.*, 46, 77–103, 1988.

83. Dillon, P. J. and Rigler, F. H., A simple method for predicting the capacity of a lake for development based on lake trophic status, *J. Fish. Res. Board Can.*, 32, 1519–1531, 1975.

84. Diner, D. J. and Martonchik, J. V., Atmospheric transfer of radiation above an inhomogeneous, non-Lambertian reflective ground—Theory, *J. Quant. Spectrosc. Radiat. Transfer*, 31, 97–125, 1984.

85. DiToro, D. M., Optics of turbid estuarine waters: Approximations and applications, *Water Res.*, 12, 1059–1068, 1978.

86. Dobson, H. F. H., Trophic conditions and trends in the Laurentian Great Lakes, *W. H. O. Water Qual. Bull.*, 6, 146–151 and 158–160, 1981.

87. Dobson, H. F. H., Some observations of (non-toxic) chemicals in water samples from the St. Lawrence River from Kingston to Quebec City, in *Collection Environnement et Géologie*, Vol. II, Legendre, P. and Delisle, C. E., Eds., Association des Biologistes du Quebec, 19–51, 1990.

88. Dorsey, N. E., *Properties of Ordinary Water-Substance*, Rheinhold Publishing Co., New York, 1940.

89. Drabkova, V. G., *Zonal Variation of the Intensity of Microbial Processes in Lakes* (Russian), Nauka Publishing Co., Leningrad, 210 pp, 1981.

90. Duntley, S. Q., *The Visibility of Submerged Objects*, Mass. Inst. Tech. Visibility Laboratory, Cambridge, 74 pp, 1952.

91. Duntley, S. Q., *Improved Nomographs for Calculating Visibility by Swimmers (Natural Light)*, Scripps Inst. of Oceanography Rep. 5-3, Task 5, Contract 72039, Univ. of Calif., San Diego, 1960.

92. Duntley, S. Q., Light in the sea, *J. Opt. Soc. Am.*, 53, 214–233, 1963.

93. Duntley, S. Q. and Preisendorfer, R. W., *The Visibility of Submerged Objects*, Mass. Inst. Tech. Visibility Lab. Final Rep. N5ori-07864, Cambridge, 1952.

94. Duntley, S. Q., Tyler, J. E., and Taylor, J. H., *Field Test of a System for Predicting Visibility by Swimmers from Measurements of the Clarity of Natural Waters*, Scripps Inst. of Oceanogr. Ref. 59-39, Univ. of Calif., San Diego, 1959.

95. Duntley, S. Q., Austin, R. W., Wilson, W. H., Edgerton, C. F., and Moran, S. E., *Ocean Color Analysis*, Scripps Institution of Oceanography, San Diego, SIO Ref 74-10, 67 pp, 1974.

96. Duysens, L. N. M., Photosynthesis, *Progr. Biophys. Molecul. Biol.*, 14, 1, 1964.

97. Duysens, L. N. M., Photobiological principles and methods, in *Photobiology of Microorganisms*, Halldal, P., Ed., Wiley-Interscience, London and New York, 1–16, 1970.

98. Eppley, R. W. and Peterson, B. J., Particulate organic matter flux and planktonic new production in the deep ocean, *Nature*, 282, 677–680, 1979.

99. Esaias, W. E., Feldman, G. C., McClain, C. R., and Elrod, J. A., Monthly satellite-derived phytoplankton pigment distribution for the North Atlantic ocean basin, *Eos*, 67, 835–837, 1986.

100. Ettreim, S., Ewing, M., and Thorndike, E. M., Suspended matter along the continental margin of the North American Basin, *Deep-Sea Res.*, 16, 613–624, 1969.

101. Ewing, M. amd Connary, S. D., Nepheloid layer in the North Pacific, in *Geol. Soc. Amer. Mem. 126*, Hays, J. D., Ed., Geologic Society of America, Boulder, 41–82, 1970.

102. Falkowski, P. G., Kim, Y., Kolber, Z., Wilson, C., Wirick, C., and Cress, R., Natural versus anthropogenic factors affecting low-level cloud over the North Atlantic, *Science*, 256, 1311–1313, 1992.

103. Farman, J. C., Gardiner, B. G., and Shanklin, J. D., Large losses of total ozone in Antarctica reveal seasonal ClO_x/NO_x interaction, *Nature*, 315, 207–210, 1985.

104. Fee, E. J., A numerical model for the estimation of photosynthetic production, integrated over time and depth, in natural waters, *Limnol. Oceanogr.*, 14, 906–911, 1969.

105. Fee, E. J., The importance of diurnal variation of photosynthesis vs. light curves to estimates of integral primary production, *Int. Assoc. Appl. Limnol. Verh.*, 19, 39–46, 1975.

106. Fee, E. J., Important factors for estimating annual phytoplankton production in the Experimental Lakes Area, *Can. J. Fish. Aquat. Sci.*, 37, 513–522, 1980.

107. Fee, E. J., *Computer Programs for Calculating in Situ Phytoplankton Photosynthesis*, Can. Tech. Rep. Fish. Aquat. Sci. No. 1740, 27 pp, 1990.

108. Fee, E. J., Hecky, R. E., Stainton, M. P., Sandberg, P., Hendzel, L. L., Guildford, S. J., Kling, H. J., McCullough, G. K., Anema, C., and Salki, A., *Lake Variability and Climate Research in Northwest Ontario: Study, Design, and 1985–1986 Data from the Red Lake District*, Can. Tech. Rep. Fish. Aquat. Sci. No. 1662, 39 pp, 1989.

109. Fournier, G. R., Forand, L., Pelletier, G., and Pace, P., NEARSCAT full spectrum narrow forward angle transmissometer-nephelometer, *SPIE Vol. 1750 Ocean Optics XI*, 114–125, 1992.

110. Fournier, G. R., Forand, L., Pelletier, G., and Pace, P., Ground-truthing the oceanic light field using an in-situ tunable transmissometer-nephelometer, *Proc. 16th Can. Symp. Rem. Sens.*, Sherbrooke, QU, 173–175, 1993.

111. Fraser, R. S., Kaufman, Y. J., and Mahoney, R. L., Satellite measurement of aerosol mass and transport, *Atmos. Environ.*, 18, 2577–2584, 1984.

112. Fraser, R. S., Ferrare, R. A., Kaufman, Y. J., and Mattoo, S., Algorithm for atmospheric corrections of aircraft and satellite imagery, *Int. J. Rem. Sens.*, 13, 541–557, 1992.

113. Freemantle, J. R., Pu, R., and Miller, J. R., Calibration of imaging spectrometer data to reflectance using pseudo-invariant features, *Proc. 15th Can. Symp. Rem. Sens.*, Toronto, 452–457, 1992.

114. French, C. S., Various forms of chlorophyll *a* in plants, in *The Photochemical Apparatus: Its Structure and Function*, Brookhaven Symposia in Biology No. 11, Brookhaven National Laboratory, Upton, New York, 52–64, 1959.

115. French, C. S., Myers, J., and McLeod, G. C., Automatic recording of photosynthesis action spectra used to measure the Emerson enhancement effect, in *Comparative Biochemistry of Photoreactive Systems*, Allen, M. B., Ed., Academic Press, London and New York, 361–365, 1960.

116. Frouin, R., Lingner, D. W., Gautier, C., Baker, K. S., and Smith, R. C., A simple analytical formula to compute clear sky total and photosynthetically available solar irradiance at the ocean surface, *J. Geophys. Res.*, 94, 9731–9742, 1989.

117. Gale, R. M., Water quality surveys, *IHD-WHO Studies and Reports in Hydrology*, UNESCO, Paris, 350 pp, 1978.

118. Gallegos, C. L., Correll, D. L., and Pierce, J. W., Modelling spectral diffuse attenuation, absorption, and scattering coefficients in a turbid estuary, *Limnol. Oceanogr.*, 35, 1486–1502, 1990.

119. Gallie, A. E., Calibrating optical models of lake water colour using lab measurements—preliminary results, *Proc. 16th Can. Symp. Rem. Sens.*, Sherbrooke, Quebec, Gagnon, P. and O'Neill, N., Eds., 119–123, 1993.

120. Gallie, E. A. and Murtha, P. A., Specific absorption and backscattering spectra for suspended minerals and chlorophyll *a* in Chilko Lake, British Columbia, *Rem. Sens. Environ.*, 39, 103–118, 1992.

121. Gallie, E. A. and Murtha, P. A., A modification of chromaticity analysis to separate the effects of water quality variables, *Rem. Sens. Environ.*, 44, 47–65, 1993.

122. Gallo, K. P. and Daughtry, C. S. T., Differences in vegetation indices for simulated LANDSAT-5 MSS and TM, NOAA-9 AVHRR, and SPOT-1 sensor systems, *Rem. Sens. Environ.*, 23, 439–452, 1987.

123. Gershun, A. A., To the problem of diffuse light transmission (Russian), *Trans. Opt. Instr.*, 8, 35–47, 1928.

124. Gershun, A., The light field, *J. Math. Phys.*, 18, 51–151, 1939.

125. Ghassemi, M. and Christman, R. F., Properties of the yellow organic acids of natural waters, *Limnol. Oceanogr.*, 13, 583–597, 1968.

126. Glover, R. M., *Diatom Fragmentation in Grand Traverse Bay, Lake Michigan, and Its Implications for Silica Recycling*, Ph.D. Thesis, University of Michigan, Ann Arbor, 1982.

127. Goodison, B. E., Whiting, J. M., Wiebe, K., and Cihar, J., *Operational Requirements for Water Resources Remote Sensing in Canada*, IUGG, IAHS Publication IAHS/WMO No. HS1, 1983.

128. Goodwin, T. W., *Chemistry and Biochemistry of Plant Pigments*, Academic Press, London and New York, 1965.

129. Goodwin, T. W., *Biochemistry of Chloroplasts*, Vol. 1, Academic Press, London and New York, 1966.

130. Goodwin, T. W., *Biochemistry of Chloroplasts*, Vol. 2, Academic Press, London and New York, 1967.

131. Gordon, H. R., Simple calculation of the diffuse reflectance of the ocean, *Appl. Opt.*, 2803–2804, 1973.

132. Gordon, H. R., Removal of atmospheric effects from satellite imagery of the oceans, *Appl. Opt.*, 17, 1631–1636, 1978.

133. Gordon, H. R., Dependence of the diffuse reflectance of natural waters on the sun angle, *Limnol. Oceanogr.*, 34, 1484–1489, 1989.

134. Gordon, H. R. and Brown, O. B., Irradiance reflectivity of a flat ocean as a function of its optical properties, *Appl. Opt.*, 12, 1549–1551, 1973.

135. Gordon, H. R., Brown, O. B., and Jacobs, M. M., Computed relationships between the inherent and apparent optical properties of a flat homogeneous ocean, *Appl. Opt.*, 14, 417–427, 1975.

136. Gordon, H. R. and McCluney, W. R., Estimation of the depth of sunlight penetration in the sea for remote sensing, *Appl. Opt.*, 14, 413–416, 1975.

137. Gordon, H. R. and Wouters, A. W., Some relationships between Secchi depth and inherent optical properties of natural waters, *Appl. Opt.*, 17, 3341–3343, 1978.

138. Gordon, H. R. and Clark, D. K., Remote sensing optical properties of a stratified ocean: An improved interpretation, *Appl. Opt.*, 19, 3428–3430, 1980.

139. Gordon, H. R. and Clark, D. K. Atmospheric effects in the remote sensing of phytoplankton pigments, *Boundary-Layer Meteorol.*, 18, 299–313, 1980.

140. Gordon, H. R., Clark, D. K., Mueller, J. L., and Hovis, W. A., Phytoplankton pigments from the Nimbus-7 Coastal Zone Color Scanner: Comparisons with surface measurements, *Science*, 210, 63–66, 1980.

141. Gordon, H. R., Clark, D. K., Brown, J. W., Brown, O. B., Evans, R. H., and Broenkow, W. W., Phytoplankton pigment concentrations in the Middle Atlantic Bight: Comparison of ship determinations and CZCS estimates, *Appl. Opt.*, 22, 20–36, 1983.

142. Gordon, J. I., *Directional Radiance (Luminance) of the Sea Surface*, Visibility Laboratory, San Diego, SIO Ref. 69–20, 50 pp, 1969.

143. Gordon, J. I., *Daytime Visibility, A Conceptual Review*, Air Force Geophysics Lab., Hanscom AFB Research Report AFGL-TR-70-0257, AD A085451, 1970.

144. Gorlenko, V. M., Dubinina, G. A., and Kuznetsov, S. I., *Ecology of Aquatic Microorganisms* (Russian), Nauka Publishing Co., Moscow, 289 pp, 1974.

145. Gower, J. F. R., Low-cost satellite sensor image reception for NOAA HRPT and other compatible data, *Int. J. Rem. Sens.*, 177–181, 1993.

146. Green, A. E. S., Cross, K. R., and Smith, L. A., Improved analytic characterization of ultraviolet skylight, *Photochem. and Photobiol.*, 31, 59–65, 1980.

147. Grishchenko, D. L., Dependence of the albedo of the sea upon the solar height and roughness of the sea surface (Russian), *Transactions Main Geophysical Observations*, 80, 75–83, 1959.

148. Gushchin, G. K., Sea surface albedo: Ocean and sea meteorology (Russian), *Trans. DVNIGMI, Far-East Institute of Hydrometeorology*, 30, 244–251, 1970.

149. Guzzi, R., Rizzi, R., and Vindigni, S., Reconstruction of the solar spectrum at the ground using measured turbidity and precipitable water as input parameters, in *Proceedings of the 2nd International Solar Forum*, Hamburg, 35–41, 1978.

150. Guzzi, R., Rizzi, R., and Zibordi, G., Atmospheric correction of data measured by a flying platform over the sea: Elements of a model and its experimental verification, *Appl. Opt.*, 26, 3043–3051, 1987.

151. Häder, D.-P. and Worrest, R. C., Effects of enhanced solar ultraviolet radiation on aquatic ecosystems, *Photochem. Photobiol.*, 53, 717–725, 1991.

152. Haffner, G. D., Poulton, D. J., and Kohli, B., Physical processes and eutrophication, *Water Res. Bull.*, 18, 457–463, 1982.

153. Hall, F. G., Strebel, D. E., Goetz, S. J., and Nickeson, N. E., *Radiometric Rectification of Multi-date, Multi-sensor Satellite Images*, Proc. 10th Annual IEEE Int. Geosci. Rem. Sens. Symp., New York, 165–168, 1990.

154. Halldal, P., Pigment formation and growth in blue-green algae in crossed gradients of light intensity and temperature, *Physiol. Plant.*, 11, 401–420, 1958.

155. Halldal, P., Automatic recording of action spectra of photobiological processes, spectro-photometric analyses, fluorescence measurements and recording of the first derivative of the absorption curve in one simple unit, *Photochem. Photobiol.*, 10, 23–34, 1969.

156. Halldal, P., The photosynthetic apparatus of microalgae and its adaptation to environmental factors, in *Photobiology of Microorganisms*, Halldal, P., Ed., Wiley-Interscience, London and New York, 17–55, 1970.

157. Halldal, P., Light and photosynthesis of different marine algal groups, in *Optical Aspects of Oceanography*, Jerlov, N. G. and Steemann Nielsen, E., Eds., Academic Press, London and New York, 345–360, 1974.

158. Harris, G. P., The relationship between chlorophyll *a* fluorescence, diffuse attenuation changes and photosynthesis in natural phytoplankton populations, *J. Plankton Res.*, 2, 109–127, 1980.

159. Harris, G. P., Piccinin, B. B., Haffner, G. D., Snodgrass, W., and Polak, J., Physical variability and phytoplankton communities: I. The descriptive limnology of Hamilton Harbour, *Arch. Hydrobiol.*, 88, 303–327, 1980.

160. Harris, G. P., Haffner, G. D., and Piccinin, B. B., Physical variability and phytoplankton communities: II. Primary productivity by phytoplankton in a physically variable environment, *Arch. Hydrobiol.*, 88, 447–473, 1980.

161. Harris, G. P. and Piccinin, B. B., Physical variability and phytoplankton communities: IV. Temporal changes in the phytoplankton community of a physically variable lake, *Arch. Hydrobiol.*, 89, 447–473, 1980.

162. Harron, J. W., Freemantle, J. R., Hollinger, A. B., and Miller, J. R., Methodologies and errors in the calibration of a compact airborne imaging spectrometer, in *A World of Applications*, 391–396, 1992.

163. Hasler, A. D., Eutrophication of lakes by domestic drainage, *Ecology*, 28, 383–395, 1947.

164. Hatfield, J. L., Giorgis, R. B., Jr., and Flocchini, R. G., A simple solar radiation model for computing direct and diffuse spectral fluxes, *Solar Energy*, 27, 323–329, 1981.

165. Haxo, F. T. and Blinks, L. R., Photosynthetic action spectra of marine algae, *J. Gen. Physiol.*, 33, 389–422, 1950.

166. Hebert, P. D. N., Muncaster, B. W., and Mackie, G. L., Ecological and genetic studies on *Dreissena polymorpha* (Pallas): A new mollusc in the Great Lakes, *Can. J. Fish Aquat. Sci.*, 46, 1587–1592, 1989.

167. Herman, B. M. and Browning, S. R., The effect of aerosols on the earth-atmosphere albedo, *Atmos. Sci.*, 32, 158–165, 1975.

168. Hishida, K. and Kishino, M., On the albedo of radiation of the sea surface, *J. Oceanogr. Soc. Japan*, 21, 148–153, 1965.

169. Hoepffner, N. and Sathyendranath, S., Determination of major groups of phytoplankton pigments from the absorption spectra of total particulate matter, *J. Geophys. Res.*, 98, 22,789–22,803, 1993.

170. Højerslev, N. K., *Daylight Measurements for Photosynthetic Studies in the Western Mediteranean*, Univ. Copenhagen, Inst. Phys. Oceanogr. Report No. 26, 1979.

171. Højerslev, N. K., Visibility of the sea with special reference to the Secchi disc, *Proc. Soc. Photo-opt. Instr. Eng.*, 637, 294–305, 1986.

172. Højerslev, N. K., *Natural Occurrences and Optical Effects of Gelbstoff*, Geophs. Inst., Dept. Phys. Oceanogr., University of Copenhagen, Denmark, 29 pp, 1988.

173. Holm, R. G., Moran, M. S., Jackson, R. D., Slater, P. N., Yuan, B., and Biggar, S. F., Surface reflectance factor retrieval from Thematic Mapper data, *Rem. Sens. Environ.*, 27, 47–57, 1989.

174. Holmes, R. W., The Secchi disk in turbid coastal waters, *Limnol. Oceanogr.*, 15, 688–694, 1970.

175. Hovis, W. A., Clark, D. K., Anderson, F., Austin, R. W., Wilson, W. H., Baker, E. T., Ball, D., Gordon, H. R., Mueller, J. L., El Sayed, S. Y., Sturm, B., Wrigley, R. C., and Yentsch, C. S., Nimbus-7 Coastal Zone Color Scanner, system description and initial imagery, *Science*, 210, 60–63, 1980.

176. Hovis, W., Szajna, E. F., and Bohan, W. A., *Nimbus-7 CZCS Coastal Zone Color Scanner Imagery for Selected Coastal Regions*, NASA Goddard Space Flight Center Atlas of CZCS imagery, 1988.

177. Hulbert, E. O., Optics of distilled and natural water, *J. Opt. Soc. Amer.*, 35, 698–705, 1945.

178. Hunkins, K., Thorndike, E. M., and Mathieu, G., Nepheloid layers and bottom circulation in the Arctic Ocean, *J. Geophys. Res.*, 74, 6995–7008, 1969.

179. Hurley, J. P. and Watras, C. J., Identification of bacteriophylls in lakes via reverse-phase HPCL, *Limnol. Oceanogr.*, 36, 307–315, 1991.

180. Hutchinson, G. E., On the relation between oxygen deficit and the productivity and typology of lakes, *Int. Rev. Hydrobiol.*, 36, 336–355, 1938.

181. Hutchinson, G. E., *Treatise on Limnology*, Vol. 1, John Wiley and Sons, New York, 1957.

182. Hutchinson, G. E., Eutrophication: The scientific background of a contemporary practical problem, *Amer. Scientist*, 61, 269–279, 1973.

183. IMSL, *The IMSL Library Reference Manual*, Edition 8 (revised June 1980), International Mathematics and Statistics Libraries, Houston, 1980.

184. Ivanoff, A., Au sujet de la teneure de l'eau en particules en suspension au voisinage immédiat du fond de la mer, *C. R.*, 250, 1881–1883, 1960.

185. Ivanov, A. P., *Physical Background of Hydrooptics* (Russian), Nauka Technika Publishers, Minsk, USSR, 500 pp, 1975.

186. Jackson, R. D., Spectral indices in n-space, *Rem. Sens. Environ.*, 13, 409–421, 1983.

187. Janus, L. L., *Hamilton Harbour in Relation to OECD Standards*, Report to Remedial Action Plan Team, National Water Research Institute internal document, 1987.

188. Jaquet, J.-M. and Zand, B., Colour analysis of inland waters using Landsat TM data, *ESA*, SP-1102, 57–67, 1989.

189. Jaquet, J.-M., Schanz, F., Bossard, P., Hanselmann, K., and Gendre, F., Measurements and

significance of bio-optical parameters in two subalpine lakes of different trophic state, *Aquat. Sci.*, 56, 263–305, 1994.

190. Jaquet, J.-M., Strauffacher, M., Lehmann, A., and Nakayama, M., Water quality assessment and management by GIS and remote sensing: The GISWAQ project, in *Advances in Water Quality Monitoring*, Report of a World Meteorological Organization Regional Workshop, 7–11 March, Vienna Austria, Tech. Rep. in Hydrology and Water Resources No. 42, 249–268, 1994.

191. Jassby, A. D. and Platt, T., Mathematical formulation of the relationship between photosynthesis and light for phytoplankton, *Limnol. Oceanogr.*, 21, 540–547, 1976.

192. Jerlov, N. G., Optical studies of ocean water, *Rep. Swedish Deep-Sea Exped.*, 3, 1–59, 1951.

193. Jerlov, N. G., Influence of suspended and dissolved matter on the transparency of sea water, *Tellus*, 5, 59–65, 1953.

194. Jerlov, N. G., *Optical Measurements in the Eastern North Atlantic*, Medd. Oceanographic Inst. Goteborg, Series B, 8, 40 pp, 1961.

195. Jerlov, N. G., Optical classification of ocean water, in *Physical Aspects of Light in the Sea*, Univ. Hawaii Press, Honolulu, 44–49, 1964.

196. Jerlov, N. G., *Optical Oceanography, Elsevier Oceanography Series 5*, Elsevier Publishing Co., Amsterdam, 194 pp, 1968.

197. Jerlov, N. G., Significant relationships between optical properties of the sea, in *Optical Aspects of Oceanography*, Jerlov, N. G. and Steeman Nielsen, E., Eds., Academic Press, London, 77–94, 1974.

198. Jerlov, N. G., *Marine Optics, Elsevier Oceanography Series 14*, Elsevier Publishing Co., Amsterdam, 231 pp, 1976.

199. Jerlov, N. G. and Steeman Nielson, E., Eds., *Optical Aspects of Oceanography*, Academic Press, London, 494 pp, 1974.

200. Jerlov, N. G. and Nygård, K., *Influence of Solar Elevation on Attenuation of Underwater Irradiance*, Univ. Copenhagen, Inst. Phys. Oceanogr. Report No. 4, 9 pp, 1969.

201. Jerome, J. H., McNeil, W. R., and Elder, F. C., *Optical Indices for in Situ and Remote Classification of Lakes*, Abstracts of Eighteenth Conference on Great Lakes Research, Albany, 1975.

202. Jerome, J. H., Bruton, J. E., and Bukata, R. P., Influence of scattering phenomena on the solar zenith angle dependence of in-water irradiance levels, *Appl. Opt.*, 21, 642–647, 1982.

203. Jerome, J. H., Bukata, R. P., and Bruton, J. E., Spectral attenuation and irradiance in the Laurentian Great Lakes, *J. Great Lakes Res.*, 9, 60–68, 1983.

204. Jerome, J. H., Bukata, R. P., and Bruton, J. E., Utilizing the components of vector irradiance to estimate the scalar irradiance in natural waters, *Appl. Opt.*, 27, 4012–4018, 1988.

205. Jerome, J. H., Bukata, R. P., and Bruton, J. E., Determination of available subsurface light for photochemical and photobiological activity, *J. Great Lakes Res.*, 16, 436–443, 1990.

206. Jerome, J. H., Bukata, R. P., Whitfield, P. H., and Rousseau, N., Colours of natural waters: 1. Factors controlling the dominant wavelength, *Northwest Science*, 68, 43–52, 1994.

207. Jerome, J. H., Bukata, R. P., Whitfield, P. H., and Rousseau, N., Colours of natural waters: 2. Observations of spectral variations in British Columbia rivers, *Northwest Science*, 68, 53–64, 1994.

208. Kalle, K., Zum problem der Meereswasserfarbe, *Ann. Hydrol. Mar. Mitt.*, 66, 1–13, 1938.

209. Kalle, K., *Die Farbe des Meeres*, Rapp. P.-V. Réun., Cons. Perm. Int. Explor. Mer., 109, 96–105, 1939.

210. Kalle, K., The problem of the gelbstoff in the sea, *Oceanogr. Mar. Biol. Ann. Rev.*, 4, 91–104, 1966.

211. Kattawar, G. W. and Plass, G. N., Radiative transfer in the Earth's atmosphere-ocean system: II. Radiance in the atmosphere and ocean, *J. Phys. Oceanogr.*, 2, 146–156, 1972.

212. Kattawar, G. W. and Humphreys, T. J. Remote sensing of chlorophyll in an atmosphere-ocean environment: A theoretical study, *Appl. Opt.*, 15, 273–282, 1976.

213. Kaufman, Y. J. and Sendra, C., Automatic atmospheric correction, *Int. J. Rem. Sens.*, 9, 1357–1381, 1988.

214. Kauth, R. J. and Thomas, G. S., The tasseled cap—A graphic description of the spectral-temporal development of agricultural crops as seen by LANDSAT, *Proc. Symp. Machine Processing of Rem. Sens. Data*, Purdue University, West Lafayette, IN, 41–51, 1976.

215. Kerr, J. B., Trends in total ozone at Toronto between 1960 and 1991, *J. Geophys. Res.*, 96, 20,703–20,709, 1991.

216. Kerr, J. B. and McElroy, C. T., Evidence for large upward trends of ultraviolet-B radiation linked to ozone depletion, *Science*, 262, 1032–1034, 1993.

217. Kiefer, D. A. and SooHoo, J. B., Spectral absorption by marine particles of coastal waters off Baja California, *Limol. Oceanogr.*, 27, 492–499, 1982.

218. King, M., Bryne, D., Herman, B., and Reagan, J., Aerosol size distributions obtained by inversion of spectral optical depth measurements, *J. Atmos. Sci.*, 35, 2153–2167, 1978.

219. Kirk, J. T. O., A theoretical analysis of the contribution of algal cells to the attenuation of light within natural waters. I. General treatment of suspensions of pigmented cells, *New Phytol.*, 75, 11–20, 1975.

220. Kirk, J. T. O., A theoretical analysis of the contribution of algal cells to the attenuation of light within natural waters. II. Spherical cells, *New Phytol.*, 75, 21–36, 1975.

221. Kirk, J. T. O., Yellow substance (gelbstoff) and its contribution to the attenuation of photosynthetically active radiation in some inland and coastal southeastern Australian waters, *Aust. J. Mar. Freshwater Res.*, 27, 61–71, 1976.

222. Kirk, J. T. O., Use of a quanta meter to measure attenuation and underwater reflectance of photosynthetically active radiation in some inland and coastal south-eastern Australian waters, *Aust. J. Mar. Freshwater Res.*, 28, 9–21, 1977.

223. Kirk, J. T. O., Spectral absorption properties of natural waters: Contribution of the soluble and particulate fractions to light absorption in natural waters, *Aust. J. Mar. Freshwater Res.*, 31, 287–296, 1980.

224. Kirk, J. T. O., *Monte Carlo Procedure for Simulating the Penetration of Light into Natural Waters*, CSIRO (Australia) Division of Plant Industry Tech. Paper No. 36, 16 pp, 1981.

225. Kirk, J. T. O., Monte Carlo study of the nature of the underwater light field in, and the relationships between optical properties of, turbid yellow waters, *Aust. J. Mar. Freshwater Res.*, 32, 517–532, 1981.

226. Kirk, J. T. O., *Light and Photosynthesis in Aquatic Ecosystems*, Cambridge University Press, Melbourne, Australia, 401 pp, 1983.

227. Kirk, J. T. O., Dependence of relationship between inherent and apparent optical properties of water on solar altitude, *Limnol. Oceanogr.*, 29, 350–356, 1984.

228. Kirk, J. T. O., Volume scattering function, average cosines, and the underwater light field, *Limnol. Oceanogr.*, 36, 455–467, 1991.

229. Kirk, J. T. O. and Tyler, P. A., The spectral absorption and scattering properties of dissolved and particulate components in relation to the underwater light field of some tropical Australian freshwaters, *Freshwater Biol.*, 16, 573–583, 1986.

230. Klapwijk, A. and Snodgrass, W. J., Model for lake-bay exchange, *J. Great Lakes Res.*, 11, 43–52, 1985.

231. Kneizys, F. X., Settle, E. P., Gallery, W. O., Chetwynd, J. H., Abreu, L. W., Selby, J. E. A., Clough, S. A., and Fenn, R. W., *Atmospheric Transmittance/Radiance: Computer Code LOWTRAN6*, Air Force Geophysics Lab., Hanscom AFB Environmental Research Report AFGL-TR-83-0187, 1983.

232. Kneizys, F. X., Shettle, E. P., Abreu, L. W., Chetwynd, J. H., Anderson, G. P., Gallery, W. O., Selby, J. E. A., and Clough, S. A., *Users Guide to LOWTRAN 7*, Air Force Geophysics Lab., Hanscom AFB Environmental Research Report ERP No. 1010, 1988.

233. Koepke, P., Effective reflectance of oceanic white caps, *Appl. Opt.*, 23, 1816–1824, 1984.

234. Kohli, B., Mass exchange between Hamilton Harbour and Lake Ontario, *J. Great Lakes Res.*, 5, 36–44, 1979.

235. Kondratyev, K. Ya., *Radiation in the Atmosphere, International Geophysics Series 12*, Academic Press, New York, 912 pp, 1969.

236. Kondratyev, K. Ya. and Ter-Markariants, N. E., On the diurnal variation of albedo (Russian), *Meteorol. and Hydrol.*, 6, 71–79, 1953.

237. Kondratyev, K. Ya., Barten'eva, O. D., and Vasilyev, O. B., *GARP—Climate: Climate and Aerosol* (Russian), Gidrometeoizat Publishing Co., Leningrad, 130 pp, 1976.

238. Kondratyev, K. Ya. and Pozdnyakov, D. V., *Optical Properties of Natural Waters and Remote Sensing of Phytoplankton* (Russian), Nauka Publishing Co., Leningrad, 182 pp, 1988.

239. Kondratyev, K. Ya., Pozdnyakov, D. V., and Isakov, V. U., *Hydrooptical Radiation Experiments on Lakes* (Russian), Nauka Publishing Co., Leningrad, 1990.

240. Kullenberg, G., Scattering of light by Sargasso Sea water, *Deep-Sea Res.*, 15, 423–432, 1968.

241. Kuzmin, P. P., *Physical Properties of Snow Cover* (Russian), Gidrometeoizdat Publishers, Leningrad, 320 pp, 1957.

242. Lauscher, F., Optik der gewässer: Sonnen und himmelstrahlung im meer und im gewässen, in *Handbook der Geophysik*, Springer, Berlin, 7, 723–763, 1955.

243. Lean, D. R. S., Cooper, W. J., and Pick, F. R., Hydrogen peroxide formation and decay in lakewaters, in *Aquatic and Surface Photochemistry*, Helz, G. R., Zepp, R. G., and Crosby, D. G., Eds., Lewis Publishers, Ann Arbor, MI, 207–214, 1994.

244. Lenoble, J., Remarque sur la couleur de la mer, *C. R.*, 242, 662–664, 1956.

245. Levenburg, K., A method for the solution of certain non-linear problems in least squares, *Quant. Appl. Math.*, 2, 164–168, 1944.

246. Lewis, M. R., Cullen, J. J., and Platt, T., Phytoplankton and thermal structure in the upper ocean: Consequences of nonuniformity in chlorophyll profile, *J. Geophys. Res.*, 88, 2565–2570, 1983.

247. Lindell, T., Karlsson, B., Rosengren, M., and Alföldi, T., A further development of the chromaticity technique for satellite mapping of suspended sediment load, *Photogramm. Eng. Rem. Sens.*, 52, 1521–1530, 1986.

248. Lindeman, R. L., The trophic-dynamic aspect of ecology, *Ecology*, 23, 399–418, 1942.

249. Longuet-Higgins, M. S., On the skewness of sea-surface slopes, *J. Phys. Oceanogr.*, 12, 1283–1291, 1982.

250. Lovelock, J. E., Gaia as seen through the atmosphere, *Atmos. Environ.*, 6, 579–580, 1972.

251. Lovelock, J. E. and Margulis, L., Atmospheric homeostasis by and for the biosphere: the Gaia hypothesis, *Tellus*, 26, 2–10, 1974.

252. Lyon, K. G. and Willard, M. R., SeaStar (SeaWiFS) ocean color remote sensing data for the 1990s, *Proc. 16th Can. Symp. Rem. Sens.*, 137–142, 1993.

253. Mackie, G. L., Gibbons, W. N., Muncaster, B. W., and Grey, I. M., *The zebra mussel, Dreissena polymorpha: A synthesis of European experiences and a preview for North America*, Report of Water Resources Branch, Ontario Ministry of the Environment, Toronto, Ontario, 1989.

254. Madronich, S., Photodissociation in the atmosphere. I. Actinic flux and the effects of ground reflections and clouds, *J. Geophys. Res.*, 92, 9740–9752, 1987.

255. Makhotkin, L. G., Regularites in the variation of diffuse radiation under a cloudless sky (Russian), *Transactions Main Geophysical Observations*, 80, 35–46, 1959.

256. Markham, B. L. and Barker, J. L., Spectral characteristics of the Landsat Thematic Mapper sensors, *Int. J. Rem. Sens.*, 6, 697–716, 1985.

257. Markham, B. L. and Barker, J. L., *LANDSAT MSS and TM Post-calibration Dynamic Ranges, Exoatmospheric Reflectances, and At-satellite Temperatures*, EOSAT LANDSAT Tech. Notes 1, 3–7, Landham, MD, 1986.

258. Markham, B. L. and Barker, J. L., Thematic Mapper bandpass solar exoatmospheric radiances, *Int. J. Rem. Sens.*, 8, 517–523, 1987.

259. Marquardt, D. W., An algorithm for least squares estimation of nonlinear parameters, *J. Int. Soc. Appl. Math.*, 11, 36–48, 1963.

260. Marra, L., Effect of short term variations in light intensity on photosynthesis of a marine phytoplankter—laboratory simulation study, *Mar. Biol. (N.Y.)*, 46, 191–202, 1978.

261. Matvejev, L. T., *Physics of the Atmosphere* (Russian), Gidrometeoizat Publishing Co., Leningrad, 750 pp, 1984.

262. Maul, G. A., *Introduction to Satellite Oceanography*, Martinus Nijhoff Publishers, Dordrecht, The Netherlands, 606 pp, 1985.

263. Meszaros, E., On the possible role of the biosphere in the control of atmospheric clouds and precipitation, *Atmos. Environ.*, 22, 423–424, 1988.

264. Michaels, P. J., Singer, S. F., and Knappenberger, P. C., Analyzing ultraviolet-B radiation: Is there a trend? *Science*, 264, 1341–1342, 1994.

265. Mitchell, B. G., Algorithms for determining the absorption coefficient of aquatic particulates using the quantitative filter technique (QFT), in *Ocean Optics X*, SPIE, 1302, 137–148, 1990.

266. Mitchell, B. G. and Kiefer, D. A., Variability in pigment specific particulate fluorescence and absorption spectra in the northeastern Pacific Ocean, *Deep-Sea Res.*, 35, 665–689, 1988.

267. Monteith, J. L., *Principles of Environmental Physics*, Edward Arnold Pub., London, 1973.

268. Moon, P. and Spencer, D. E., Illumination from a non-uniform sky, *Illum. Eng.*, 37, 707–726, 1942.

269. Morel, A., Diffusion de la luminière par les eaux de mer: Résultats expérimentaux et approche théorique, in *Optics of the Sea*, AGARD Lecture Series 61, 31.1–31.74, 1973.

270. Morel, A., Optical properties of pure water and sea water, in *Optical Aspects of Oceanography*, Jerlov, N. G. and Steeman Nielsen, E., Eds., Academic Press, London, 1–24, 1974.

271. Morel, A., Inwater and remote measurements of ocean color, *Boundary-Layer Meteorol.*, 18, 177–201, 1980.

272. Morel, A., Optical modeling of the upper ocean in relation to its biogenous matter content (Case I waters), *J. Geophys. Res.*, 93, 10, 749–10, 768, 1988.

273. Morel, A. and Prieur, L., Analysis of variations in ocean color, *Limnol. Oceanogr.*, 22, 709–722, 1977.

274. Morel, A. and Bricaud, A., Theoretical results concerning light absorption in a discrete medium, and application to specific absorption of phytoplankton, *Deep-Sea Res.*, 11, 1375–1393, 1981.

275. Morel, A. and Berthon, J.-F., Surface pigments, algal biomass profiles, and potential production of the euphotic layer: Relationships reinvestigated in view of remote sensing applications, *Limnol. Oceanogr.*, 34, 1545–1562, 1989.

276. Mudroch, A. and Mudroch, P., Geochemical composition of the nepheloid layer in Lake Ontario, *J. Great Lakes Res.*, 18, 132–153, 1992.

277. Mueller, J. L., *Influence of Phytoplankton on Ocean Color*, Ph.D. thesis, Oregon State University, 1973.

278. Müller-Karger, F. E., McClain, C. R., Fisher, T. R., Esaias, W. E., and Varela, R., Pigment distribution in the Caribbean Sea: Observations from space, *Prog. Oceanogr.*, 23, 23–64, 1989.

279. Multi-author, *Optics of the Ocean*, Vol. 1. (Russian), Nauka Publishing Co., 360 pp, 1983.

280. Multi-author, *Present State of the Ladoga Lake Ecosystem* (Russian), Nauka Publishing Co., Leningrad, Russia, 210 pp, 1987.

281. Munawar, M. and Munawar, G. F., A lakewide study of phytoplankton biomass and its species composition in Lake Erie, April-December, 1970, *J. Fish. Res. Board Can.*, 33, 581–600, 1976.

282. Munday, J. C., Jr., Lake Ontario water mass delineation from ERTS-1, *Proc. 9th Int. Symp. Rem. Sens. Environ.*, Ann Arbor, MI, 1355–1368, 1974.

283. Munday, J. C., Jr. and Alföldi, T. T., Chromaticity changes from isoluminous techniques used to enhance multispectral remote sensing data, Rem. Sens. Environ., 4, 221–236, 1975.

284. Munday, J. C., Jr. and Alföldi, T. T., LANDSAT test of of diffuse reflectance models for aquatic suspended solids measurements, Rem. Sens. Environ., 8, 169–183, 1979.

285. Munday, J. C., Jr., Alföldi, T. T., and Amos, C. L., Verification and application of a system for automated multidate LANDSAT measurement of suspended sediment, Proc. 5th Ann. William T. Pecora Mem. Symp., Sioux Falls, S. D., Am. Soc. Photogramm. Rem. Sens., Falls Church, VA, 622–640, 1979.

286. Murrell, J. N., The Theory of the Electronic Spectra of Organic Molecules, Methuen and Co., London, and Wiley and Sons, New York, 1963.

287. NASA, Global Change: Impacts on Habitablity: A Scientific Basis for Assessment, Report of Executive Committee of Workshop, Woods Hole, Mass., 21–26 June, NASA Document, JPL, D-85, Washington, DC, 1982.

288. NASA, Global Biology Research Program, NASA Tech. Memo. 85629, Washington, DC, 1983.

289. NASA, Landsat Data Users Handbook, Goddard Space Flight Center, Greenbelt, MD, 1983.

290. NASA, Earth Orbiting System: Science and Mission Requirements, Working Group Report Vol. 1, NASA Tech. Memo. 86129, Goddard Space Flight Center, Greenbelt, MD, 1984.

291. NASA, From Pattern to Process: The Strategy of the Earth Observing System, Vols. I and II, Reports of the EOS Science Steering Committee, Washington, DC, 1985.

292. NASA, Earth Observing System (Eos) Background Information Package (BIP) Part Five: Research Facility and Operational Facility Instrument Descriptions, NASA Announcement of Opportunity No. OSSA-1-88, January 19, 1988.

293. NASA Advisory Council, Earth System Science—Overview: A Program for Global Change, Report of the Earth System Science Committee, Washington, DC, 1986.

294. NASA Advisory Council, Earth System Science, a Closer View, Report of the Earth Science Committee, Washington, DC, 1988.

295. National Research Council, Toward an International Geosphere-Biosphere Program: A Study of Global Change, Report of the National Research Council Workshop, Woods Hole, Mass., 25–29 July, National Academy Press, Washington, DC, 1983.

296. National Research Council, Global Change in the Geosphere-Biosphere: Initial Priorities for an IGBP, U. S. Committee for an International Geosphere-Biosphere Program, National Academy of Sciences, Washington, DC, 1986.

297. Neale, P. J. and Richerson, P. J., Photoinhibition and the diurnal variation of phytoplankton photosynthesis—I. Development of a photosynthesis-irradiance model from studies of in situ responses, J. Plankton Res., 9, 167–193, 1987.

298. NOAA, NOAA Polar Orbiter Data User's Guide, Kidwell, K. B., Ed., NOAA/NESDIS, Washington, DC, 1991.

299. Nixon, S. W., Ed., Comparative ecology of freshwater and marine ecosystems, Limnol. Oceanogr., Special issue, Vol. 33, 1988.

300. Ontario Ministry of the Environment (MOE), Hamilton Harbour Study 1977, Vol. 1, Water Resources Branch, Ontario Ministry of the Environment, Toronto, 1981.

301. Ontario Ministry of the Environment (MOE), Hamilton Harbour Technical Summary and General Management Options, Water Resources Branch, Ontario Ministry of the Environment, Toronto, 1985.

302. Paerl, H. W., Tucker, J., and Bland, P. T., Carotenoid enhancement and its role in maintaining blue-green (Microcystis aeruginosa) surface blooms, Limnol. Oceanogr., 28, 847–857, 1983.

303. Pak, H. and Plank, W. S., Some applications of the optical tracer method, in Symposium on Optical Aspects of Oceanography, Jerlov, N. G. and Steeman Nielsen, E., Eds., Academic Press, London, 221–235, 1972.

304. Patten, B. C., Mathematical models of plankton production, Int. Rev. Gesamten Hydrobiol., 53, 357–408, 1968.

305. Payne, R. E., Albedo of the sea surface, *J. Atmos. Sci.*, 29, 959–970, 1972.

306. Pelevin, V. N. and Rutkovskaya, V. A., On the optical classification of ocean waters from the spectral attenuation of solar radiation, *Oceanology*, 18, 278–282, 1977.

307. Petrova, N. A., *Successions of Phytoplankton with the Anthropogenic Eutrophication of Large Lakes* (Russian), Nauka Publishing Co., Leningrad, 200 pp, 1990.

308. Petzold, T. J., *Volume Scattering Functions for Selected Ocean Waters*, Scripps Institute of Oceanography Ref. 72–28, Univ. of Calif., San Diego, 79 pp, 1972.

309. Plass, G. N. and Kattawar, G. W., Monte Carlo calculations of light scattering from clouds, *Appl. Opt.*, 7, 415–419, 1968.

310. Plass, G. N. and Kattawar, G. W., Radiative transfer in an atmosphere-ocean system, *Appl. Opt.*, 8, 455–466, 1969.

311. Plass, G. N. and Kattawar, G. W., Monte Carlo calculations of radiative transfer in the Earth's atmosphere-ocean system: I. Flux in the atmosphere and ocean, *J. Phys. Oceanogr.*, 2, 139–145, 1972.

312. Plass, G. N., Kattawar, G. W., and Catchings, F. E., Matrix operator theory of radiative transfer, *Appl. Opt.*, 12, 314–329, 1973.

313. Plass, G. N., Humphreys, T. J., and Kattawar, G. W., Color of the ocean, *Appl. Opt.*, 17, 1432–1446, 1978.

314. Plass, G. N., Humphreys, T. J., and Kattawar, G. W., Ocean-atmosphere interface: Its influence on radiation, *Appl. Opt.*, 20, 917–931, 1981.

315. Platt, T. and Irwin, B., Caloric content of phytoplankton, *Limnol. Oceanogr.*, 18, 306–310, 1973.

316. Platt, T., Sathyendranath, S., Caverhill, C. M., and Lewis, M. R., Ocean primary production and available light: Further algorithms for remote sensing, *Deep-Sea Res.*, 35, 855–879, 1988.

317. Platt, T. and Sathyendranath, S., Oceanic primary production: Estimation by remote sensing at local and regional scales, *Science*, 241, 1613–1620, 1988.

318. Platt, T. and Sathyendranath, S., Biological production models as elements of coupled, atmosphere-ocean models for climate research, *J. Geophys. Res.*, 96, 2585–2592, 1991.

319. Platt, T., Caverhill, C., and Sathyendraneth, S., Basin-scale estimates of oceanic primary production by remote sensing: The North Atlantic, *J. Geophys. Res.*, 96, 15,147–15,159, 1991.

320. Pokrovski, G. I., Über einen scheinbaren Mie-Effekt und seine möglishe rolle in der atmosphärenoptik, *Z. Phys.*, 53, 67–71, 1929.

321. Polak, J. and Haffner, G. D., Oxygen depletion of Hamilton Harbour, *Water Res.*, 12, 205–215, 1978.

322. Poliakova, E. A., Spectrographic study of atmospheric transparency for ultraviolet radiation (Russian), *Transactions Main Geophysical Observations*, 26, 81–89, 1950.

323. Poole, H. H. and Adkins, W. R. G., Photoelectric measurements of submarine illumination throughout the year, *J. Mar. Biol. Assoc. U. K.*, 16, 297–324, 1929.

324. Poulton, D. J., Trace contaminant status of Hamilton Harbour, *J. Great Lakes Res.*, 13, 193–201, 1987.

325. Preisendorfer, R. W., Exact reflectance under a cardioidal luminance distribution, *Quart. J. Royal Meteorol. Soc.*, 83, 540–551, 1957.

326. Preisendorfer, R. W., Application of radiative transfer theory to light measurements in the sea, *Union Geod. Geophys. Instr. Monogr.*, 10, 11–30, 1961.

327. Preisendorfer, R. W., *Lectures on Photometry, Hydrologic Optics, Atmospheric Optics*, Lecture notes, Vol. 1, Scripps Inst. of Oceanogr., Univ. of Calif., San Diego, 1953.

328. Preisendorfer, R. W., *Hydrologic Optics, Vol. I: Introduction*, U.S. Dept. of Commerce, Washington, 1976.

329. Preisendorfer, R. W., *Hydrologic Optics, Vol. II: Foundations*, U.S. Dept. of Commerce, Washington, 1976.

330. Preisendorfer, R. W., *Hydrologic Optics, Vol. III: Solutions*, U.S. Dept. of Commerce, Washington, 1976.

331. Preisendorfer, R. W., *Hydrologic Optics, Vol. IV: Imbeddings*, U.S. Dept. of Commerce, Washington, 1976.

332. Preisendorfer, R. W., *Hydrologic Optics, Vol. V: Properties*, U.S. Dept. of Commerce, Washington, 1976.

333. Preisendorfer, R. W., Secchi disk science: Visual optics of natural waters, *Limnol. Oceanogr.*, 31, 909–926, 1986.

334. Price, J. C., Calibration of satellite radiometers and the comparison of vegetation indices, *Rem. Sens. Environ.*, 21, 15–27, 1987.

335. Prieur, L., *Transfer radiatif dans les eaux des mer*, D.Sc. thesis, Univ. Pierre et Marie Curie, Paris, 243 pp, 1976.

336. Prieur, L. and Morel, A., *Apercu sur les théories du transfert radiatif applicables a la propagation dans la mer*, AGARD lect. Series No. 61, 1.3-1 to 1.3-25, NATO, Neuilly-sur-Seine, 1973.

337. Prieur, L. and Sathyendranath, S., An optical classification of coastal and oceanic waters based on the specific spectral absorption curves of phytoplankton pigments, dissolved organic matter, and other particulate materials, *Limnol. Oceanogr.*, 26, 671–689, 1981.

338. Putnam, E. S., Ed., *System Concept for Wide-Field-of-View Observations of Ocean Phenomena from Space*, Report of the SeaWiFS (Sea viewing, wide-field-of-view sensor) Working Group, NASA, Greenbelt, MD, 1987.

339. Raschke, E., Multiple-scattering calculation of the transfer of solar radiation in an atmosphere-ocean system, *Beitr. Phys. Atmos.*, 45, 1–19, 1972.

340. Raman, C. V., On the molecular scattering of light in water and the colour of the sea, *Proc. R. Soc. London, Ser. A*, 101, 64–79, 1922.

341. Reynolds, C. S., The ecology of freshwater phytoplankton, in *Cambridge Studies in Ecology*, Cambridge University Press, London, 1–150, 1984.

342. Richardson, A. J. and Wiegand, C. L., Distinguishing vegetation from soil background information, *Photogramm. Eng. Rem. Sens.*, 43, 1541–1552, 1977.

343. Rodgers, C. F., *The Radiative Heat Budget of the Troposphere and Lower Stratosphere*, Report A2, Planetary Circulation Project, MI Dept. of Meteorology, 197 pp, 1967.

344. Roesler, C. S., Perry, M. J., and Carder, K. L., Modeling *in situ* phytoplankton absorption from total absorption spectra in productive inland marine waters, *Limnol. Oceanogr.*, 34, 1510–1523, 1989.

345. Romankevich, E. A., *Geochemistry of Organic Matter in the Sea* (Russian), Nauka Publishing Co., Moscow, 908 pp, 1977.

346. Ryther, J. H. and Dunstan, W. M., Nitrogen, phosphorus, and eutrophication in the coastal marine environment, *Science*, 171, 1008–1013, 1971.

347. Sager, G., Zur refraktion von licht im meerwässer, *Beitr. Meeresk.*, 33, 63–72, 1974.

348. Sandilands, R. G. and Mudroch, A., Nepheloid layer in Lake Ontario, *J. Great Lakes Res.*, 9, 190–200, 1983.

349. Sathyendranath, S. and Platt, T., The spectral irradiance field at the surface and in the interior of the ocean: A model for applications in oceanography and remote sensing, *J. Geophys. Res.*, 93, 9270–9280, 1988.

350. Sathyendranath, S. and Platt, T., Remote sensing of ocean chlorophyll: Consequence of nonuniform pigment profile, *Appl. Opt.*, 28, 490–495, 1989.

351. Sathyendranath, S., Gouveia, A. D., Shetye, S. R., Ravindran, P., and Platt, T., Biological control of surface temperature in the Arabian Sea, *Nature*, 349, 54–56, 1991.

352. Schelske, C. L., Comment on small particles of amorphous silica in the nepheloid layer, *J. Great Lakes Res.*, 10, 94–95, 1984.

353. Schindler, D. W., Evolution of phosphorus limitation in lakes, *Science*, 195, 260–262, 1977.

354. Schindler, D. W., Beaty, K., Fee, E. J., Cruikshank, D. R., DeBruyn, E. R., Findlay, D. L., Linsey, G. A., Shearer, J. A., Stainton, M. P., and Turner, M. A., Effects of climatic warming on the lakes of the central boreal forest, *Science*, 250, 967–970, 1990.

355. Schindler, D. W., Bayley, S. E., Curtis, P. J., Parker, B. R., Stainton, M. P., and Kelly, C.

A., Natural and man-caused factors affecting the abundance and cycling of dissolved organic substances in precambrian shield lakes, *Hydrobiol.*, 229, 1–21, 1992.

356. Schlesinger, W. H., *Biochemistry: An Analysis of Global Change*, Academic Press, New York, 1991.

357. Schneider, K., Mauser, W., and Bach, H., Water quality parameters derived from remote sensing data, in *Advances in Water Quality Monitoring*, Report of a World Meteorological Organization Regional Workshop, 7–11 March, Vienna, Austria, Tech. Rep. in Hydrology and Water Resources No. 42, 269–286, 1994.

358. Schneider, S. H., Debating Gaia, *Environment*, 32, 5–9 and 29–32, 1990.

359. Schwartz, S. E., Are cloud albedo and climate controlled by marine phytoplankton? *Nature*, 336, 441–445, 1988.

360. Scully, N. H. and Lean, D. R. S., The attenuation of ultraviolet radiation in temporate lakes, *Arch. Hydobiol.*, 43, 135–144, 1995.

361. Secchi, A., Relazione della esperienze fatta a bordo della Pirocorvetta L'Immacolata Concezione per determinaire la transparenza del mare, in Cialdi, A., *Sul moto ondoso del mare e su le correnti di esso specialment auquelle littorali*, 2nd ed., 1866, 255–288. (Translation available, Dept. of the Navy, Office of Chief of Naval Operations, O.N.I. Trans. No. A-655, Op-923M4B, Dec. 1955).

362. Selby, J. E. and McClatchey, R. A., *Atmospheric Transmittance from 0.25–28.5 μm: Computer Code LOWTRAN 3*, Air Force Geophysics Lab., Hanscom AFB Environmental Research Paper No. 513, 75 pp, 1975.

363. Setlow, R. B., The wavelengths in sunlight effective in producing skin cancer, *Proc. Natl. Acad. Sci. USA*, 71, 3363–3366, 1974.

364. Shaw, G. E., Aerosols as climate regulators: A climate-biosphere linkage? *Atmos. Environ.*, 985–986, 1987.

365. Shearer, J. A., DeBruyn, E. R., DeClercq, D. R., Schindler, D. W., and Fee, E. J., *Manual of Phytoplankton Primary Production Methodology*, Can. Tech. Rep. Fish. Aquat. Sci. No. 1341, 58 pp, 1985.

366. Shifrin, K. S., *Scattering of Light in a Turbid Medium* (Russian), Gidrometeoizat Publishing Co., Moscow, 565 pp, 1951.

367. Shifrin, K. S., *Introduction to the Optics of the Ocean* (Russian), Gidrometeoizdat Publishing Co., Leningrad, 278 pp, 1983.

368. Shuleikin, V. V., On the colour of the sea, *Phys. Rev.*, 22, 86–100, 1923.

369. Skopintsev, B. A., *Organic Matter in Oceanic Waters in Progress in Soviet Oceanology* (Russian), Nauka Publishing Co., Moscow, 64–89, 1979.

370. Slater, P. N., Biggar, S. F., Holm, R. G., Jackson, R. D., Mao, Y., Moran, M. S., Palmer, J. M., and Yuan, B., Absolute radiometric calibration of the Thematic Mapper, *SPIE*, 660, 2–8, 1986.

371. Slater, P. N., Biggar, S. F., Holm, R. G., Jackson, R. D., Mao, Y., Moran, M. S., Palmer, J. M., and Yuan, B., Reflectance- and radiance-based methods for the in-flight absolute calibration of multi-spectral sensors, *Rem. Sens. Environ.*, 22, 11–37, 1987.

372. Smith, E. L., Photosynthesis in relation to light and carbon dioxide, *Proc. Nat. Acad. Sci. U.S.A*, 22, 504–511, 1936.

373. Smith, R. C. and Baker, K. S., Optical classification of natural waters, *Limnol. Oceanogr.*, 23, 260–267, 1978.

374. Smith, R. C. and Baker, K. S., Penetration of UV-B and biologically effective dose-rates in natural waters, *Photochem. Photobiol.*, 29, 311–323, 1979.

375. Smith, R. C. and Baker, K. S., Optical properties of the clearest natural waters (200–800 nm), *Appl. Opt.*, 20, 177–184, 1981.

376. Smith, R. C. and Baker, K. S., Assessment of the influence of enhanced UV-B on marine productivity, in *The Role of Solar Ultraviolet Radiation in Marine Ecosystems*, Calkins, J., Ed., Plenum Press, New York, 509–537, 1982.

377. Smith, R. C. and Baker, K. S., Stratospheric ozone, middle ultraviolet radiation, and phytoplankton productivity, *Oceanogr. Mag.*, 2, 4–10, 1989.

378. Smith, R. C., Baker, K. S., Holm-Olson, O., and Olson, R. S., Photoinhibition of photosynthesis in natural waters, *Photochem. Photobiol.*, 31, 585–592, 1980.

379. Smith, R. C. and Wilson, W. H., Ship and satellite bio-optical research in the California Bight, in *Oceanography from Space*, Gower, J. F. R., Ed., Plenum Press, New York, 281–294, 1981.

380. Smith, R. C., Prézelin, B. B., Baker, K. S., Bidigare, R. R., Boucher, N. P., Coley, T., Karentz, D., MacIntyre, S., Matlick, H. A., Menzies, D., Ondrusek, M., Wan, Z., and Waters, K. J., Ozone depletion: Ultraviolet radiation and phytoplankton biology in Antarctic waters, *Science*, 255, 952–959, 1992.

381. Sobolov, V. V., *Scattering of Light in the Atmosphere of Planets* (Russian), Nauka Publishing Co., Moscow, 375 pp, 1972.

382. Spanner, M., Wrigley, R., Pueschel, R., Livingston, J., and Colburn, D., Determination of atmospheric optical properties for the First ISCLSCP Field Experiment (FIFE), *J. Spacecr. Rockets*, 27, 373–379, 1990.

383. Spencer, J. W., Fourier series representation of the position of the sun, *Search*, 2, 172, 1971.

384. Stainton, M. P., Capel, M. J., and Armstrong, F. A. J., *The Chemical Analysis of Fresh Water*, 2nd ed., Can. Fish. Mar. Serv. Misc. Spec. Publication No. 25, 180 pp, 1977.

385. Steeman Nielsen, E., Light and primary production, in *Optical Aspects of Oceanography*, Jerlov, N. G. and Steeman Neilsen, E., Eds., Academic Press, London and New York, 361–388, 1974.

386. Steven, M. D., Standard distribution of clear sky radiance, *Quart. J. Roy. Meteorol. Soc.*, 103, 457–465, 1977.

387. Stoermer, E. F. and Kopczinska, E., Phytoplankton populations in the extreme southern basin of Lake Michigan, *Proc. 15th Conf. Great Lakes Res.*, 181–191, 1972.

388. Stravinskaya, E. A. and Kulisheva, Yu. I., The role of bottom sediments in the cycles of nitrogen and phosphorus, in *Preservation of Natural Water Basin Ecosystems in Urbanized Landscapes* (Russian), Nauka Publishing Co., Leningrad, 39–43, 1984.

389. Strickland, J. D. H., Production of organic matter in primary stages of the marine food chain, in *Chemical Oceanography*, Riley, J. P. and Skirrow, G., Eds., Academic Press, London, 477–610, 1965.

390. Strickland, J. D. H. and Parsons, T. R., *A Practical Handbook of Seawater Analysis*, Bull. Fish. Res. Board Can., 167, 310 pp, 1972.

391. Strong, A. E., Chemical whitings and chlorophyll distributions in the Great Lakes as viewed by Landsat, *Rem. Sens. Environ.*, 7, 61–72, 1978.

392. Talling, J. F., The phytoplankton population as a compound photosynthetic system, *New Phytol.*, 56, 287–295, 1957.

393. Tam, C. K. N., and Patel, A. C., Optical absorption coefficients of water, *Nature*, 280, 302–304, 1979.

394. Ter-Markariants, N. E., On the calculation of water surface albedo (Russian), *Proc. Acad. Sci. USSR, Ser. Geophys.*, 8, 32–34, 1959.

395. Thekaekara, M. P. and Drummond, A. J., Standard values for the solar constant and its spectral components, *Nature*, 229, 6–9, 1971.

396. Thomas, J. B., *Primary Photoprocesses in Biology*, North Holl. Publ. Cy., Amsterdam, and Wiley and Sons, New York, 1965.

397. Thomson, K. B. P., Jerome, J. H., and McNeil, W. R., Optical properties of the Great Lakes (IFYGL), in *Proc. 17th Conf. Great Lakes Res.*, 811–822, 1974.

398. Thomson, K. P. B. and Jerome, J. H., *Transmissometer Measurements of the Great Lakes*, Environment Canada Inland Waters Directorate Scientific Series No. 53, 1975.

399. Twomey, S., *Atmospheric Aerosols*, Elsevier Scientific Publishing Co., Amsterdam, 1977.

400. Twomey, S., Piepgras, M., and Wolfe, T. L., Am assessmnent of the impact of pollution on global cloud albedo, *Tellus*, 5, 356–366, 1984.

401. Tyler, J. E. The Secchi disc, *Limnol. Oceanogr.*, 13, 1–6, 1968.

402. Tyler, J. E. and Richardson, W. H., Nephelometer for the measurement of volume scattering function *in situ*, *J. Opt. Soc. Amer.*, 48, 354–357, 1958.

403. Ulianova, D. S., Conditions for carbonate formation in the Lake Sevan waters, in *Limnology of Mountain Water Basins* (Russian), Armenian Academy of Sciences Publication, 320–321, 1984.

404. United Nations Environment Programme, *Environmental Effects of Ozone Depletion: 1991 Update*, United Nations Environmental Programme, Nairobi, 1991.

405. Unoki, S., Okami, N., Kishino, M., and Sugihara, S., *Optical Characteristics of Sea Water at Tokyo Bay*, Science and Technology Agency Report, Japan, 1978.

406. U. S. JGOFS, *The Role of Ocean Biogeochemical Cycles in Climate Change*, U. S. Joint Global Ocean Flux Planning Report No. 11, National Academy Press, Washington, DC, 1990.

407. Vallentyne, J. R., The Process of Eutrophication and Criteria for Trophic State Determination, in *Proceedings of the Workshop on Modeling the Eutrophication Process*, St. Petersburg, FL, 19–21 Nov., U. of Florida Dept. of Environ. Engin. and U.S. Dept. of Interior, 57–67, 1969.

408. Valley, S. E. (Ed.), *Handbook of Geophysics and Space Environments*, Air Force Cambridge Research Laboratories, McGraw-Hill Book Co., New York, 16–2, 1965.

409. Varotsos, C. and Kondratyev, K. Ya., Changes in solar ultraviolet radiation reaching the ground due to tropospheric and stratospheric ozone variations (in press), 1995.

410. Varotsos, C. and Kondratyev, K. Ya., Athens environmental dynamics: From a rural to an urban region, *Opt. Atmos. Ocean* (in press), 1995.

411. Vincent, W. F., Neale, P. J., and Richerson, P. J., Photoinhibition: Algal responses to bright light during diel stratification and mixing in a tropical alpine lake, *J. Phycol.*, 20, 201–211, 1984.

412. Vincent, W. F. and Roy, S., Solar ultraviolet-B radiation and aquatic primary production: Damage, protection and recovery, *Environ. Rev.*, 1, 1–12, 1993.

413. Visser, S. A., Seasonal changes in the concentration and colour of humic substances in some aquatic environments, *Freshwater Biol.*, 14, 79–87, 1984.

414. Vollenweider, R. A., Some observations on the C 14 method for measuring primary production, *Verh. Internat. Verein. Limnol.*, 14, 134–139, 1961.

415. Vollenweider, R. A., Calculation models of photosynthesis-depth curves and some implications regarding day rate estimates in primary production measurements, *Mern. Inst. Ital. Idrobiol.*, 18(suppl), 425–457, 1965.

416. Vollenweider, R. A., *Scientific Fundamentals of the Eutrophication of Lakes and Flowing Waters, with Particular Reference to Nitrogen and Phosphorus as Factors in Eutrophication*, Tech. Rep. No. DAS/DSI/68, Organization for Economic Cooperation and Development, Paris, 1968.

417. Vollenweider, R. A., Advances in defining critical loading levels for phosphorus in lake eutrophication, *Mem. 1st Ital. Idrobiol.*, 33, 53–83, 1976.

418. Vollenweider, R. A., *Global Problems of Eutrophication and Its Control*, Symp. Biol. Hung. 38, Akadémiai Kiadó, Budapest, 19–41, 1989.

419. Vollenweider, R. A., Munawar, M., and Stadelmann, M., A comparative review of phytoplankton and primary production in the Laurentian Great Lakes, *J. Fish. Res. Board Can.*, 31, 739–762, 1974.

420. Walsh, J. J., Importance of continental margins in the marine biogeochemical cycling of carbon and nitrogen, *Nature*, 350, 53–55, 1991.

421. Walsh, J. W. T., *The Science of Daylight*, Macdonald Press, London, 369 pp, 1961.

422. Weaver, E. C. and Wrigley, R., *Factors Affecting the Identification of Phytoplankton Groups by Means of Remote Sensing*, NASA Tech. Mem. No. 108799, Ames Research Center, Moffett Field, CA, 117 pp, 1994.

423. Welschmeyer, J. A. and Lorenzen, C. J., Chlorophyll-specific photosynthesis and quantum-efficiency at saturating light intensities, *J. Phycol.*, 17, 283–293, 1981.

424. Whiting, J. and Bukata, R. P. An examination of the role of satellites in monitoring the impact of climate change on the hydrological cycle, *Proceedings of International Society for Photogrammetry and Remote Sensing Commission, VII Symposium on Global and Environmental Monitoring*, Vol. 2, 17–21 Sept., Victoria, B C, 10 pp, 1990.

425. Whitlock, C. H., Poole, L. R., Ursy, J. W., Houghton, W. M., Witte, W. G., Morris, W. D., and Gurganus, E. A., Comparison of reflectance with backscatter for turbid waters, *Appl. Opt.*, 20, 517–522, 1981.

426. Whitney, L. V., Measurement of continuous solar radiation in Wisconsin lakes, *Trans. Wisc. Acad. Sci. Arts Lett.*, 31, 177, 1938.

427. Williamson, C. E., Zagarese, H. E., Schulze, P. C., Hargreaves, B. R., and Seva, J., The impact of short-term exposure to UV-B radiation on zooplankton communities in north temperate lakes, *J. Plankton Res.*, 16, 205–218, 1994.

428. Witte, W. G., Whitlock, C. H., Harriss, R. C., Ursy, J. W., Poole, L. R., Houghton, W. M., Morris, W. D., and Gurganus, E. A., Influence of dissolved organic materials on turbid water optical properties and remote sensing reflectance, *J. Geophys. Res.*, 87, 441–446, 1982.

429. Wollast, R., The coastal carbon cycle: Fluxes, sources, and sinks, in *Ocean Margin Processes in Global Change*, Manoura, R. C. F., Martin, J. M., and Wollast, R., Eds., John Wiley and Sons, New York, 365–381, 1991.

430. Wrigley, R. C. and Klooster, S. A., *Coastal Zone Color Scanner Data of Rich Coastal Waters*, Proc. International Geoscience and Remote Sensing Symp., San Francisco, California, FA-6, 2.1–2.5, 1983.

431. Wrigley, R. C., Spanner, M. A., Slye, R. E., Pueschel, R. F., and Aggarwal, H. R., Atmospheric correction of remotely sensed image data by a simplified model, *J. Geophys. Res.*, Special FIFE issue, 97, 18,797–18,814, 1992.

432. Wrigley, R. C., Klooster, S. A., Freedman, R. S., Carle, M., Slye, R. E., and McGregor, L. F., The Airborne Ocean Color Imager: System description and image processing, *J. Imag. Sci. Tech.*, 35, 423–430, 1992.

433. Wu, J., Wind stress and surface roughness at air-sea interface, *J. Geophys. Res.*, 74, 444–455, 1969.

434. Yentsch, C. S., The influence of phytoplankton pigments on the color of sea water, *Deep-Sea Res.*, 7, 1–9, 1960.

435. Yocum, C. S. and Blinks, L. R., Light induced efficiency and pigment alterations in red algae, *J. Gen. Physiol.*, 41, 1113–1117, 1958.

436. Zepp, R. G. and Cline, D. M., Rates of photolysis in aquatic environment, *Environ. Sci. Technol.*, 11, 359–366, 1977.

437. Zepp, R. G. and Schlotzhauer, P. F., Comparison of photochemical behaviour of various humic substances in water. III. Spectroscopic properties of humic substances, *Chemosphere*, 10, 479–486, 1981.

GLOSSARY

Absorption Coefficient $a(\lambda)$:   The fraction of radiant energy absorbed from an incident light beam as it traverses an infinitesimal distance ∂r divided by ∂r.

Absorption Cross Section $a_i(\lambda)$:   See **Specific Absorption Coefficient $a_i(\lambda)$**

Achromatic Color S: See **White Point S**.

Action Spectrum: An organism's spectral response function to an incident irradiance spectrum.

Active Remote Sensing Device: A remote sensor that transmits a signal to the target being monitored and then measures the signal that is returned from that target. Active remote sensing devices interact with the target oftimes stimulating an optical transition within the target resulting in an optical return that would not have been observed from a non-stimulated target. Active remote sensing devices include fluoro-sensors, scatterometers, and synthetic aperture radar systems.

Aerosols: Suspended atmospheric matter and liquid particles that may exist in a myriad of diverse forms and shapes from a myriad of diverse sources. Aerosols include smoke, water and H_2SO_4 droplets, dust, ashes, pollen, spores, and other forms of atmospheric suspensions.

Aerosol Optical Depth $\tau(\lambda)$:   An estimate of atmospheric total aerosol concentration, defined as the integration of the atmospheric beam attenuation coefficient due to aerosols over altitude h.

Aerosol Phase Function $P(\theta,\phi)$:   A measure of light scattered by an atmospheric aerosol particle as a function of angle relative to the direction of the incident photon beam, defined as the ratio of 4π times the volume scattering function to the atmospheric scattering coefficient.

Albedo $A(\lambda)$:   The ratio of the energy returning from a point or surface to the energy incident upon that point or surface for a particular wavelength λ. Also referred to as **spectral albedo**.

Algal Fungi: Colorless, chlorophyll-free lower aquatic plant forms.

Algae: The most familiar of the genus of **phytoplankton**. Algae are a high-variety group of plant structures that may be one-celled, colonial, or filamentous, and include such diverse members as seaweed and pond scum.

Allochthonic: An adjective referring to non-indigenous plant, animal, and microbial populations within a water column, the consequence of external inputs.

Apparent Optical Property: An optical property of a water body that is dependent upon the spatial distribution of the incident radiation. Apparent optical properties include the irradiance attenuation coefficient and the volume reflectance.

Apparent Radiance: A term referring to remotely-sensed data in its non-radiometrically calibrated form, for example, "raw" satellite data recorded and stored in a convenient computer code comprising integers proportional to, but not converted to physical units of radiance.

Aquatic Brightness: A term originally used to refer to the total spectral radiance recorded over a water body by the Landsat MSS (the sum of Bands 1, 2, and 3). It is now generally used to refer to the total spectral upwelling radiance from the water column irrespective of the recording sensor.

Aquatic Humic Matter: See **Yellow Substance**.

Atmospheric Attenuation Coefficient $c_s(\lambda)$: The atmospheric counterpart of **attenuation coefficient $c(\lambda)$**. The fraction of radiant energy removed from the solar beam as it traverses an infinitesimal distance (due to the combined processes of scattering and absorption within the atmosphere) divided by that infinitesimal distance.

Atmospheric Transparency Coefficient $T(\lambda)$: The atmospheric counterpart of **transmittance $T(\lambda)$**. The exponential of the negative of the **optical thickness $\tau(\lambda)$** which when multiplied by the solar constant $E_{solar}(\lambda)$ yields the direct solar irradiance $E_{sun}(\lambda)$ [uncorrected for distance from sun].

Attenuance $C(\lambda)$: The ratio of radiant flux lost from a beam within an infinitesimal attenuating layer, $(\Phi_{inc} - \Phi_{trans})$, to the radiant flux within the beam impinging upon that infinitesimal layer, Φ_{inc}.

Attenuation Coefficient $c(\lambda)$: The fraction of radiant energy removed from an incident light beam as it traverses an infinitesimal distance ∂r (due to the combined processes of absorption and scattering) divided by ∂r.

Attenuation Length $\tau(\lambda)$: The path length in an attenuating medium required to reduce the energy of a light beam by a factor of $1/e$. Determined by the reciprocal of the beam attenuation coefficient.

Autochthonous: An adjective referring to indigenous plant, animal, and microbial populations within a water column.

Backscattering Coefficient $b_B(\lambda)$: The integral of the volume scattering function $\beta(\theta)$ over the hemisphere trailing the incident flux [defined by the angular ranges $(\pi/2 \leq \theta \leq \pi)$ and $(0 \leq \phi \leq 2\pi)$]. b_B is the product of $B(\lambda)$ and $b(\lambda)$.

Backscattering Cross Section $B_i(\lambda)b_i(\lambda)$ or $(b_B)_i(\lambda)$: See **Specific Backscattering Coefficient $B_i(\lambda)b_i(\lambda)$ or $(b_B)_i(\lambda)$**.

Backscattering Probability B(λ): The ratio of the scattering into the hemisphere trailing the incident flux to the total scattering into all directions. Determined as the ratio of the backscattering coefficient $b_B(\lambda)$ to the scattering coefficient $b(\lambda)$.

Bacterioplankton: The microbial organism component of **plankton**.

Beam Attenuance C(λ): See **Attenuance C(λ)**.

Beam Attenuation Coefficient c(λ): See **Attenuation Coefficient c(λ)**.

Beam Transmittance T(λ): See **Transmittance T(λ)**.

Beer's Law: A statistical relationship, expressed in exponential form, between radiation incident upon an attenuating medium and radiation transmitted through that attenuating medium.

Bio-optical Models: Models that relate the optical return from a natural water body to the biological aquatic components and processes responsible for that optical return.

Blackbody: An object which absorbs all radiative energy incident upon it and which itself radiates in accordance with **Planck's Law**.

Boundary Layer: See **Free Surface Layer**.

Brewster's Angle θ$_B$: The angle of incidence upon the interface between two attenuating media for which the angle of reflection and the angle of refraction sum to 90°. It is the incident angle for which there is no reflected parallel polarized electromagnetic radiation from a plane interface.

Bulk Inherent Optical Property: An inherent optical property of a water column wherein the water column is considered as a composite entity with no regard as to the specific components contributing to that property. $a(\lambda)$, $b(\lambda)$, $c(\lambda)$, and $b_B(\lambda)$ are examples of bulk inherent optical properties.

Carotenoids: One of the three basic types of phytoplankton photosynthizing agents. The other two are **chlorophylls** and **phycobilins**.

Chlorophylls: One of the three basic types of phytoplankton photosynthizing agents. The other two are **carotenoids** and **phycobilins**.

Chloroplast: Subcellular plant structure housing the pigmentation in photosynthetic organisms.

Chromaticity: The apportionment of white light into its tristimulus red, green, and blue components.

Chromaticity Coordinates X (red), Y (green), Z (blue): The fractions of a visible irradiance spectrum that define each of the tristimulus red, green, and blue apportionments. The sum of X, Y, and Z is unity.

Chromaticity Diagram: A diagram in which two of the **chromaticity coordinates** are generally plotted as corresponding pairs in order to illustrate the color of a given spectrum. The loci of all possible pairs are surrounded by an envelope, that consists of the coordinate pairs pertinent to monchromatic spectra for all the visible wavelengths.

Chromatophore: Bacterial substructure housing the pigmentation in photosynthetic organisms.

Chromosphere: The layer of transparent glowing gas above the **photosphere** in the solar atmosphere.

CIE Color Mixture $x(\lambda)$, $y(\lambda)$, and $z(\lambda)$: Hypothetical standards set by the Commission Internationale de l'Éclairage to represent the sensitivity of the human eye to stimuli by red, green, and blue wavelengths, respectively.

Color Index $L(\lambda_1)/L(\lambda_2)$: The ratio, at depth z, of upwelling radiance values at two different wavelengths λ_1 and λ_2.

Contrast $C_T(\lambda)$: An indication of the optical distinctiveness of an object from its surroundings, given as the difference between the reflectivity of the object's surroundings and the reflectivity of the object normalized to the reflectivity of the object.

Corona: The faint white halo in the region above the **chromosphere** at the maximum extent of the solar atmosphere. The corona is also referred to as the **solar crown.**

Critical Angle of Incidence: That angle of incidence, measured in the denser of two adjacent attenuating media, for which the the angle of refraction is 90°. For incident angles larger than the critical angle of incidence, photons cannot escape their confining medium and **total internal reflection** occurs.

Daily Irradiation $\Gamma(\lambda,z)$: See **Irradiation $\Gamma(\lambda,z,t)$**.

Deep Chlorophyll Maximum (DCM): The peak concentration in the subsurface aquatic chlorophyll pigment depth profile in natural waters. It is generally located within the photic zone and deeper than 1 optical depth.

Detritus: Small particles of organic and partially mineralized matter formed from decayed plants and animals and their excretions.

Diffuse Light $E_{sky}(\lambda)$: The spectral distribution of sky irradiance. Also referred to as **diffuse sky radiation** and **skylight.**

Diffuse Sky Radiation $E_{sky}(\lambda)$: See **Diffuse Light $E_{sky}(\lambda)$**.

Direct (Collimated) Light $E_{sun}(\lambda)$: The spectral distribution of solar irradiance. Also referred to as **direct solar radiation.**

Direct Solar Radiation $E_{sun}(\lambda)$: See **Direct (Collimated) Light $E_{sun}(\lambda)$**.

Dominant Wavelength: The colorimetric definition of the perceived color associated with a given spectrum in terms of the wavelength most representative of that spectrum. It is obtained in a geometrical manner from an operative pair of **chromaticity coordinates** and the envelope encompassing all such possible pairs in the chromaticity diagram.

Downwelling Irradiance $E_d(\lambda)$: The irradiance at a point in an attenuating medium due to the stream of downwelling light.

Downwelling Scalar Irradiance $E_{Od}(\lambda)$: The integrated downwelling radiance distribution over the hemisphere above the horizontal plane containing a selected point in an attenuating medium. The $\cos\theta$ term which incorporates directionality into the determination of non-scalar irradiance is excluded from the 2π integration.

Epilimnion: The warmer upper layer of a lake that is seasonally thermally stratified.

Energy Flux $\Phi_N(\lambda)$:   The product of the individual energies of the photons comprising a photon beam and the number of such photons per unit area and time.

Euphotic Depth: See **Photic Depth.**

Euphotic Zone: See **Photic Zone.**

Eutrophication: The enrichment of waters with plant nutrients, generally phosphorus and/or nitrogen, resulting in enhanced aquatic plant growth. See **Trophic Status.**

Eutrophy: See **Trophic Status.**

Exosphere: That region above the Earth's surface (height > 200 km) for which temperature is solely under solar control.

Extraterrestrial Radiation $E_{\text{solar}}(\lambda)$:   See **Solar Constant $E_{\text{solar}}(\lambda)$.**

Fluorescence: The release by a substance of electromagnetic radiation of a particular wavelength as a consequence of that substance absorbing electromagnetic radiation of a different wavelength. This stimulated release of electromagnetic radiation ceases immediately upon cessation of input energy.

Forwardscattering Coefficient $b_F(\lambda)$:   The integral of the volume scattering function $\beta(\theta)$ over the hemisphere preceding the incident flux [defined by the angular ranges $(0 \leq \theta \leq \pi/2)$ and $(0 \leq \phi \leq 2\pi)$].

Forwardscattering Probability $F(\lambda)$:   The ratio of the scattering into the hemisphere preceding the incident flux to the total scattering into all directions. Determined as the ratio of the forwardscattering coefficient $b_F(\lambda)$ to the scattering coefficient $b(\lambda)$.

Free Surface Layer: The first subsurface millimeter or so just beneath the air-water interface. Also referred to as the **Boundary Layer.**

Fresnel Reflectance Formulae: Formulae which enable the computation of reflectance from the interface between two attenuating media in terms of the angles of incidence and refraction appropriate to that interface. The Fresnel reflectance formulae may be applied to **polarized** or **nonpolarized** incident light.

Fulvic Acid: A water-soluble fraction of **humic substances** present in the water column as dissolved organic carbon, differing in molecular weight from **humic acid.**

Gelbstoff: See **Yellow Substance.**

Gilvin: See **Yellow Substance.**

Global Radiation $E(\lambda)$:   The total irradiance reaching a point on the Earth's surface. It is the sum of the **direct solar radiation $E_{\text{sun}}(\lambda)$** and the **diffuse sky radiation $E_{\text{sky}}(\lambda)$.**

Humic Substances: A variety of complex polymers (comprised of a water-soluble and a water-insoluble fraction) resulting from the decomposition of phytoplankton cells. The water-soluble fraction gives rise to **yellow substance**, imparting a yellow hue to natural waters.

Humic Acid: A water-soluble fraction of **humic substances** present in the water column as dissolved organic carbon, differing in molecular weight from **fulvic acid**.

Humin: The water-insoluble fraction of **humic substances** present in the water column as suspended matter.

Humolimnic Acid: See **Yellow Substance**.

Hyphae: Assemblages of cellular filaments containing the spores of algal fungi.

Hypolimnion: The cooler bottom layer of a lake that is seasonally thermally stratified.

Hypertrophy: See **Trophic Status**.

Illuminance: The photometric parameter counterpart of the physical parameter **irradiance**.

Inherent Optical Property: An optical property of a water body which is totally independent of the spatial distribution of the impinging radiation. Inherent optical properties include the attenuation coefficient, the absorption coefficient, the scattering coefficient, the forwardscattering probability, the backscattering probability, and the scattering albedo.

Irradiance $E(\lambda)$: The radiant flux per unit area at a point within a radiative field or at a point on an extended surface.

Irradiance Attenuation Coefficient $K_d(\lambda,z)$: The logarithmic depth derivative of the spectral irradiance $E(\lambda,z)$ at depth z.

Irradiance Attenuation Coefficient for PAR K_{PAR}: The logarithmic depth derivative of the spectral irradiance $E_{PAR}(z)$. $E_{PAR}(z)$ is the downwelling irradiance in the PAR broadband range 400 nm to 700 nm at depth z.

Irradiance Reflectance $R(\lambda,z)$: The ratio of the upwelling irradiance at a point in an attenuating medium to the downwelling irradiance at that point. In water this ratio is termed the **volume reflectance $R(\lambda,z)$**

Irradiation $\Gamma(\lambda,z)$: The integration of the irradiance $E(\lambda,z,t)$ at depth z over time. The integration of the subsurface irradiance $E(\lambda,z,t)$ at depth z over the daylight period is termed the **daily irradiation**.

Light: A general term usually referring to radiation in that portion of the electromagnetic spectrum to which the human eye is sensitive (about 390 nm to about 740 nm). See **Visible Wavelengths**.

Luminance $\mathscr{L}(\theta,\phi,\lambda)$: The photometric parameter counterpart of the physical parameter **radiance $L(\theta,\phi,\lambda)$**.

Luminosity Function $\ell(\lambda)$: A statistically derived factor relating a human observer's pyschophysical sensation to light of wavelength λ to the observer's pyschophysical sensation to light of wavelength 555 nm. See **Standard Luminosity Curve**.

Luminous Energy: The photometric parameter counterpart of the physical parameter **radiant energy.**

Luminous Intensity: The photometric parameter counterpart of the physical parameter **radiant intensity.**

Mass Attenuation Coefficient $\beta_i(\lambda)$**:**   The specific attenuation that can be attributed to a unit density of atmospheric component i as solar radiation propagates through the atmosphere.

Mean Sunspot Number: See **Wolf Sunspot Number.**

Mesopause: The surface (at a height of ~90 km) of minimum temperature in the Earth's atmosphere (around $-90°$C).

Mesosphere: The layer of the earth's atmosphere (roughly between heights of 50 km and 90 km) in which atmospheric temperature decreases with increasing height.

Mesotrophy: See **Trophic Status.**

Meteorological Range V_M**:**   A quantification of the qualitative term **visibility,** defined by a relationship between the atmospheric extinction coefficient (due to combined molecular and aerosol extinction) and an object's contrast to its surroundings.

Mie Scattering: The relationship between scattering intensity and wavelength for the situation in which the wavelength of the impinging radiation is comparable to or smaller than the diameter of the scattering centre. Mie scattering describes atmospheric aerosol scattering phenomena.

Nepheloid Layer: A region of low transmission frequently observed near the bottom of a water body and generally attributed to the resuspension of bottom sediments.

Net Downward Irradiance $E\downarrow(\lambda)$**:**   The difference between the downwelling irradiance $E_d(\lambda)$ and the upwelling irradiance $E_u(\lambda)$ at any point in the radiative field. The net downward irradiance is obtained by integrating the radiance distribution multiplied by $\cos\theta$ over 4π space.

Net Downwelling Scalar Irradiance $E_0\downarrow(\lambda)$**:**   The difference between the downwelling scalar irradiance $E_{0d}(\lambda)$ and the upwelling scalar irradiance $E_{0u}(\lambda)$ at any point in the radiative field. The net downwelling scalar irradiance is obtained by integrating the radiance distribution over the upper and lower 2π hemispheres individually and then subtracting them.

Oligotrophy: See **Trophic Status.**

Optical Air Mass m**:**   The ratio of the mass of an air column encountered by solar radiation emanating from the sun located at solar zenith angle θ to the mass of an air column encountered by solar radiation emanating from the sun located directly overhead ($\theta = 0°$).

Optical Attenuation $C(\lambda)$**:**   See **Attenuance** $C(\lambda)$.

Optical Cross Sections: The specific amount of absorption and scattering, as a function of wavelength λ, that may be attributed to a unit concentration of each organic and inorganic component of a natural water body. The **absorption, scattering,** and **backscattering cross sections** [$a_i(\lambda)$, $b_i(\lambda)$, and $B_i(\lambda)$, $b_i(\lambda)$] are interchangeable with the terms **specific absorption, specific scattering,** and **specific backscattering coefficients.**

Optical Depth $\zeta(\lambda)$:   The integration of the downwelling irradiance attenuation coefficient $K_d(\lambda,z)$ over depth z.

Optical Range: A collective term referring to the region of the electromagnetic spectrum that includes the ultraviolet, visible, and near-infrared wavelengths.

Optical Thickness $\tau(\lambda)$:   The integral, from ground-level to the top of the atmosphere, of the product of the atmospheric **mass attenuation coefficients** $\beta_i(\lambda)$ and the densities $\rho_i(z)$ of the principal atmospheric attenuating components.

Partial Thickness: The height of the atmosphere (at normal temperature 15°C and pressure 1010 mb) if the atmosphere were comprised solely of a designated gas.

Passive Remote Sensing Device: A remote sensor that faithfully responds to the optical field emanating from a target whether that optical field be reflected or self-generated. The most familiar passive remote sensing device is the multispectral radiometer.

Penetration Depth $z_p(\lambda)$:   The vertical aquatic depth from which 90% of the signal recorded by a remote sensing device is considered to originate. $z_p(\lambda)$ corresponds to an **optical depth** of unity, or alternatively, to a subsurface irradiance level of ~37%.

Phase Function $P(\theta)$:   The volume scattering function $\beta(\theta)$ normalized to the scattering coefficient $b(\lambda)$, and given as 4π times $\beta(\theta)$ divided by $b(\lambda)$.

Phaeoforbid: A product of the decomposition of aquatic chlorophyll due to oxygen hydrolysis. Phaeoforbid is a chlorophyll derivative that represents a portion of the non-living phytoplankton within the water column. See **Phaeophytin.**

Phaeophytin: A product of the decomposition of aquatic chlorophyll due to oxygen hydrolysis. Phaeophytin is a chlorophyll derivative that represents a portion of the non-living phytoplankton within the water column. See **Phaeoforbid.**

Photic Depth: The vertical distance from the air-water interface to the 1% subsurface irradiance level. Also referred to as the **euphotic depth.**

Photic Zone: The aquatic region in which maximum photosynthesis occurs (generally taken to be the upper aquatic layer bounded by the 100% and 1% subsurface irradiance levels). Also referred to as the **euphotic zone.**

Photon: A quantum of energy displaying both a particulate and a wavelike character. The energy of a photon, ξ, is given by the product of the speed of light in a vacuum c (3.00×10^8 m s^{-1}), Planck's Constant h (6.625×10^{-34} joules s), and the inverse of its associated wavelength . The total energy of a beam of photons is designated by ξ_T. See **Radiant Energy Q.**

Photon Flux *N:* The number of photons incident upon a unit area per unit time.

Photosphere: The apparent solar surface including the lowest layer of the solar atmosphere. Most of the solar mass resides within the photosphere.

Photosynthesis: The biological combination of chemical compounds in the presence of light. It generally refers to the production of organic substances from carbon dioxide and water residing in green plant cells provided the cells are sufficiently irradiated to allow chlorophyllous pigments to convert radiant energy into a chemical form.

Photosynthetic Available Radiation (PAR): The total radiation in the wavelength interval 400 nm to 700 nm. It is the wavelength interval of greatest significance to aquatic photosynthesis and primary productivity.

Photosynthetic Efficiency: The ratio of the energy stored as chemical energy through photosynthesis to the total energy absorbed by the photosynthetic pigments.

Photosynthetic Usable Radiation (PUR): The amount of the Photosynthetic Available Radiation (PAR) which is pertinent to the local photosynthetic process. PUR is the value of PAR as weighted by the absorption capabilities of indigenous aquatic algal species.

Phycobilins: One of the three basic types of phytoplankton photosynthesizing agents. The other two are **chlorophylls** and **carotenoids**.

Phytoplankton: The plant organism component of **plankton**. See **Algae** and **Algal Fungi**.

Phytoplankton Biomass: The organic carbon content of the phytoplankton population.

Planck's Law: An expression of the relationship between the radiation emitted by a perfect radiator (**blackbody**), and the temperature of that perfect radiator.

Plankton: A collective term representing the principal living organisms (plant, animal, microbial) present within a natural water column. Plankton include **phytoplankton** (plant), **zooplankton** (animal), and **bacterioplankton** (microbial).

Polarization: The orientation of the electric field vector in the plane perpendicular to the direction of electromagnetic propagation. Depending upon this orientation, the traveling wave may be **plane, circularly, elliptically,** or **randomly (non)** polarized.

Poynting Vector *S:* The vector product of the electric field density vector **E** and the magnetic field density vector **B** associated with a traveling electromagnetic wave.

Primary Production: The chemical energy contained within an ecosystem as a direct result of photosynthesis.

Primary Productivity: The sum of all photosynthetic rates within an ecosystem.

Quantum: Quanta: See **Photon**.

Quiescent Sun: The steady-state dynamical equilibrium of solar processes that enable solar behavior to be described as a statistically predictable series of recurrent physical phenomena.

Radiance $L(\theta,\phi,\lambda)$: The radiant flux per unit solid angle $d\Omega$ (the solid angle lying along a specified direction) per unit area dA (the area lying at right angles to the specified direction) at any point in a radiative field.

Radiance Attenuation Coefficient $K(\lambda,z,\theta,\phi)$: The logarithmic depth derivative of the radiance at depth z.

Radiant Energy Q: The quantity of energy in joules transferred by radiation. It is interchangeable with the symbol ξ_T (see **Photon**) that represents the total energy contained within a beam of photons.

Radiant Flux $\Phi(\lambda)$: The time rate of flow of radiant energy Q.

Radiant Intensity $I(\lambda)$: The radiant flux $\Phi(\lambda)$ per unit solid angle $d\Omega$ in a particular direction.

Radiative Transfer: The dynamics of the energy changes associated with the propagation of radiation through media which absorb, scatter, and/or emit photons. The **radiative transfer equation** is the mathematical formulism describing the transfer of energy between an electromagnetic field and a physical containment medium.

Rayleigh Scattering: The relationship between scattering intensity and wavelength for the situation in which the wavelength of the impinging radiation is considerably larger than the diameter of the scattering centre. Rayleigh scattering describes molecular scattering phenomena.

Reflectance $R(\lambda)$: The ratio of the measured upwelling irradiance to the downwelling irradiance at a given wavelength. For the special case of underwater reflectance, this reflectance is termed the **subsurface irradiance reflectance** or the **volume reflectance $R(\lambda,z)$**.

Reflectivity $R(\lambda)$: The ratio of the radiation returning from a surface to the radiation impinging on that surface for a given wavelength. See **Reflectance $R(\lambda)$**.

Scalar Irradiance $E_0(\lambda)$: The measure of irradiance at a point in the radiative field in which the radiation from all directions are treated equally. Scalar irradiance is determined as the integral of the radiance distribution over 4π space without the $\cos\theta$ term that weights the directionality defining the non-scalar irradiance at that point.

Scalar Irradiance Attenuation Coefficient $K_0(\lambda,z)$: The logarithmic depth derivative of the scalar irradiance at depth z.

Scattering Albedo $\omega_0(\lambda)$: The number of scattering interactions that occur within a fixed volume of an attenuating medium expressed as a fraction of the total number of interactions (both scattering and absorption) that occur within that fixed volume.

Scattering Coefficient $b(\lambda)$: The fraction of radiant energy scattered from a light beam as it traverses an infinitesimal distance ∂r divided by ∂r.

Scattering Cross Section $b_i(\lambda)$: See **Specific Scattering Coefficient $b_i(\lambda)$**.

Scattering Phase Function $P(\theta,\phi)$: See **Phase Function** $P(\theta,\phi)$.

Secchi Depth S: The depth at which a lowered disk vanishes from the view of an observer situated above the air-water interface.

Seston: A collective term referring to all organic and inorganic matter suspended within the water column.

Single-Scattering Albedo $\omega_o(\lambda)$: Defined for the atmosphere as the ratio of the atmospheric scattering coefficient to the atmospheric beam attenuation coefficient. See **Scattering Albedo** $\omega_o(\lambda)$.

Skylight: See **Diffuse Light** $E_{sky}(\lambda)$.

Snell's Law: The relationship among the angle of incidence, the angle of refraction, and the indices of refraction, as light crosses the plane interface separating two attenuating media.

Solar Activity: See **Sunspots**.

Solar Constant $E_{solar}(\lambda)$: The total electromagnetic irradiance impinging at the top of the Earth's atmosphere. Also referred to as the **extraterrestrial radiation**.

Solar Crown: See **Corona**.

Solar Zenith Angle θ: The angle, as measured positively from the vertical, that locates the position of the sun. The zenith angle is a function of local time.

Specific Absorption Coefficient $a_i(\lambda)$: The absorption at wavelength λ that may be attributed to a unit concentration of aquatic component i. Also referred to as the **absorption cross section** of component i.

Specific Backscattering Coefficient $B_i(\lambda)b_i(\lambda)$ or $(b_B)_i(\lambda)$: The backscattering at wavelength λ that may be attributed to a unit concentration of aquatic component i. Also referred to as the **backscattering cross section** of component i.

Specific Inherent Optical Properties: The inherent optical properties of a water column that can be attributed to the individual scattering and absorption centers comprising the water column. The **optical cross sections** are specific inherent optical properties.

Specific Scattering Coefficient $b_i(\lambda)$: The scattering at wavelength λ that may be attributed to a unit concentration of aquatic component i. Also referred to as the **scattering cross section** of component i.

Spectral Albedo $A(\lambda)$: See **Albedo** $A(\lambda)$.

Spectral Purity: An indication (on a scale of 0 to 1) of the distinctiveness of the dominant wavelength within the colorimetric definition of color associated with a given spectrum. As is the case for dominant wavelength, the spectral purity is geometrically determined from an operative pair of **chromaticity coordinates** and the envelope encompassing all such possible pairs in the chromaticity diagram.

Standard Luminosity Curve: A Gaussian-type curve illustrating a **luminosity function** $\ell(\lambda)$ as a function of λ throughout the visible spectrum. The standard luminosity curve is considered to represent a comparison of the energy possessed by photons of a given wavelength to their pyschophysical effects on a human observer with normal vision.

Stratopause: The upper boundary of the **stratosphere,** the height (~50 km) at which the maximum high altitude atmospheric temperature (just under 0°C) is observed.

Stratosphere: The layer of the Earth's atmosphere (roughly between heights of 15 km and 50 km) in which atmospheric temperature increases with increasing height. The stratosphere contains the maximum concentration of atmospheric ozone.

Subsurface Irradiance Reflectance $R(\lambda,z)$: See **Volume Reflectance $R(\lambda,z)$.**

Subsurface Sighting Range: A subjective term representing the maximum distance at which an object may be detected underwater by the human eye.

Sunspots: Relatively cold regions in the photosphere (solar surface). Sunspots are transient, non-stationary, but statistically predictable phenomena that produce relatively minor disturbances on the solar surface. The totality of the sunspot disturbance is termed **solar activity,** and while sunspots do not dramatically impact the electromagnetic solar spectrum observed at the Earth, they have a profound impact upon the corpuscular radiation arriving at the Earth (through the class of short-lived disruptive phenomena known as solar flares) and the magnetic field configurations comprising the inner solar cavity. Sunspots follow a roughly 28-day repetition cycle as observed from Earth, and are directly responsible for the 11-year solar activity cycle.

Thermal Bar: The demarcation between the elevated nearshore aquatic temperature and the as-yet-to-be-elevated offshore colder aquatic temperature as seasonal thermal lake stratification develops.

Thermocline: The region of demarcation between the distinctive warmer upper thermal regime **(epilimnion)** and the distinctive cooler lower thermal regime **(hypolimnion)** comprising a thermally stratified lake.

Thermosphere: An ill-defined layer above the **mesopause** that is considered to define the maximum distance above the Earth's surface (perhaps as high as 150 km to 200 km) at which the atmosphere can still influence temperature.

Tolerance Ellipse dA_t: The range (i.e., tolerance) of possible upwind and crosswind projections of wave slope (onto the plane defining the untilted air-water interface) for which a wave facet will reflect a highlight of the sun directly towards a remote sensing device.

Total Attenuation Coefficient $c(\lambda)$: See **Attenuation Coefficient $c(\lambda)$.**

Total Internal Reflection: The elimination of transmission through the interface between two attenuating media occurring within the denser of two attenuating media for angles of incidence exceeding the **critical angle of incidence.**

Transmittance $T(\lambda)$: The ratio of the radiant flux within a beam emerging from an infinitesimal attenuating layer, Φ_{trans}, to the incident radiant flux within the beam impinging upon that infinitesimal layer, Φ_{inc}.

Tristimulus Values X' (red), Y' (green), Z' (blue): The integrations of an impinging spectral irradiance spectrum with the red, green, and blue spectral sensitivities (**CIE color mixture**) of the human eye.

Trophic Levels: The succession of living plant and animal species that constitute the life-cycle dynamics (or food chain) of a natural water column. Trophic levels represent the nutrient-life-death-decomposition relationships of the water column.

Trophic Status: A qualitative term defining the nutrient and plant growth rate conditions of natural water bodies. A water body displaying enhanced nutrient concentrations and plant growth is termed **eutrophic**, a water body displaying low nutrient concentrations and plant growth is termed **oligotrophic**, and a water body displaying intermediate nutrient concentrations and plant growth is termed **mesotrophic**. **Hypertrophy** and **ultra-oligotrophy** refer to extended boundary conditions of eutrophy and oligotrophy, respectively.

Tropopause: The upper limit of the **troposphere**. It is the height (~8 km to 18 km) above the Earth's surface at which decrease in atmospheric temperature with height abruptly stops.

Troposphere: The lowest 8 to 18 km of the Earth's atmosphere in which the temperature decreases with increasing altitude.

Turbidity: A rather loosely defined term describing aquatic composition. It is generally considered to refer to the totality of suspended organic and inorganic matter present in the water column. It does not include matter in solution.

Ultra-oligotrophy: See **Trophic Status**.

Upwelling Irradiance $E_u(\lambda)$: The irradiance at a point in an attenuating medium due to the stream of upwelling light.

Upwelling Scalar Irradiance $E_{0u}(\lambda)$: The integrated upwelling radiance distribution over the hemisphere below the horizontal plane containing a selected point in an attenuating medium. The $\cos\theta$ term which incorporates directionality into the determination of non-scalar irradiance is excluded from the 2π integration.

Visibility V: A qualitative term representing the greatest horizontal distance at which the unaided human eye can see and identify an object. An atmospheric equivalent of **Secchi depth**.

Visible Wavelengths: That region of the electromagnetic spectrum to which the human eye is considered to be responsive. Generally taken to encompass the wavelength interval from about 390 nm to about 740 nm.

Volume Reflectance $R(\lambda,z)$: The ratio of the upwelling irradiance $E_u(\lambda)$ at a depth z in the water column to the downwelling irradiance $E_d(\lambda)$ at that depth. See **Irradiance Reflectance $R(\lambda,z)$.**

Volume Scattering Function $\beta(\theta,\phi)$: The scattered radiant intensity dI in a direction (θ,ϕ) per unit scattering volume dV normalized to the value of the incident irradiance E_{inc}.

Water Humus: See **Humic Substances.**

Water Quality: The state or condition of a water body in relation to a set of criteria established for its designated use. As considered in this book, water quality refers to the coexisting concentrations of chlorophyll a, suspended minerals, and dissolved organic carbon.

White Point S: The point on a **chromaticity diagram** representing a "white" spectrum. The white point is located by the **chromaticity coordinates** $X = Y = Z = 0.333$.

Whitings: Extended regions of near-surface milky-white calcium carbonate precipitation resulting from supersaturated solution conditions in natural waters.

Wolf Sunspot Number: A characterization of **solar activity** as a number that equals the sum of the total number of **sunspots** visible from Earth at a particular time plus ten times the number of sunspot groups. Also referred to as the **Zurich sunspot number** or the **mean sunspot number.**

Yellow Organic Acids: See **Yellow Substance.**

Yellow Substance : Dissolved aquatic humus, which, due to the presence of yellow and brown melanoids (the products of the temperature-dependent Maillard reaction) impart a yellow hue to the aquatic column. Also referred to as **gelbstoff, aquatic humic matter, yellow organic acids, humolimnic acid,** and **gilvin.**

Zooplankton: The animal organism component of **plankton.**

Zurich Sunspot Number: See **Wolf Sunspot Number.**

INDEX

relations with curvature of the Earth,
76, 78–81
relations with sensor viewing angles,
76, 81–85
relations with wave slope, 73–75
Sunspots, 39–40

Thermal bar, 197
Thermocline, 197
Thermosphere, 41
TM
See Landsat Thematic Mapper
Tolerance ellipse, 75
Total attenuation coefficient
See Attenuation coefficient
Total internal reflection, 70, 94
Total scattering coefficient
See Scattering coefficient
Transmissivity, 69
Transmittance, 43, 65, 69, 185, 190, 233–234
Tristimulus values, 170–171
X', 171
Y', 171
Z', 171
Trophic levels, 115
Trophic status, 127
boundary value quantification for
natural waters, 214–217
definition, 115
Tropopause, 41
Troposphere, 41, 46
Turbidity, 169, 236

Ultra-oligotrophy, 214, 217
Ultraviolet radiation, 38, 316
and action spectra, 312
groundlevel measurements, 50–51, 295
impacts on ecosystems, 48–52, 302–303
and ozone depletion, 47–51, 124, 311
photoinhibition, 312
UV-A, UV-B, UV-C, 47, 295, 311–313
Upwelling irradiance, 11–12, 58, 94, 171, 226
Upwelling radiance spectra, 221–222, 266
CASI-type spectra, 255

conversion to volume reflectance, 223,
229, 278–279
Landsat Thematic Mapper, 232
spatial sensitivity requirements, 300
spectral sensitivity requirements, 300
Upwelling scalar irradiance, 14, 25, 283–284

Visibility, 236
Volume reflectance, 12, 58–59, 90, 94
deconvolving aquatic component
concentrations from, 159–162
definition, 12
dependence on solar zenith angle, 112,
243, 279
impact of chlorophyll on, 145–147, 155
impact of dissolved organic carbon on,
149–151, 155
impact of suspended minerals on,
147–149
impact on color of, 167–169
at a single wavelength, 156–159
transference through the air-water
interface, 33–35, 141, 228
Volume scattering function, 19–20, 90–91,
95

Water humus, 122
Water quality, 3
definition, 318
parameters, 318–319, 320
remote sensing of, 318–322
Waves
dependence on wind, 60–63, 72–76
impact on surface reflection, 72–76
White point, 172
Whitings, 126–127
Wolf sunspot number, 40

Yellow organic acids, 123
Yellow substance, 51, 123–124, 271

Zooplankton, 115–116, 126–127
Zurich sunspot number, 40